Riti Thapar Kapoor, Mohd Rafatullah and Aseem Vashisht (Eds.)

Electronic Wastes

Also of interest

Electronic Wastes.
Biological Treatments, Valorization, Circular Economy
Edited by Riti Thapar Kapoor, Mohd Rafatullah and Aseem Vashisht, 2026
ISBN 978-3-11-222549-3, e-ISBN (PDF) 978-3-11-222550-9

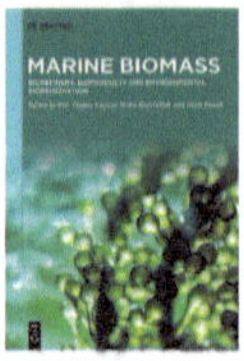

Marine Biomass.
Biorefinery, Bioproducts and Environmental Bioremediation
Edited by Riti Thapar Kapoor, Mohd Rafatullah, Norli Ismail, 2024
ISBN 978-3-11-135358-6, e-ISBN (PDF) 978-3-11-135395-1

Bioremediation Technologies.
For Wastewater and Sustainable Circular Bioeconomy
Edited by Riti Thapar Kapoor, Mohd Rafatullah, 2023
ISBN 978-3-11-101665-8, e-ISBN (PDF) 978-3-11-101682-5

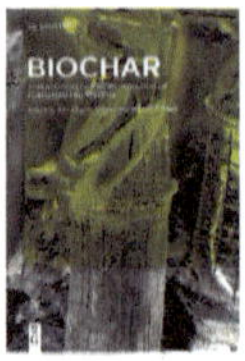

BioChar.
Applications for Bioremediation of Contaminated Systems
Edited by Riti Thapar Kapoor, Maulin P. Shah, 2022
ISBN 978-3-11-073858-2, e-ISBN (PDF) 978-3-11-073400-3

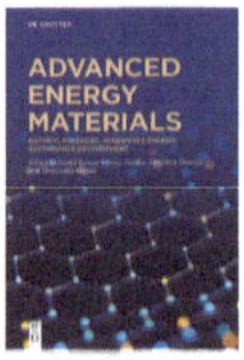

Advanced Energy Materials.
Natural Polymers, Renewable Energy, Sustainable Development
Edited by Sushil Kumar Verma, Sonika Gupta, Abhishek Sharma, Shailendra Rajput, 2025
ISBN 978-3-11-128890-1, e-ISBN (PDF) 978-3-11-128907-6

Biomedical Waste Management.
Bioremediation and Recycling
Edited by Neha Suyal, Prasenjit Debbarma, Manali Singh, Anita Mishra, Roop S. Bora, 2024
ISBN 978-3-11-130316-1, e-ISBN (PDF) 978-3-11-130528-8

Electronic Wastes

Pollution, Management Strategies, Policy Frameworks

Edited by
Riti Thapar Kapoor, Mohd Rafatullah and Aseem Vashisht

DE GRUYTER

Editors
Riti Thapar Kapoor
Professor
Centre for Plant and Environmental Biotechnology,
Amity Institute of Biotechnology,
Amity University Uttar Pradesh,
Noida - 201313,
Uttar Pradesh, India

Mohd Rafatullah
Associate Professor
Department of Mathematics and Natural Sciences,
Prince Mohammad Bin Fahd University,
P.O. Box, 1664, Al-Khobar,
31952, Saudi Arabia

Aseem Vashisht
Associate Professor
University Centre for Research and Development,
Chandigarh University, Mohali,
Punjab, 140413, India

ISBN 978-3-11-224806-5
ISBN 978-3-11-224807-2 (PDF)
ISBN 978-3-11-224808-9 (EPUB)
DOI https://doi.org/10.1515/9783112248072

Library of Congress Control Number: 2026934616

Bibliographic information published by the Deutsche Nationalbibliothek
The Deutsche Nationalbibliothek lists this publication in the Deutsche Nationalbibliografie; detailed bibliographic data are available on the internet at http://dnb.dnb.de.

De Gruyter and Walter de Gruyter GmbH are part of De Gruyter Brill.
www.degruyterbrill.com

Questions about General Product Safety Regulation:
productsafety@degruyterbrill.com

Cover illustration: photoguns/iStock/Getty Images Plus
Printing: CPI books GmbH, Leck

MIX
Papier | Fördert gute Waldnutzung
FSC® C083411

Preface

The rapid pace of digital transformation has reshaped worldwide economy, but it has also led to a significant environmental challenges. Approximately sixty million tons of electronic wastes (e-wastes) are generated globally due to increased demand and technological advancements. Electronic wastes have become one of the main sources of pollution globally, due to application of electronic and smart devices aimed for daily life convenience. Electrical and electronic equipments play pivotal role for upscaling life standards and societal status. As per the report of United Nations, around 54 million metric tons of e-wastes were produced in 2019, which is expected to reach 120 million metric tons by 2050. The main sources of e-wastes include computers, laptops, televisions, home appliances, mobile phones, medical and industrial equipments. The e-waste composition is quite complex and heterogenous in nature, as it contains plastics, heavy metals, azo dyes, and xenobiotic substances. Electronic goods are designed with short lifecycle, which makes their fast disposal and finally leads to unprecedented rise in production of e-wastes. Most of the e-wastes generated in developed nations are shipped to developing countries every year in the form of donations or due to the low cost of recycling. Application of improper procedures for e-waste management like open burning generates dioxins, acids, furans and acid wash release the hazardous substances which show negative impacts on environment and human health. Exposure of the e-wastes toxins through inhalation, ingestion, or dermal absorption or their entry into human body via food chain lead to chronic health consequences such as damage to nervous and respiratory system, heart, brain and liver disorders etc. Consumers should be informed about the lifespan of electronic devices and encouraged to choose items based on quality ratings. However, e-waste reduction is a challenging task, but it can be achieved through improved recycling infrastructure, updated policies, and skilled workforce. There is an urgent need for effective implementation of e-waste regulations at all levels which include extended producer responsibility, effective segregation procedures, proper recycling practices, awareness campaigns, and technological solutions. Public-private partnerships can foster management of e-wastes by boosting collaborations between government and private organizations. By closing regulatory gaps, improving informal collection networks, and strengthening post-processing practices, policymakers can prioritize robust tracking system and material flow monitoring to ensure the sustainable management of e-waste.

The implementation of a single strategy is not enough to address this problem effectively. Instead, combined efforts that integrate technological advancements, policy reforms, and planning at national and global level are crucial for mitigation of e-waste pollution. This book focuses on adverse effects of electronic wastes on ecosystem, health of human-beings as well as recent developments in strategies and policies for management of e-wastes and environmental clean-up. Thus, this book serves as an essential resource for researchers, scientists, policymakers, environmentalists, in-

 | https://doi.org/10.1515/9783112248072-202

dustry professionals, and wider public, offering understanding on e-waste crisis and highlighting significance of effective policies for its management in the present context. We are thankful for the authors and learned reviewers for their valuable feedback for improvement in the content of the book chapters. We would also like to convey our sincere appreciation to De Gruyter Brill Editorial Office, especially Ms Helene Chavaroche for bringing us this opportunity and guiding us throughout the book publication process. We would also like to express our heartfelt thanks to Ms Ute Skambraks, Ms Adelinde Luyken, Ms Sathiyavathi Pajaniradj and other publication team members for their assistance and unwavering support throughout the book publication journey.

Riti Thapar Kapoor
Professor
Centre for Plant and Environmental Biotechnology
Amity Institute of Biotechnology
Amity University Uttar Pradesh
Noida - 201 313, Uttar Pradesh, India

Mohd Rafatullah
Associate Professor
Department of Mathematics and Natural Sciences
Prince Mohammad Bin Fahd University
P.O. Box 1664, Al-Khobar 31952, Saudi Arabia

Aseem Vashisht
Associate Professor
University Centre for Research and Development
Chandigarh University, Mohali, Punjab
140413, India

Contents

Lincy Sara Varghese*

Electronic waste: sources, composition, environmental, and public health consequences

Abstract: Our everyday lives now involve EEE (electrical and electronic equipment) due to the ongoing advancement of technology. These EEE are eventually disposed of in landfills as e-waste, or waste EEE, once their practical uses are over. Due to the rising manufacturing and use of EEE, a massive amount of e-waste is produced globally. These e-waste materials include a number of dangerous substances that can harm people, the environment, and other living things. This chapter covers the origin, categorization, constitution, and environmental and public health consequences of e-wastes.

Keywords: E-waste, composition, classification, ecological and health impacts

1 Introduction

Electronic waste, or e-waste, refers to discarded electrical and electronic equipment (EEE) and their parts that have reached the end of their useful life. The term is often used interchangeably with waste electrical and electronic equipment (WEEE) or e-scrap [1, 2]. It includes electronic items that are no longer functional, obsolete, or unwanted and encompasses a diverse range of electronic devices, such as telecommunications and information technology equipment, large household items, lighting equipment, automatic dispensers, medical devices, monitoring and control devices, as well as consumer electronics, including electronic and electrical tools, sports and leisure equipment, toys, mobile phones, and computers [3, 4].

Rapid industrialization, technological advancement, and changing consumer lifestyles have resulted in a dramatic rise in e-waste generation worldwide, leading to a significant environmental impact and health concern [5, 6]. With the growing dependence on electronic devices, e-waste has become one of the fastest-growing solid waste streams globally [7]. In 2022 alone, approximately 62 million tons of e-waste were produced, of which only 22.3% was formally collected and recycled.

*Corresponding author: Dr. Lincy Sara Varghese, Associate Professor, Department of Microbiology, Bishop Kurialacherry College for Women, Amalagiri, Kottayam 686561, Kerala, India, e-mail: svlincy45@gmail.com

 | https://doi.org/10.1515/9783112248072-001

Recycling e-waste is a valuable source of raw material as it comprises a wide range of materials like metals, glass, and plastics. However, only 17.4% of global e-waste is collected and recycled due to the complexity of collection and recycling [8]. Recycling e-waste is a difficult task because e-waste consists of metals with different physicochemical properties, including hazardous halogen metals [9]. If the substances containing these hazardous compounds are not properly disposed of, they may cause an environmental threat [10, 11]. Proper waste management techniques and the safe disposal of e-waste have become a global priority to mitigate human health risks and environmental degradation caused by landfilling [12].

2 Global origin and generation of e-waste

The origin of e-waste is closely linked to global patterns of production, consumption, and disposal of electronic goods. Developed nations are the largest producers of e-waste, while developing countries are the major recipients through illegal transboundary movement of discarded electronics. Nearly 80% of e-waste from developed regions is exported to low- and middle-income countries such as India, China, Ghana, Nigeria, and Pakistan, where environmental regulations are weak and disposal is cheaper [13].

In 2019, the world generated 53.6 million tons of e-waste. Asia contributed the highest share (46.4%), followed by the Americas (24.4%), Europe (22.4%), Africa (5.4%), and Oceania (1.3%) [8]. Only about 17.4% of global e-waste was formally recycled, with the rest being burned or dumped illegally – activities that release toxic substances into the environment. While resource recovery through e-waste can provide commercial prospects, its management may be hindered by inadequate infrastructure and efficient scientific technologies [14]. As a result, e-waste management in the twenty-first century is regarded as one of the biggest global challenges, posing serious threats to the environment and human health [15–17].

2.1 The birth of e-waste

The origin of e-waste can be traced to the late twentieth century, particularly during the 1980s and 1990s, when personal computers and consumer electronics became widespread. Several key developments contributed to this:

(i) **Consumer electronics boom:** The affordability and accessibility of gadgets increased globally.

(ii) **Planned obsolescence:** Manufacturers began designing products with shorter life spans to encourage frequent replacements.

(iii) **Technological advancements:** Rapid innovation and technological advancement accelerated the introduction of new models, faster processors, and enhanced features at a pace that quickly made earlier devices and older models obsolete and outdated. As consumers embraced modern technology and shifted towards newer gadgets, older electronics were discarded more often rather than repaired or reused. Over time, this pattern of consumption gave rise to a completely new category of waste, the electronic waste (e-waste), marking one of the most significant environmental challenges of the digital age.

(iv) **Globalization and outsourcing of e-waste:** In the early 2000s, developed countries started exporting e-waste to developing nations under the pretext of donation or second-hand use. However, much of this exported equipment was nonfunctional and ended up in informal recycling sectors in Asia and Africa, especially in countries like India, China, Ghana, and Nigeria [18–20]. This outsourcing of e-waste contributed to the environmental degradation and health hazards in these regions due to unsafe recycling practices such as open burning of cables to extract copper, acid baths to retrieve gold from circuit boards, and landfilling of toxic remnants.

3 Classification of e-waste

3.1 Classification based on patterns of use, consumer demand, and product function

WEEE are sometimes informally grouped into three major categories based on patterns of use, consumer demand, and product function, and are described as white waste, brown waste, and gray waste (Figure 1). These terms are not formal legal categories, but they are commonly used in waste-management and consumer-electronics contexts to group devices by type and typical material composition [21, 22]:

- **White goods** include large household appliances such as refrigerators, washing machines, air conditioners, ovens, and freezers. These are typically bulky domestic devices used for routine household activities.
- **Gray goods** cover information and communication technology (ICT) equipment. Examples include desktop computers, laptops, CD-ROM drives, LCD monitors, keyboards, computer mice, typewriters, printers, telecommunication devices, fax machines, modems, and pagers.
- **Brown goods** consist mainly of entertainment and audio-visual equipment such as televisions, cameras, mobile phones, sound systems, projectors, and recording devices.

This classification is an industry-style grouping for e-waste used in many regions, including Europe, the UK, Australia, and India, that helps in understanding patterns of consumption, disposal behavior, and recycling needs for different types of electronic products.

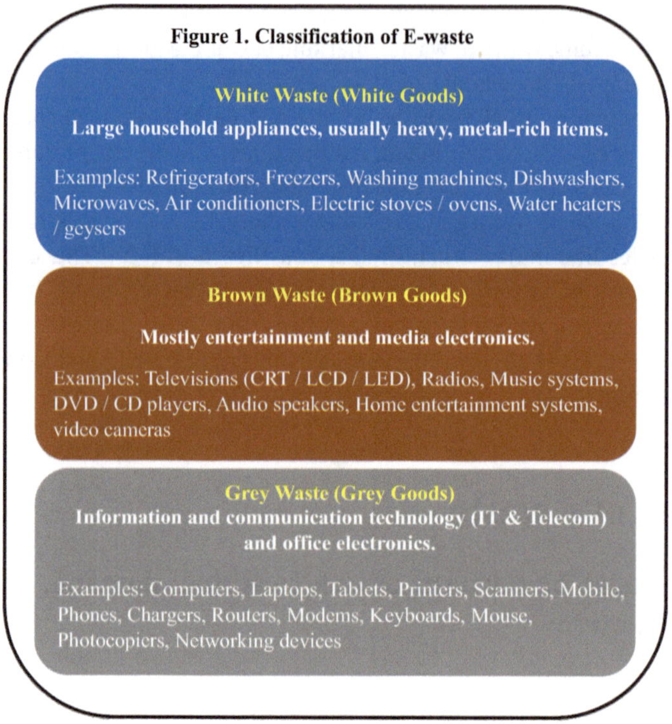

Figure 1: Classification of electronic wastes.

3.2 Classification based on type and function of equipment from which it originates

E-waste can be classified based on the type and function of equipment from which it originates. Globally, 10 main categories are recognized as given in Table 1 [23].

Large household items are the largest contributor (≈42%), followed by IT and telecommunications equipment (≈34%) and consumer electronics (≈14%). The specific composition and proportions vary across regions depending on income levels, population size, and technological penetration.

Table 1: Types of e-waste and common examples [23].

Categories	Examples
Large household appliances	Fridges and freezers, washing machines, cookers, air conditioners, dishwashers, microwaves, electric radiators, etc.
Small household appliances	Toasters, irons, clocks, vacuum cleaners, electric kettles, grinder-mixture blenders, etc.
IT and telecommunication equipment	Laptop/desktop computers, phones, calculating devices, printers and scanners, copying machines, network cables, Wi-Fi modem, communication satellite, etc.
Consumer equipment	TVs, radios, musical equipment, hi-fi equipment, camcorders, digital cameras, audio players, etc.
Lighting equipment	Fluorescent tubes (straight and compact), halogen bulbs, LED bulbs, sodium bulbs, high-intensity discharge lamps, etc.
Electrical and electronic tools	Lawn mowers, drills, sewing machines, electrical saws, etc.
Toys, leisure, and sports equipment	Game consoles, running machines, electric trains, racing cars, etc.
Medical devices	Dialysis machines, medical fridges and freezers, ultrasound machines, cardiology equipment, analyzers, respiratory ventilators, electronic thermometers, etc.
Monitoring and control instruments	Data acquisition systems, microcontrollers, sensors, valves, smoke detectors, thermostats, heating regulators, etc.
Automatic dispensers	Hot and cold drink dispensers, spray dispensers, money dispensers, etc.

4 Composition of e-waste

E-waste is a complex mixture of valuable, base, and hazardous materials. On average, it contains 40% metals, 30% plastics, and 30% oxides and other compounds [24].

4.1 Valuable and base materials

- **Precious metals:** Gold (Au), silver (Ag), palladium (Pd)
- **Inorganic elements:** Approximately 90%, confirming that a substantial portion of the WEEE particles is composed of these inorganic elements. This includes sodium (Na), calcium (Ca), potassium (K), chlorine (Cl), sulfur (S), along with nonmetallic elements such as phosphorus (P), bromine (Br), and fluorine (F)

- **Metal(oids):** Silicon (Si), iron (Fe), aluminum (Al), zinc (Zn), chromium (Cr), copper (Cu), lead (Pb)
- **Trace elements:** Cadmium (Cd), nickel (Ni), tin (Sn), and magnesium (Mg).

The distribution of these elements is consistent with their roles in the components of WEEE, such as solder joints (Pb, Sn), printed circuit boards (Si, Fe, Al), and cathode ray tubes (Pb, Cr, Ni). These materials can be recovered through urban mining, which offers a sustainable alternative to conventional mining. For instance, 1 ton of circuit boards may yield up to 1.5 kg of gold and 210 kg of copper, concentrations many times higher than in natural ores [25].

4.2 Hazardous substances

E-waste also contains toxic metals such as lead (Pb), mercury (Hg), cadmium (Cd), arsenic (As), and beryllium (Be), along with halogenated compounds (bromine, chlorine, fluorine) found in plastics and flame retardants [26]. When improperly handled – through open burning, acid leaching, or incineration – these substances can contaminate air, soil, and water, posing risks to both the environment and human health.

5 Mechanisms of toxicity and exposure

E-waste exposure occurs through multiple environmental and occupational pathways, particularly in informal recycling and disposal settings [27]:
- **Inhalation** is one of the most significant routes, as open burning, shredding, and manual dismantling generate fine particulate matter, metal fumes, and toxic gases that can penetrate deep into the respiratory tract.
- **Ingestion** may occur indirectly through the consumption of crops, fish, or drinking water contaminated with heavy metals and persistent organic pollutants that bioaccumulate in soil and aquatic systems surrounding e-waste sites.
- **Dermal contact** is another relevant pathway, especially among workers who handle circuit boards, plastics, and soldering materials without protective equipment; many e-waste chemicals can be absorbed through the skin or enter via cuts and abrasions.

Although the exact mechanisms of toxicity are still being understood, current research shows that these exposure pathways allow harmful metals and chemicals from e-waste to enter and build up in the body, leading to a range of health effects across different organs. Inhalation of particulates and metal fumes is particularly associated with respiratory inflammation, oxidative stress, and impaired lung function, while ingested contaminants may contribute to gastrointestinal absorption of toxicants, hepatic stress, and bioaccumulation in internal organs. Dermal exposure, especially among workers handling e-waste, can also increase chemical exposure through absorption and minor skin injuries. Over time, these combined effects may contribute to hormone imbalance, damage to the nervous system and cognitive development, DNA and cellular changes, and weakened immune responses.

5.1 Ecological impacts of e-waste exposure

5.1.1 Airborne and particulate pollution

Mechanical processing of WEEE, including shredding, drilling, grinding, and manual dismantling, generates large quantities of airborne particulate matter enriched with toxic metals such as Pb, Cd, Cr, Cu, and Zn. Studies report particularly high levels of respirable dust fractions (<1 µm) containing elemental carbon, metal oxides, brominated flame retardants, and halogenated compounds [28]. These ultrafine particles are capable of remaining suspended in air for extended periods and dispersing over long distances from processing sites. Due to their small aerodynamic diameter, they can be deeply inhaled and deposited in the alveolar regions of the lungs, where they may enter systemic circulation. Prolonged exposure to such emissions has been associated with respiratory irritation, oxidative stress, impaired lung function, and elevated risk of chronic respiratory and cardiovascular disorders in nearby populations and workers.

5.1.2 Soil and water contamination

Improper disposal and primitive recycling practices such as open dumping, acid leaching of circuit boards, and open burning of plastics facilitate the release and migration of hazardous chemicals into surrounding soil and groundwater systems [27, 29]. Heavy metals and persistent organic pollutants adsorb strongly to soil particles, where they remain environmentally stable for decades. Over time, these contaminants are taken up by crops and soil organisms, gradually entering terrestrial food chains. In groundwater-dependent communities, contaminated water sources pose additional risks through drinking, irrigation, and livestock exposure. The long-term accumulation of toxic substances such as Pb, Cd, and As in soils not only degrades land quality but also increases the likelihood of chronic human exposure, particularly in regions where informal e-waste recycling is located near residential and agricultural areas.

5.1.3 Aquatic and ecotoxic effects

Discharge of e-waste residues, leachates, and processing effluents into surface waters introduces a complex mixture of metals and organic pollutants into aquatic ecosystems. Toxicity assessments using model organisms such as *Daphnia magna* have demonstrated significant mortality and sub-lethal effects at elevated concentrations of WEEE-derived particles, indicating strong acute and chronic aquatic toxicity [28]. Metals, including Cd, Pb, and Zn, interfere with essential biochemical pathways, disrupt ion regulation, inhibit enzyme activity, and impair growth and reproduction in fish and invertebrates. Bioaccumulation and biomagnification further amplify these effects across trophic levels, threatening aquatic biodiversity and ecosystem stability. In communities reliant on contaminated rivers and wetlands for fishing or domestic water needs, these ecological impacts also translate into important human health concerns.

5.2 Human health impacts of e-waste exposure

E-waste exposure occurs through inhalation, ingestion, or dermal absorption of contaminated air, dust, food, and water. Informal recycling, which involves open burning and manual dismantling, is the major source of exposure [29–33]. According to the WHO (2024), millions of workers, including women and children, are engaged in these unsafe recycling practices worldwide [7].

5.2.1 Major toxicological effects

i. **Neurotoxicity:** Metals such as Pb, Hg, and polychlorinated biphenyls (PCBs) are well-established neurotoxicants that interfere with brain development and neuronal signaling, particularly in fetuses and young children. These substances disrupt synapse formation, neurotransmitter balance, and myelination – processes that are critical for learning and memory. Chronic exposure has been linked to reduced IQ, attention and behavioral disorders, and impaired motor coordination, with children in e-waste regions considered highly vulnerable.
ii. **Hepatotoxicity:** Exposure to Pb, Cd, and organic pollutants from e-waste recycling alters liver enzyme activity, particularly alanine aminotransferase and alkaline phosphatase, indicating hepatocellular injury. These contaminants promote oxidative stress, lipid peroxidation, and inflammatory responses within liver tissues. Over time, such changes may contribute to metabolic disturbances, impaired detoxification capacity, and increased susceptibility to chronic liver disease.

iii. **Nephrotoxicity:** Metals such as Cd and As accumulate in renal tissue, where they damage proximal tubules and impair filtration and reabsorption functions. Biomarkers, including elevated serum creatinine and β2-microglobulin, reflect early kidney dysfunction among exposed populations. Prolonged exposure increases the risk of chronic kidney disease, especially among workers involved in manual dismantling and metal recovery.
iv. **Endocrine disruption:** Brominated flame retardants (polybrominated diphenyl ethers), PCBs, and bisphenols act as endocrine-disrupting chemicals by mimicking or blocking natural hormone signaling pathways. They interfere particularly with thyroid function, reproductive hormones, and metabolic regulation. Disruption during critical developmental windows may influence growth, puberty timing, fertility, and neurodevelopmental outcomes.
v. **Reproductive and developmental toxicity:** Prenatal and early-life exposure to e-waste-related contaminants has been associated with adverse birth outcomes, including low birth weight and preterm delivery. In adults, metals and organic pollutants may impair reproductive function by altering hormone levels, damaging germ cells, and reducing sperm quality and motility. These effects highlight the intergenerational implications of chronic environmental exposure in e-waste communities.
vi. **Cardiovascular and hematological effects:** Heavy metals disrupt calcium signaling, vascular tone, and oxidative balance, contributing to endothelial dysfunction and increased risk of hypertension. Lead and other metals also interfere with hemoglobin synthesis and iron metabolism, leading to anemia and reduced oxygen-carrying capacity. Such effects may be compounded in populations with nutritional deficiencies or prolonged occupational exposure.
vii. **Genotoxicity and carcinogenicity:** Chromium, As, and polycyclic aromatic hydrocarbons induce DNA strand breaks, oxidative DNA lesions, and chromosomal instability. These genotoxic effects are accompanied by epigenetic alterations that may affect gene expression and cell cycle regulation. Long-term exposure is therefore associated with elevated cancer risk, particularly for lung, skin, and gastrointestinal malignancies in high-exposure environments.

5.2.2 Vulnerable populations

Children and pregnant women are especially vulnerable to the health effects of e-waste exposure because their bodies are still developing and their physiological defense mechanisms are not fully mature. Children breathe more air, drink more water, and consume more food per unit body weight than adults, which means they absorb proportionally higher levels of environmental contaminants when exposed. Toxicants such as Pb and Hg can cross the placental barrier during pregnancy and may also be transferred through breast milk, exposing infants during critical periods of brain and organ development. These early-life exposures can have lifelong consequences, including learning and behavioral difficulties, growth delays, and increased susceptibility to chronic disease later in life, making protection of these populations a major public-health priority.

6 Policy, prevention, and management

Despite its hazards, e-waste also represents a resource opportunity. Efficient recycling and recovery of valuable materials can support a circular economy [33]. However, major challenges include:
- Lack of universal e-waste legislation
- Limited recycling infrastructure
- High costs of environmentally safe technologies
- Informal and unregulated processing sectors

To address these challenges, international frameworks like the Basel Convention and the WHO–ILO E-waste and Child Health Initiative emphasize [34]:
- Adoption of national e-waste laws
- Formal recycling and safe material recovery
- Public health monitoring and worker protection
- Eco-design and extended producer responsibility

7 Conclusion

E-waste is both a resource and a risk. While it contains valuable metals, improper recycling releases toxic pollutants that threaten human health and ecosystems. Asia remains the largest generator of e-waste, yet much of it is recycled under unsafe conditions. The health burden – especially among vulnerable populations – underscores the need for stronger policy frameworks, safer technologies, and public awareness. A sustainable, circular approach integrating eco-design, green chemistry, and responsible recycling is essential for minimizing the toxic legacy of e-waste in the modern world.

References

[1] Baldé CP, Wang F, Kuehr R, Transboundary Movements of Used and Waste Electronic and Electrical Equipment. United Nations University, ViE – SCYCLE, Bonn, Germany, 2016.
[2] Mary JS, Meenambal T, Inventorisation of E-waste and developing a policy – bulk consumer perspective. Procedia Environ Sci, 2016, 35, 643–655, https://doi.org/10.1016/j.proenv.2016.07.058.

[3] Huang K, Guo J, Xu Z, Recycling of waste printed circuit boards: A review of current technologies and treatment status in China. J Hazard Mater, 2009, 164, 399–408, doi: 10.1016/j.jhazmat.2008.08.051.
[4] Chen M, Huang J, Ogunseitan OA, Zhu N, Wang Y, Comparative study on copper leaching from waste printed circuit boards by typical ionic liquid acids. Waste Manag, 2015a, 41, 142–147, https://doi.org/10.1016/j.wasman.2015.03.037.
[5] Matsakas L, Gao Q, Jansson S, Rova U, Christakopoulos P, Green conversion of municipal solid wastes into fuels and chemicals. Electron J Biotechnol, 2017, 26, 69–83, https://doi.org/10.1016/j.ejbt.2017.01.004.
[6] Bhuiya MMK, Rasul M, Khan M, Ashwath N, Mofijur M, Comparison of oil extraction between screw press and solvent (N-Hexane) extraction technique from beauty leaf (*Calophyllum inophyllum L.*) feedstock. Ind Crops Prod, 2020, 144, 112024, https://doi.org/10.1016/j.indcrop.2019.112024.
[7] World Health Organization. Electronic Waste (E-waste), 1 Oct. World Health Organization, 2024, www.who.int/news-room/fact-sheets/detail/electronic-waste-%28e-waste%29.
[8] Forti V, Baldé CP, Kuehr R, Bel G The Global E-Waste Monitor 2020: Quantities, Flows and the Circular Economy Potential. United Nations University (UNU)/United Nations Institute for Training and Research (UNITAR) – Co-Hosted SCYCLE Programme, International Telecommunication Union (ITU) & International Solid Waste Association (ISWA), Bonn/Geneva/Rotterdam, 2020.
[9] Ilankoon IMSK, Ghorbani Y, Chong MN, Herath G, Moyo T, Petersen J, E-Waste in the international context – a review of trade flows, regulations, hazards, waste management strategies and technologies for value recovery. Waste Manag, 2018, 82, 258–275, https://doi.org/10.1016/j.wasman.2018.10.018.
[10] Chen M, Huang J, Chen H, Lead during the leaching process of copper from waste printed circuit boards by five typical ionic liquid acids. J Clean Prod, 2015b, 95, 142–147.
[11] Lu C, Zhang L, Zhing Y, Ren W, Tobias M, Mu Z, Ma Z, Geng Y, Xue B, An overview of e-waste management in China. J Mater Cycles Waste Manag, 2015, 17(1), 1–12, 10.1007/S10163-014-0256-8.
[12] Song Q, Li J, A review on human health consequences of metals exposure to e-waste in China. Environ Pollut, 2015, 196, 450–461, https://doi.org/10.1016/j.envpol.2014.11.004.
[13] Sthiannopkao S, Wong MH, Handling e-waste in developed and developing countries: Initiatives, practices, and consequences. Sci Total Environ, 2013, 463–464, 1147–1153, http://dx.doi.org/10.1016/j.scitotenv.2012.06.088.
[14] Lakshmi S, Raj A, A review study of e-waste management in India. Asian J Appl Sci Technol, 2017, 1, 33–36.
[15] Bhutta MKS, Omar A, Yang X, Electronic waste: A growing concern in today's environment. Econ Res Int, 2011, 2011, 1–8, https://doi.org/10.1155/2011/474230.
[16] Moletsane R, Venter C, The 21st century learning environment tools as electronic waste, In Proceedings of The World Congress on Engineering and Computer Science. 1, 2018.
[17] Magalini F, Global challenges for e-waste management: The societal implications. Rev Environ Health, 2016, 31(1), 137–140, https://doi:10.1515/reveh-2015-0035.
[18] Needhidasan S, Samuel M, Chidambaram R, Electronic waste – an emerging threat to the environment of urban India. J Environ Health Sci Eng, 2014, 12(1), 36, doi: 10.1186/2052-336X-12-36.
[19] Nnorom IC, Osibanjo O, Overview of electronic waste (e-waste) management practices and legislations, and their poor applications in the developing countries. Resour Conserv Recycl, 2008, 52(6), 843–858, 10.1016/j.resconrec.2008.01.004.
[20] Puckett J, Byster L, Westervelt S, Gutierrez R, Davis S, Hussain A, Dutta M, Exporting Harm: The High-Tech Trashing of Asia. Basel Action Network & Silicon Valley Toxics Coalition, 2002.
[21] Arya S, Kumar S, E-waste in India at a glance: Current trends, regulations, challenges and management strategies. J Clean Prod, 2020, 271, 122707, https://doi.org/10.1016/j.jclepro.2020.122707.

[22] Saini A, Taneja A, Managing e-waste in India – a review. Int J Appl Eng Res, 2012, 7(11), 1–6.
[23] Buczkó NA, Papp M, Maróti B, Kis Z., Szentmiklósi L.. Classification of electronic waste components through x-ray and neutron-based imaging techniques. Materials 2024, 17(19), 4707.
[24] Sahajwalla V, Hossain R, The science of microrecycling: A review of selective synthesis of materials from electronic waste. Mater Today Sustain, 2020, https://doi.org/10.1016/j.mtsust.2020.100040.
[25] Bazargan A, Lam KF, McKay G, Challenges and Opportunities in E-Waste Management. Nova Science Publishers, United States, 2012, 39–66.
[26] Goyal S, Gupta S, A comprehensive review of current techniques, issues, and technological advancements in sustainable e-waste management. Adv Electr Electron Eng, 2024, 9, 100702, https://doi.org/10.1016/j.prime.2024.100702.
[27] Eze C, Vinken M, E-waste mechanisms of toxicity and safety testing. FEBS Open Bio, 2024, 14(9), 1420–1440, https://doi.org/10.1002/2211-5463.13863.
[28] Odnevall I, Brookman-Amissah M, Stábile F, Ekvall MT, Herting G, Vargas MB, Messing ME, Sturve J, Hansson LA, Isaxon C, Rissler J, Characterization and toxic potency of airborne particles formed upon waste from electrical and electronic equipment waste recycling: A case study. ACS Environ, 2023, 3(6), 370–382, https://doi.org/10.1021/acsenvironau.3c00034.
[29] Zhang T, Ruan J, Zhang B, Lu S, Gao C, Huang L, Bai X, Xie L, Gui M, Qiu RL, Heavy metals in human urine, foods and drinking water from an e-waste dismantling area: Identification of exposure sources and metal-induced health risk. Ecotoxicol Environ Saf, 2019, 169, 707–713, 10.1016/j.ecoenv.2018.10.039.
[30] Lin S, Ali MU, Zheng C, Cai Z, Wong MH, Toxic chemicals from uncontrolled e-waste recycling: Exposure, body burden. Health Impact J Hazard Mater, 2022, 426, 127792, https://doi.org/10.1016/j.jhazmat.2021.127792.
[31] Parvez SM, Jahan F, Brune MN, Gorman JF, Rahman MJ, Carpenter D, Islam Z, Rahman M, Aich N, Knibbs LD, Sly PD, Health consequences of exposure to e-waste: An updated systematic review. Lancet Planet Health, 2021, 5(12), e905–e920, https://doi.org/10.1016/S2542-5196(21)00263-1.
[32] Rehman B, Iqbal M, Nawaz I, Toxicity, eco-toxicity, and phytoremediation of e-waste, In: Hashmi M, Varma A, Eds, Electronic Waste Pollution. Soil Biology. Springer Cham, 2019, ,57 https://doi.org/10.1007/978-3-030-26615-815.
[33] Joon V, Shahrawat R, Meena K, The emerging environmental and public health problem of electronic waste in India. J Health Pollut, 2017, 7(15), 1–7, https://doi.org/10.5696/2156-9614-7.15.1.
[34] Basel Convention. The Basel Convention on the Control of Transboundary Movements of Hazardous Wastes and Their Disposal. UNEP 1989. Retrieved from: https://www.basel.int.

Pooja Bhardwaj*, Aaliya Ali, Reena Sharma

From innovation to contamination: impacts of electronic wastes on people and planet

Abstract: The fast development of the technological world has revolutionized the contemporary lifestyle, but it has also increased the creation of electronic waste at an alarming rate, which has shown detrimental effects on the population and the environment in the twenty-first century. In this chapter, the author explores the two sides of the technological advancement that is bringing incredible benefits yet leaving a legacy of dangerous wastes. E-waste has useful materials that are reusable and a broad assortment of toxic chemicals, including heavy metals, flame retardants that are brominated, and persistent organic pollutants. Unless these chemicals are properly handled, they can prove to be hazardous. This chapter explores the magnitude of e-waste production in the world, the pathways through which the contaminants enter the ecosystem, and the direct and indirect health impact of the same, such as neurotoxicity, endocrine disruption, and the development of cancer. The discussion mentions the disproportionate contribution of informal recycling to the low- and middle-income nations by using case studies of the global hotspots in e-waste. It also evaluates the effectiveness of policy responses, including the Basel Convention, extended producer responsibility (EPR) structures, and circular economy policies in minimizing risks. Connecting the trail of innovation to its less-than-notorious toxic history, this research demands quick concerted action to halt the chain of pollution and move toward a safer and more sustainable technological future.

Keywords: Electronic waste, health effects, electronic garbage, environmental contamination, e-waste governance

1 Introduction

The world technological environment has undergone unprecedented expansions in the last 40 years. The shrinking of the size of parts, growing consumer electronics, and the accelerated novelty cycles have all had an impact on human life that cannot be envisioned a generation ago. From the era of smartphones that create continuous and immediate communication around the globe to the tools of medical imaging that transform the

*Corresponding author: **Pooja Bhardwaj,** University Center for Research and Development, Chandigarh University, Mohali 140413, Punjab, India, e-mail: poojaphysics76@gmail.com
Aaliya Ali, Reena Sharma, University Center for Research and Development, Chandigarh University, Mohali 140413, Punjab

 | https://doi.org/10.1515/9783112248072-002

world of diagnostics, electronic technologies have become an inseparable part of societies in all countries of the world. But, on the one hand, in the context of such incredible developments, there is a latent and increasing crisis, the accumulation of discarded electronic products or electronic waste (e-waste), which has become one of the most rapidly increasing waste streams in the world [1]. The broad definition of e-waste is any disposed electrical or electronic device, all its parts, assemblies, and consumables used in the product during the time of disposal (Widmer et al., 2005). The duality of this shape of waste stream is the presence, on the one hand, of valuable secondary raw materials, which include gold, copper, palladium, and rare earth elements that are essential in producing new technologies [2]. Conversely, it contains dangerous chemicals such as lead, cadmium, mercury, brominated flame retardants (BFRs), and persistent organic pollutants (POPs), which, when not properly handled, can run into the environment and settle into the biological systems, causing deplorable effects on human health as well as the entire ecosystem [3]. Statistics of the world contribute to the severity of the problem. The global production of e-waste was estimated to be 53.6 million metric tons (Mt) in 2019 alone, and is expected to be more than 74 Mt as of 2030 unless the trend is reversed [1]. Various intertwined factors can explain this fast growth: the reduction of product lifespan through planned or perceived obsolescence, the growing demand by consumers in the fast-growing economies of the world, and the spread of low-cost electronic gadgets at the hands of globalized production chains (Parajuly et al., 2019). To the contrary of more traditional waste streams, e.g., municipal solid waste, e-waste management presents unique challenges due to the complex material composition, the constantly changing designs of products, and the substantial participation of the informal sector, especially in developing nations. The consequences of improperly disposed e-waste are much further than local pollution. Recycling activities like open burning of circuit-boards, washing of gold through acid use, and manual dismantling without protective measures are prevalent in most of Africa, South Asia, and Southeast Asia [4]. Through such methods, toxic fumes and leachates are emitted into the air, water, and the soil, which puts workers at risk of occupational hazards and contaminates the environment in the long run. The released pollutants may travel both via atmospheric and aquatic routes, and therefore, e-waste is a transboundary environmental hazard even to countries that do not produce a large volume of it [5]. The increasing trend in global e-waste generation over recent years is illustrated in Figure 1. Besides, e-waste is not just an environmental and public health concern, but it is also a matter of fairness and justice. Most developed nations sell their e-waste in large quantities to the low- and middle-income nations in the name of used electronics to reuse [6]. This equipment is, however, nonfunctional or at the end of life when it arrives in the recipient countries, which are charged with the responsibility of disposal and the environmental hazards. This dynamic brings about asymmetry, where the fruits of innovation get concentrated into specific regions, whereas the negative effects of discarded technology are overly borne by other folks. The fact that e-waste can negatively affect sustainable development goals (SDGs), especially the health (SDG 3), clean water and sanitation (SDG 6), sustainable cities (SDG 11), responsible consumption and production (SDG 12), and climate action (SDG 13), justifies the urgency of the issue of e-waste. The e-waste crisis is going to keep increasing in size and complexity unless systemic changes are made to production, consumption, and waste management. This chapter focuses on the complex effects of e-waste on individuals and the environment as it explores the routes of contamination in the environment, the direct and indirect health impacts of e-waste, as well as the socioeconomic and regulatory aspects of the problem. It also explains the new measures of reducing risks, such as reforming policies, innovating technology, and intervening at the community level. The chapter follows the path of technological innovation to environmental pollution to shed light upon the necessity of a concerted approach to the problem on the global level to stop the vicious cycle of toxic concentration and transition to a more environmentally friendly technological future.

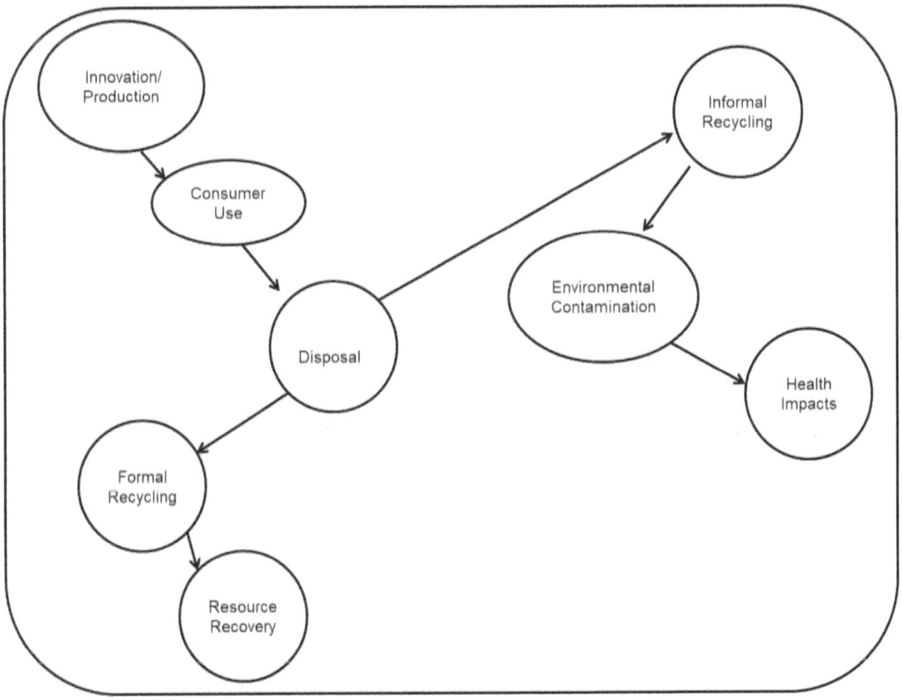

Figure 1: E-waste lifecycle flowchart.

2 Sources and composition of electronic waste

The path of innovation to pollution starts with the things we glorify as technological wonders, like smartphones, laptops, flat-screen televisions, and renewable energy components, which find their way to the e-waste stream. One must know the materials and the chemical structure of e-waste in order to set up methods of alleviating these effects.

2.1 Sources of E-Waste

Electronic waste has various sources depending on the diversity of the electronic technology in our everyday life and manufacturing processes. Generally, such sources are: (1) Consumer electronics: The largest and most visible stream of e-waste consists of mobile phones, tablets, personal computers, televisions, and home appliances, including refrigerators, washing machines, and others. The aggressive marketing and planned obsolescence, which lead to rapid product turnover, are contributing factors to the high rate of growth in the discarded devices. (2) ICT equipment: network servers, data centers, routers, and other telecommunications infrastructure are all being phased out due to the increasing demand for digital connectivity. (3) Industrial and medical equipment: diagnostic imaging and laboratory analyzers, industrial control systems, and automated machinery are also sources of a large quantity of e-waste when they are upgraded or replaced. (4) Components of the energy sector: Renewable energy technologies, including solar panels, wind turbine electronics, and battery storage units, will substantially contribute to the e-waste in the future as massive installations date out of service in the early 2000s. (5) Military and aerospace equipment: These highly specialized electronics are less likely to be discussed in the common sphere, but they also get into the waste, and they have rare, valuable, and potentially dangerous materials.

Global statistics on electronic waste generation across different regions are summarized in Table 1.

Table 1: Sources of e-waste.

Source category	Examples	Notes
Consumer electronics	Mobile phones, tablets, and PCs, TVs, refrigerators, and washing machines	Largest and fastest-growing contributor; driven by short product lifespans
ICT equipment	Network servers, routers, and data centers	Replaced frequently as digital infrastructure scales
Industrial/ medical	Diagnostic imaging machines, lab analyzers, and industrial control systems	Significant waste when upgraded or replaced
Energy sector	Solar panels, wind turbine electronics, and battery storage systems	Growing source as renewable installations reach end of life
Military/ aerospace	Radars, avionics, and defense electronics	Contain rare and hazardous materials

2.2 The role of product life cycle in e-waste generation

Electronic products have had a significantly reduced lifespan in the last 20 years. Whereas electronic devices such as mobile phones previously remained in use for five to seven years, the average replacement cycle has now reduced to less than three years. The same trends are observed in televisions, computers, and home appliances. This has not only been caused by technological advancement but also by design decisions, planned obsolescence, in which parts are hard to repair, software updates slow down computers, or replacement is being sold as a better option than fixing.

2.3 Composition of e-waste

E-waste is a complex mixture of materials, making its management challenging. Broadly, its composition includes:

- **Valuable metals** – Gold, silver, palladium, and copper are present in small but economically significant quantities, making recovery an attractive economic activity.
- **Common metals** – Aluminum, iron, and tin are used in structural components and soldering.
- **Hazardous metals** – Lead (in solder and cathode ray tubes), mercury (in switches and flat-screen backlights), cadmium (in batteries and semiconductors), and hexavalent chromium (in anticorrosion coatings) pose serious risks to human health and ecosystems.
- **Plastics** – Used in casings, insulation, and internal components, plastics in electronics often contain brominated flame retardants (BFRs), which are POPs.
- **Glass and ceramics** – Found in display screens, resistors, and insulating components.
- **Rare earth elements** – Neodymium, dysprosium, and yttrium are critical for high-performance magnets, displays, and batteries, but their recovery is technologically challenging.

Table 2: Composition of e-waste.

Material type	Examples	Environmental/health risks
Valuable metals	Gold, silver, palladium, and copper	High recycling value and economic driver
Common metals	Aluminum, iron, and tin	Structural components and solder
Hazardous metals	Lead, mercury, cadmium, and chromium(VI)	Toxic to nervous, renal, and reproductive systems; persistent pollutants
Plastics	Casings and insulation with BFRs	Release dioxins/furans on burning; persistent organic pollutants

Table 2 (continued)

Material type	Examples	Environmental/health risks
Glass and ceramics	Display glass, resistors, and insulators	Nonbiodegradable and accumulate in landfill
Rare earth elements	Neodymium, dysprosium, and yttrium	Critical for magnets and displays; difficult recovery

2.4 The hidden toxic cocktail

Although valuable metals are the main topic of most e-waste recycling, the most hazardous aspect of e-waste recycling is the toxic portion, which causes most harm to the surroundings and human health. As an illustration, inappropriate dismantling or open burning of BFR containing plastics releases dioxins and furans, whereas acid leaching to extract gold yields wastewater which contains heavy metals. These chemicals may remain in soil, water, and body tissues of living organisms decades later and accumulate damage.

2.5 The socioeconomic dimension of composition

In Ghana, India, and Nigeria, in the informal recycling industries, workers sacrifice their lives by dismantling electronics by hand, and in some cases, without protective gear. Precious metals have a high value, which motivates the recovery even in risky exposure. This is an economic motivation, coupled with a lack of proper infrastructure, that drives the cycle where communities rely on unsafe recycling to make a living, which would continue health hazards as well as environmental destruction.

The classification of electronic waste into major material categories is presented in Table 2.

3 Global landscape of e-waste generation and management

3.1 Introduction to the global landscape

The production of electronic waste (e-waste) is a very global phenomenon. Modern life generates a continuous flow of discarded equipment, with smartphones and laptops through to industrial electronics and medical devices. Contrary to other waste [26–29] streams, e-waste is the only waste stream that has been influenced by the globalized production, consumption, and disposal systems. The good side of electronics is shared globally but the disposal side of electronics is frequently rather unevenly distributed – draining disproportionately between the high and low-income countries.

The Global E-Waste Monitor 2024 [7] estimates that the global e-waste creation in 2022 was about 62 million metric tons (Mt) of e-waste, which is more than 80% higher than in 2010. It is estimated that the total volumes of waste generated in the world will grow to more than 75 Mt by 2030 due to the rapid urbanization process, dynamic development of the emerging economies, and the increasing rates of obsolescence.

3.2 Distribution of e-waste generation by regions

Production of e-waste differs radically in terms of region, as this is a difference in consumption levels, penetration of technology, and availability of waste management infrastructure. A regional comparison of per capita and total e-waste generation is presented in Table 3.

Europe boasts of the greatest per capita e-waste generation (about 17 kg per person/year) and the most advanced collection and recycling strategies.

North America produces approximately 16 kg per capita per annum, with exports from the region to Latin America and Asia in large quantities despite the regulations.

Asia is the largest absolute volume producer of around 24 Mt per year, with its huge population and growing consumer uptake of electronics.

Africa produces comparatively low amounts per head (~2.5 kg) but is disproportionately affected by imported e-waste through the informal sector.

The disparity is immense: advanced economies prefer to outsource the most dangerous side of e-waste processing to the areas that have lower environmental and occupational standards.

Table 3: Regional e-waste generation.

Region	Per capita (kg/year)	Total volume (Mt/year)	Key notes
Europe	~17	~12	High per capita generation; advanced collection and recycling systems
North America	~16	~7	High volumes; significant exports to Latin America and Asia
Asia	~5	~24	Largest total due to population growth and rising consumption
Africa	~2.5	~2.9	Low per capita but burdened with imported e-waste via informal channels

3.3 Transboundary movement of e-waste

One of the most actively traded waste streams in the world is e-waste, both on a legal and an illegal basis. In the Basel Convention on the Control of Transboundary Movements of Hazardous Waste (in force since 1992), the export of hazardous waste to countries wastes and their Disposal, without adequate treatment facilities is restricted. Nonetheless, loopholes exist – especially via the rebranding of garbage as second-hand electronics to sell.

For example:
Massive European and North American export traffic is diverted to West Africa (notably Ghana and Nigeria) and South Asia (India, Pakistan, and Bangladesh). In most instances, most of the imported second-hand electronics are hardly functional; the rest is disposed of or disposed in open landfills.

The Basel Ban Amendment of 2019 – forbids the export of hazardous waste by OECD to non-OECD countries – has contributed to the decrease of some flows but these are still not enforced in a consistent manner.

3.4 Informal and formal recycling systems

The e-waste management market in the world is defined as being in a dual form:
1. **Formal recycling** – shredding that is being done in controlled processes, by means of a machine, automated material separation, and technology of pollution control. These amenities have the potential to be very high metal recovery of metals such as copper, gold, and palladium, and reduction of toxicity releases. There are state-of-the-art plants in Japan, Switzerland, and Belgium.
2. **Informal recycling** – usually small, unregulated recycling processes based on manual recycling, breaking down, destroying by fire, and leaching with acid to recover valuable parts. Prejudiced in such countries as Ghana (Agbogbloshie), China (Guiyu), and India (Seelampur, Moradabad). Informal recycling presents a threat to the local communities, although it is economically significant to them. Immediate dangers to health due to direct contact with lead, mercury, cadmium, and POPs.

The world records a very low percentage of e-waste generated as collected, and only 22%, which is formally recorded and recycled [7]. The remaining is either informally attended to, kept at home, or thrown away like city garbage.

3.5 Case examples and e-waste hotspots

Some places have turned into a symbol of the issue of e-waste in the world:
- **Guiyu, China** – Guiyu was once called the e-waste capital of the world, and it took huge quantities. Quantities of foreign waste prior to the 2010s government crackdown, polluted soil and water, and high levels of blood lead in children were widely documented.
- **Agbogbloshie, Ghana** – A large African recycling center of imported electronics. In this case, young employees deassemble the equipment using crude rudimentary tools and incinerate wires to salvage copper, with clouds of deadly smoke.
- **Moradabad, India** – It is a company that specializes in recovering gold in printed circuit boards by use of acid baths, which are commonly held in unprotected settings in the open air with little protective measures.

These hot spots show the meeting point between poverty, environmental degradation, and globalized waste flows.

3.6 New trends in e-waste management

The global e-waste environment is facing a number of new dynamics:
- **Reduced product life cycles** – Smartphones are now using 2–3 years of life before replacement, which was 5–6 years ten years ago.
- **Technological convergence** – More devices have common critical metals, which is rising recovery potential and also complexity of waste.
- **Relocation of geographies of recycling** – Since China has limited imports of waste, imports are flowing to the southeast of Asia (Thailand, Vietnam, Malaysia) and to Africa.
- **Extended producer responsibility (EPR)** – EPR is becoming more popular across the globe. Policies are proposing that the manufacturers are responsible for the end-of-life management of their products.

3.7 Global governance problems

Although there are international accords, the reality is that there are still a number of challenges that are impeding effective global e-waste management:

Regulatory gaps – Not every country is a signatory of the Basel Convention; some countries have weak enforcement.

Data gap – Due to a significant number of countries not having credible statistics on e-waste generation, collection, and treatment.

Economic motivation – Recycling is informal as it provides livelihoods in the regions and a low level of employment.

Technical obstacles – Recycling plants are expensive, high-technology ones, and require a constant amount of waste in order to be profitable.

3.8 Summary

E-waste processes are characterized by uneven flows, uneven burdens, and uneven distribution of e-waste throughout the world, and fragmented governance. As the rich nations absorb most of the electronics and produce most waste per capita, low- and middle-income countries tend to incur both the environmental and health costs of mismanaged recycling and disposal. It needs to correct this unfairness not only by technological innovation but also by enhanced policy congruence, international collaboration, and sustainability community alternatives to informal recycling.

4 E-waste environmental effects

The environmental impact of electronic waste is much more than localized dumping sites, and this has multi-scalar effects on the soils, water bodies, air quality, and biodiversity. As compared to general solid waste, e-waste has a distinct mixture of valuable materials (e.g., gold, copper, and palladium) and toxic substances (e.g., lead, mercury, cadmium, brominated flame retardants, and polychlorinated biphenyls) that, when improperly disposed of, end up finding their way into the ecological system in different ways. Informal recycling is also common in developing economies, and frequently in the form of crude dismantling, open burning, and leaching with acid, forming hotspots of contamination that can continue to be felt decades later. The major sources and classification of electronic waste are summarized in Figure 2.

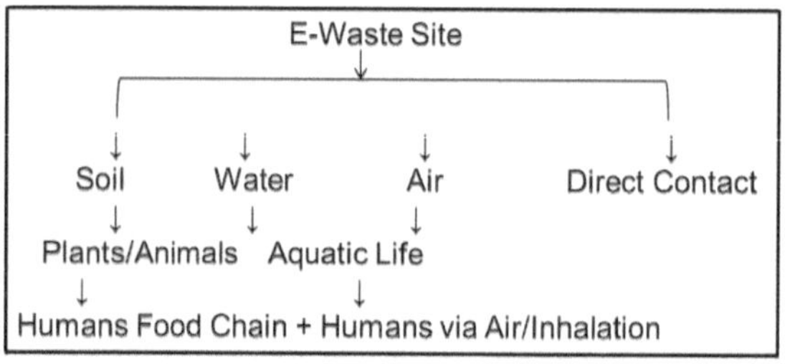

Figure 2: Pathways of contamination.

4.1 Soil contamination and heavy metal persistence

Rainfall infiltration and chemical weathering of surrounding soils release heavy metals when e-waste is ripped off and disposed of on the unlined ground. In e-waste recycling locations (e.g., Guiyu, China, and Agbogbloshie, Ghana), soils have been reported to contain very high levels of lead, up to 1,000 mg/kg, which is much higher than the safety levels in agricultural land:

- Lead (Pb) is strongly bound to the soil particles, and therefore, it is a persistent pollutant that is available to plants under acidic soils.
- Cadmium (Cd) is more mobile and leaches to lower levels of soil, causing pollution of groundwater.
- The Cr^{6+}, recently introduced as a corrosion protection compound, is toxic and has high solubility, thus becoming very difficult to remain in the water table.

This pollution affects the microbial communities in soil, diminishing the enzymatic activity and changing the nutrient cycling. The effects in the long term are that soils lose their fertility, and this would impact crop production decades after active recycling has stopped.

4.2 Water pollution and water toxicity

The leaching activities that are not properly done release effluents that are rich in heavy metals, acids, and POPs into the surface water systems. These contaminants get into the estuarine environments in the coastal recycling centers and are bioaccumulated in fish and shellfish as food consumed by the local population:

- Bioaccumulation of mercury and cadmium occurs in the benthic organisms and then magnifies in the trophic levels. Brominated flame retardants and PCBs are still present in the sediments, and they serve as long-term storage sites of pollution [31].

Research in the Pearl River Delta has indicated that there are increased levels of polybrominated diphenyl ethers (PBDEs) in sediment and living organisms, which are associated with increased levels of developmental toxicity in aquatic organisms. Water pH can also be changed by acidic effluents, affecting the survival of delicate species and the dissolution of heavy metals.

4.3 Thermal process: air pollution

Insulated wires combusted to extract copper emit polycyclic aromatic hydrocarbons (PAHs), dioxins, and furans – airborne pollutants that have respiratory and carcinogenic impacts that are severe. Plastics with brominated flame retardants are open burnt to create brominated dioxins and furans, which are more toxic than their chlorinated counterparts:

- The sizes of particulate matter ($PM_{2.5}$) that are produced due to the combustion of residues may travel hundreds of kilometers and affect the quality of the air in the region.
- Flying metals, including arsenic and lead, re-enter the soil and water surfaces, forming a feedback mechanism of contamination.

The level of atmospheric dioxin in certain e-waste centers has been reported to reach 100 times above the background level in urban areas and has caused a chronic inhalation hazard to the workers and people living around the e-waste centers.

4.4 Ecotoxicological and biodiversity effects

The health of the ecosystems is affected by the e-waste-induced contamination:

Terrestrial plants were also found to exhibit lower biomass, chlorosis, and root length inhibition in contaminated soils, which is a result of metal toxicity [37–45].

The aquatic ecosystems are characterized by a lack of species richness and changes in dominant organisms, which mostly prefer organisms that tolerate the pollution.

Top predators, such as fish-eating birds, build up large amounts of mercury and PCBs, resulting in reproductive disabilities and a decline in population.

Due to lipophilicity and persistence of most e-waste pollutants, they move across geographic borders via migratory species and ocean currents, thus expanding their ecological footprint well beyond point sources.

4.5 Long-term and legacy environmental footprint

Contrary to organic pollutants, the heavy metals contained in e-waste are not degradable but change between the environmental compartments (soil, water, sediment, biota) across decades. Although there is site remediation, the contaminants tend to rebound to deep sediments or groundwater. Such a long-term persistence presents a necessity to employ preventive measures, since remediation technologies can be expensive, slow, and ineffective only to a certain extent.

To conclude, the ecological effects of e-waste concern not only waste treatment zones but are also incorporated into the ecological cycles, impacting biodiversity, and bringing about long-term and transnational pollution threats. The solution to these issues involves reducing the sources by designing devices with innovative approaches and ensuring the waste management practices are strictly controlled to contain the level of pollutant emission of the e-waste in all stages of the e-waste life cycle.

5 Public health impacts of e-waste

E-waste behaves with human biology in several ways of exposure, such as the inhalation of polluted dust and smoke, dermal contact during dismantling, and ingestion of contaminated food and water. Although all people may be affected, informal recyclers, their families, and communities around them face the greatest dangers. To appreciate the importance of public health, it is necessary to interconnect the conditions of the workplace, family, and social determinants at large (poverty, housing, and access to healthcare) to understand their vulnerability and outcomes.

5.1 Who is exposed? The social patterning of risk

The geography of exposure mirrors global inequalities. High-income countries consume – and retire – the most devices per capita, but a substantial share of hazardous end-of-life processing occurs in **low- and middle-income countries** with limited regulatory capacity. Within those countries, informal recycling hubs frequently draw **young workers and migrants**, attracted by the relatively higher income from copper, gold, and aluminum recovery. Children and pregnant people living nearby face **secondary exposure** through contaminated air and household dust, and through food cultivated on polluted soils. In many hotspots, work and residence overlap; circuit boards and cables are dismantled in courtyards or living rooms, collapsing the boundary between occupational and domestic environments. The pathways through which e-waste affects human health and vulnerable populations are summarized in Figure 3.

5.2 Exposure profiles: what gets into the body?

E-waste releases a complex mixture of **metals** (lead, cadmium, mercury, arsenic, and nickel), **POPs** (PBDEs, PCBs, and dioxins/furans), and **volatile/semi-volatile organics** (PAHs and solvents). Biomonitoring studies of workers and residents near recycling sites consistently document:

- **Elevated blood lead** and **urinary cadmium** among dismantlers and cable burners.
- **Higher serum PBDE and PCB levels** in adults and children living near e-waste hubs compared with control populations.
- **Dioxin congeners** associated with open burning are found in breast milk and serum in communities adjacent to dump sites.

These biomarkers often correlate with environmental measurements (soil, dust, sediment), reinforcing the causal chain from activity → contamination → human uptake.

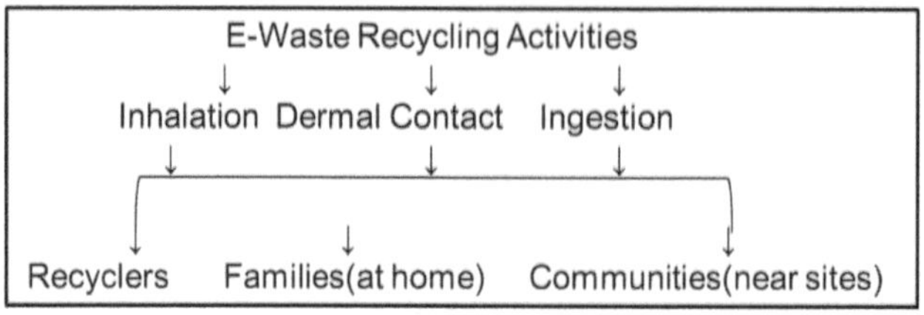

Figure 3: Exposure pathways (recyclers, families, and communities).

5.3 Health effects across the life course

5.3.1 Early life: fetal and child development

Early development is especially sensitive to toxicants. Prenatal exposure to **lead and mercury** disrupts neuronal migration, synaptogenesis, and myelination. Infants and young children in e-waste communities exhibit:

- **Reduced cognitive performance** and attention problems associated with low-to-moderate blood lead levels that would once have been dismissed as "background."
- **Altered thyroid hormones** linked to PBDE exposure, with implications for neurodevelopment and growth.
- Increased frequency of **respiratory symptoms** (persistent cough, wheeze) in neighborhoods affected by open burning and dusty dismantling yards.

Because children have **higher hand-to-mouth behavior** and higher intake per body weight, they ingest more contaminated dust and soil, compounding risk [46–52].

5.3.2 Adolescents and adults: occupational morbidity

Among workers, both **chemical** and **physical** hazards drive morbidity. Common findings include:

- **Hematologic changes** (anemia, altered hemoglobin synthesis) and **renal stress** (elevated urinary β2-microglobulin, creatinine changes) from chronic metal exposure.
- **Hepatic enzyme elevations** (ALT/AST) associated with mixed solvent and POP exposures during dismantling and crude solvent extraction.
- **Endocrine disruption**: PBDEs and other POPs are linked to altered thyroid function, reproductive hormones, and metabolic dysregulation.

- **Respiratory impairment**: Reduced pulmonary function tests among cable burners and shredder operators exposed to $PM_{2.5}$ and combustion products [33].
- **Injuries and musculoskeletal strain**: Lacerations, burns, eye injuries, and chronic back/shoulder pain arising from repetitive manual tasks and poor ergonomics.

5.3.3 Reproductive health

Emerging evidence associates e-waste exposures with **adverse reproductive outcomes**:
- In women, **subfertility** and altered menstrual cycles have been reported alongside endocrine disruption.
- In men, **reduced semen quality** (motility, morphology) and hormone perturbations correlate with elevated metals and POPs.
- **Adverse birth outcomes** (low birth weight, preterm delivery) have been observed in populations with high exposure metrics, though confounding social factors (nutrition, stress, co-pollutants) must be carefully addressed in study design.

5.3.4 Chronic disease and cancer: what we know (and don't)?

Certain PAHs and dioxin-like compounds, arsenic, cadmium, as well as some e-waste pollutants, are considered human carcinogens or probable carcinogens. Oxidative stress and DNA damage biomarkers (e.g., 8-OHdG, frequency of micronuclei) are always elevated in exposed workers. Nonetheless, there are a few longitudinal cohort studies that have associated exposure with site-specific cancer. This is a research gap of critical importance: numerous communities are highly mobile, have poor cancer registries, and are mixed, which makes it harder to draw inferences.

Table 4: Public health impacts.

Group/life stage	Key health effects
Children and infants	Cognitive decline (lead and mercury), thyroid disruption (PBDEs), respiratory illness, and low birth weight
Adolescents and adults	Anemia, renal stress, reduced lung function, endocrine disruption, and injuries from unsafe recycling
Reproductive health	Subfertility, altered hormones, reduced semen quality, and adverse pregnancy outcomes
Chronic disease and cancer	Elevated risk of cancer (arsenic, cadmium, PAHs, and dioxins); oxidative stress; DNA damage

5.4 Mixtures, modifiers, and the real-world exposure problem

E-waste is not a single chemical exposure – it is a **mixture problem**. Metals and POPs can interact additively or synergistically on shared biological pathways (oxidative stress, endocrine signaling, inflammation). **Nutritional status** (e.g., iron or calcium deficiency) modifies metal absorption; **co-exposures** (ambient air pollution, indoor biomass smoke) amplify respiratory risks; and **psychosocial stress** can potentiate neurodevelopmental impacts. Effective public health assessment must therefore shift from single-chemical risk factors to **cumulative risk frameworks** that integrate chemical and nonchemical stressors.

The major health outcomes associated with exposure to electronic waste across different population groups are summarized in Table 4.

5.5 Pathways to protection: evidence-based interventions

Although a change in the structural policies is necessary (addressed later in the chapter), practical evidence of near-term risk reduction in high-exposure settings is available:

1. **Engineering and process controls** in small enterprises: localized fume extraction, wet methods for dust suppression, enclosed shredders, acid-resistant containment with neutralization systems, and safe wastewater management.
2. **Personal protective equipment (PPE)** programs coupled with **behavioral training**: consistent glove/respirator use, dedicated work clothing, and hand-washing stations reduce dermal and ingestion pathways.
3. **Household exposure reduction**: separating work and living spaces, routine wet cleaning to lower dust, and community education on safe food preparation and storage.
4. **Health surveillance and biomonitoring**: periodic screening for blood lead, urinary cadmium, and selected POPs; integrating surveillance with **referral and treatment** (iron supplementation, chelation where clinically indicated).
5. **Child-focused protections**: banning child labor at recycling sites, creating safe childcare away from work areas, and targeted nutritional programs (iron, calcium) that reduce metal uptake.

Evidence from pilot programs shows that **multi-component interventions** – combining engineering, training, and clinical services – deliver the largest exposure reductions and are more sustainable than PPE alone.

5.6 Ethics, justice, and community participation

Ethics and justice cannot be disconnected from the public health practice in e-waste settings. Societies that are the least advantaged by the digital economy frequently assume the biggest health costs of its rubbish. In order to engage in ethics, it is necessary:

Participatory risk assessment, which incorporates the knowledge and priorities of the workers:

- Livelihood-transitions: formalization, cooperatives, and buy-back schemes, which enhance safety but do not end income.
- Clear risk information in spoken and written forms in local languages and literacy levels was developed with local leaders and health workers.

5.7 What is next on the agenda of policymakers and practitioners

To health authorities and NGOs, three priorities have been identified:

1. Construct longitudinal cohorts to measure the long-term outcomes and assess interventions.
2. Standardize biomarkers and environmental measures so that comparison across sites can occur, and meta-analysis can be done.
3. Integrate health services in e-waste management plans – primary care, reproductive health, and mental health facilities among the affected communities.

6 International response to e-waste and e-waste governance

6.1 Introduction

The management of electronic waste is a complicated issue, as well as a complicated matter of governance, besides being an environmental and public health issue. The electronic production, consumption, and disposal are global, so any solutions have to

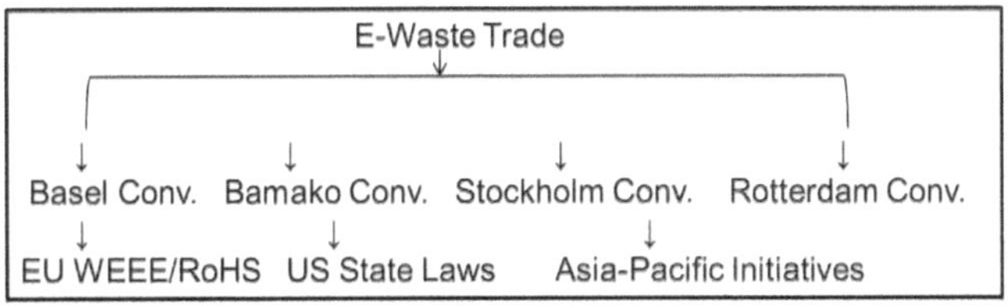

Figure 4: Global governance network.

be provided on the international, national, and local levels. Regulatory frameworks and multilateral agreements have also developed over the past twenty years to control e-waste, although the impact of their action differs greatly with the capacity of enforcement, the availability of infrastructure, and political motivation.

It is in agreement with international agreements and conventions.

6.2 International agreements and conventions

6.2.1 The Basel Convention (1989)

The best-known international treaty that controls the movement of hazardous wastes, including some of the e-wastes, is the Basel Convention on the Control of Transboundary Movements of Hazardous Wastes and their Disposal. It will strive to minimize the production of hazardous wastes and limit its export to developed countries. Nonetheless, illegal shipments continue to be prevalent, in part because of the lack of clarity of the categories of used electronics and because of the inadequate monitoring of ports.

6.2.2 The Bamako Convention (1991)

The Bamako Convention was meant to be adopted by African countries, and it outlaws the importation of hazardous wastes such as e-waste in Africa. This was embraced as a local reaction to perceived flaws in the Basel Convention.

6.2.3 The Stockholm Convention (2001)

Even though the main target of the Stockholm Convention is the POP compounds, the treaty applies to e-waste regulation since a number of electronic components in electronic devices contain POPs like brominated flame retardants, which are subject to the treaty [35].

6.2.4 The Rotterdam Convention (1998)

This agreement regulates the global commerce of dangerous chemical compounds and pesticides. Although not e-waste-specific, it facilitates the control of the toxic substances that may occur in electronics [35].

The interconnections among international treaties, regional regulations, and national policies governing e-waste are illustrated in Figure 4.

6.3 National and regional regulatory approaches

6.3.1 European Union: The WEEE and RoHS Directives

- The **Waste Electrical and Electronic Equipment (WEEE) Directive** establishes collection, recycling, and recovery targets for e-waste.
- The **Restriction of Hazardous Substances (RoHS) Directive** limits the use of hazardous substances in electronics, including lead, mercury, cadmium, and certain brominated flame retardants. Together, these directives create a producer responsibility system in which manufacturers bear part of the cost of e-waste management.

6.3.2 United States: State-Level Regulations

The US lacks a federal e-waste law, but several states, including California, New York, and Washington, have adopted Extended Producer Responsibility (EPR) programs or landfill bans for electronics.

6.3.3 Asia-Pacific Initiatives

China implemented its "Regulation on the Administration of the Recovery and Disposal of Waste Electrical and Electronic Products" in 2011, establishing a fund for re-

Table 5: Policy frameworks and governance.

Policy/region	Scope/focus	Effectiveness/challenges
Basel Convention (1989)	Restricts hazardous waste exports from developed → developing countries	Loopholes (mislabeling as "used electronics"), weak enforcement
Bamako Convention (1991)	Prohibits hazardous waste imports into Africa	Region-specific; enforcement varies
Stockholm Convention (2001)	Controls POPs (e.g., brominated flame retardants in electronics)	Relevant but limited to certain pollutants
EU WEEE Directive	Establishes collection, recycling, recovery targets	Strongest global framework; effective EPR enforcement
EU RoHS Directive	Restricts hazardous substances (Pb, Hg, Cd, BFRs) in electronics	Widely implemented; highly effective
US State Laws	State-level EPR and landfill bans (e.g., California, New York, and Washington)	No federal law; fragmented policies
Asia-Pacific Initiatives	China, Japan, India: national e-waste laws and take-back systems	Implementation uneven; informal recycling still dominant

cycling and prohibiting imports of certain e-waste streams. Japan has its *Home Appliance Recycling Law*, requiring manufacturers to take back and recycle designated products. India's E-Waste (Management) Rules (2016, amended 2022) mandate EPR compliance and encourage formal recycling channels.

Key international and regional policy frameworks governing electronic waste management are compared in Table 5.

6.4 Extended producer responsibility (EPR) and circular economy principles

EPR shifts the responsibility for postconsumer product management from municipalities to producers, creating incentives for eco-design and material recovery. EPR policies have proven effective in the EU and Japan but face enforcement challenges in countries with large informal recycling sectors. Integrating **circular economy** strategies – such as design for disassembly, modularity, and material substitution – can further reduce the environmental footprint of electronics.

6.5 Global e-waste governance problems

1. **Gaps of enforcement** – Weak monitoring and corruption are facilitating the continued existence of illegal shipments.
2. **Domination of informal sectors** – The informal recycling is readily available in most developing countries when compared to formal recycling.
3. **Technological lag** – A number of countries do not have the sophisticated recycling systems that can safely manage the streams of e-waste, which are complex.
4. **Information gap** – Data missing and discrepancies in reporting and classification do not assist in the global policy planning.

6.6 Future directions

In order to enhance governance, the experts suggest:
- Standardization of definitions and reporting of e-waste around the world.
- Increasing the formal recycling capacity of developing nations.
- Using digital tracking of supply chains (e.g., blockchain) to enhance transparency.
- Motivating product life cycle extension with repairability scoring and right-to-repair legislation.

7 E-waste management eco-friendly strategies

To address the challenge of the journey between innovation and contamination, technology, policy, markets, and community practice need to align in such a way that they use products more, repair them more readily, track them in a more transparent way, and recover them more safely [53–60]. Here, we combine scalable solutions, such as design and business models, recovery technologies, and inclusive transitions, that create a sustainable, equitable, and circular e-waste system.

7.1 Design to go circular: upstream prevention of waste

- **Design decisions entrap downstream destinies:** Design provides an extension of usable life and ease of end-of-life (EoL) recovery beginning at the drawing board.
- **Modularity and repairability:** The availability of standardized fasteners, replaceable batteries, and subassemblies minimizes early obsolescence and decreases the overall impact of the lifecycle by delaying the replacement and allowing refurbishment. Manufacturers are encouraged to include such features in their efforts by public policies like repairability scoring and the right-to-repair norms [8, 9].
- **Selection and replacement of materials:** By phasing out the use of hazardous additives (such as some brominated flame retardants) and using problem-free adhesives that can be easily broken down, mechanical recycling can be dramatically improved [10].
- **Design for disassembly:** Properly labeled materials (ishi resin name codes), disassembly instructions, and reduced material variety add value to the recovery of high-value fractions such as Cu, Au, Pd [63].
- **Design for remanufacture:** Strong housing and standard interfaces enable component harvesting and product-as-a-service (PaaS) models, in which ownership is maintained by the producer, which has a direct incentive to recuperate materials in an effective manner [9, 11].

7.2 Circular business models and market instruments

The method of selling and financing electronics can be as final as the method of designing them:

- **Product-as-a-service (leasing/subscription):** Keeping the property of assets (e.g., enterprise laptops, networking equipment) maintains the re-use value, helps to ease the collection process, and facilitates second-life markets [9].

- **Buy-back and trade-in programs:** Avoidable residual and convenient drop-off locations enhance take-back rates and reduce informal leakage.
- **Deposit-refund ystems:** Purchase refund at small deposit returned at purchase enhances the collecting of peripherals and small devices – a strategy which has significant backing in similar streams.
- **Green public procurement (GPP):** Repairability, recycled content, and verified take-back can be demanded by large institutional buyers, changing the market baselines [8].

7.3 Digital Infrastructure: traceability, passports, and metrics

Electronics circulate on complicated chains; information renders circularity quantifiable:

- **Digital product passports (DPPs)**: New EU strategies are discussed to introduce the encoding of product composition, repair records, and EoL instructions to promote safe handling and reasonable valuation.
- High-value fractions (PCBs, Li-ion batteries) can be tracked and traced using blockchain/track-and-trace to help reduce theft/leakage, as well as validate compliance, particularly across borders.
- **Standardized metrics and LCA:** Harmonized indicators (collection rate, material recovery rate, recycled content, repairability index) allow the comparability of programs and lead to constant improvement [8].

7.4 Recovery technologies, safe, and high yield

There is no single technology that can suit all fractions; hybrid flowsheets are as value and safety maximizing as possible.

- Preprocessing. Meanwhile, Metals and polymers are concentrated by manual depollution (removal of batteries, screen lamps), then shredding and advanced physical separation (magnetic, eddy current, optical) [12].
- Pyrometallurgy. The precious metals are extracted efficiently in high-grade fractions by smelting/refining and need scale and capital; high-quality facilities incorporate off-gas cleaning and slag management (Hagel, 2006; 13).
- Hydrometallurgy. Cu, Au, and Pd of PCBs with lower energy intensity are extracted selectively by acid/alkaline leaching, with precipitation/solvent extraction effluent treatment and reagent recycling being essential [13, 14].
- Bio-hydrometallurgy (bioleaching). Mobilization of metals in low-grade fines is done by microorganisms (e.g., Acidithiobacillus) to provide a lower-temperature complement to leaching; kinetics and impurity control are also an active R&D focus [15].

- Plastics & BFRs. Polymers can be recovered by clean stream segregation, solvent-based purification, and controlled thermal treatment; in situations with high levels of POP, it may be necessary to destroy them [10].

7.5 Lithium-ion batteries (LIBs) and emerging streams

The fastest-growing segment, which is the electronics and EVs LIBs, requires specialized systems:
- **Second life, then recycle:** The packs of state-of-health diagnostics can be triaged to be reused (stationary storage) before the material recovery.
- **Direct cathode recycling (relithiation, healing)** maintains structure and can reduce energy consumption compared to elemental recovery; pilots based on maturing NMC and LFP chemistries [>16].
- **Safety and collection:** Logistics that are fire-safe (UN tests, no short-circuit) and those that are paid by producers are necessary to prevent household waste streams [>17].

In addition to LIBs, PV panels (glass, Ag, Si) and wind electronics/magnets (Nd, Dy) would also need specific recovery paths; now proactive planning will curb future surges [>18].

7.6 Inclusive formalization: collaborating with the informal sector

The informal economy is the backbone of a number of regions. Sidelining it would mean failure and unfair play:
- **Cooperatives and micro-franchising:** The channeling of collectors/dismantlers into cooperatives with training, purchasing personal protective equipment, and purchase agreements feedstock to the compliant facilities and enhances incomes [>19,20].
- **Assured offtake and reasonable prices:** Sorted fraction pricing eliminates burning/acid leaching incentives.
- Social safeguards: Formalization projects can be coupled to child-labor prohibitions, available childcare, and basic health services (ILO guidance; >20).
- **Urban mining hubs:** The interface between informality and compliance is found in municipal-level projects like materials recovery parks that incorporate microenterprises with a collective management of the environment [>19].

7.7 Financing the transition

Long-term circularity requires consistent cash flows and the division of risks:

- **EPR 2.0:** Better design is encouraged with modulated prices (smaller with high-repairability/recycled material); eco-modulation is now being seen as an upstream change tool [>8,11].
- **Blended finance:** First-of-a-kind facilities in emerging markets can be de-risked using concessional loans and guarantees.
- **Results-based payments:** Remunerated achieved results (kg recycled safely, exposure reduction), which are consistent with the results of the DPP-enabled measurement.

7.8 Behavior, culture, and demand-side levers

Technology and policies are effective when individuals are able to take action:

- **Consumer nudges:** The trade-in of default devices at the point of sale, home pick-up, and visible deposit balances helps eliminate friction.
- **Repair ecosystems:** Repair is normalized through Maker spaces, certified repair networks, and open manuals (where IP allows).
- **Education and labeling:** Icons (repairability, recycled content, take-back) at purchase are simple and have an impact on decisions and literacy development.

7.9 Continuous improvement governance

A sustainable system is not a one-time construction – it is a learning and adapting structure:

- **Data feedback loops:** The enforcement and investment are led by public displays of collection/recycling rates, exposure biomarkers (where ethically gathered), and incident reports.
- **Adaptive standards:** New design and recycling requirements are regularly revised to keep pace with changing chemistries (e.g., new flame retardants, new battery chemistries).
- **International cooperation:** A harmonization of definitions, testing procedures, and customs nomenclatures minimizes the gaps and makes it possible to trade in secondary raw materials in a traceable and legal manner, which supplements formal facilities.

8 Future outlook and sustainable solutions to e-waste management

The increasing e-waste crisis is in need of novel, scalable, and sustainable methods to reduce the amount of environmental degradation and defend the health of the people. Although contemporary policies and the present-day recycling practices have improved, they are usually not in line with technological changes. This part presents prospective strategies that incorporate technological innovation, the concept of a circular economy, community involvement, and international collaboration.

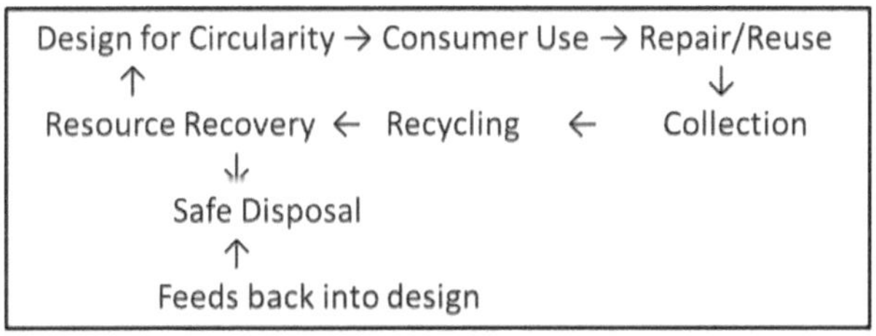

Figure 5: Circular economy approach.

8.1 Shifting to a circular economy

The notion of a circular economy stresses that materials should be in utilization as long as possible, derive the greatest value prior to recovery and regeneration [21]. This involves designing products for e-waste management in a manner that will ensure that the product is durable, modular, and repairable so that the rate of product disposal is reduced. A crucial role can be played by extended producer responsibility (EPR) as it involves causing manufacturers to be responsible for the end-of-life management of their products [2].

Key actions include:

- Design toward disassembly: Components should be designed in a way that can be disassembled easily to be reused or recycled.
- Innovative material: Biodegradable or readily recyclable electronics.
- Remanufacturing and refurbishing markets: Promoting new uses of used electronics that are their second life.

An overview of the overall framework discussed in this section is presented in Figure 5.

8.2 Green technology progress

E-waste can be recycled with a lot of technological innovation. Other, less energy-demanding and less toxic methods, such as hydrometallurgical and biohydrometallurgical processes, are also less toxic than traditional smelting [22]. The sorting and dismantling of e-waste is being more automated, AI, and robotized to enhance the recovery of useful materials such as gold, palladium, and rare earth elements [1].

Innovations in development are:

- **Bioleaching:** Recycling circuit board metals with the help of microorganisms.
- **Low-temperature processing plasma:** Reducing the number of hazardous by-products in the separation of materials.
- **Keeping track of the e-waste with blockchain:** Stakeholders should be assured of transparent flows of the waste.

8.3 Ensuring intensification of the informal sector

E-waste collection and dismantling in most of the developing nations occur in the informal sector [23]. The policies should formalize and train informal workers instead of replacing them, and this would equip informal workers with protective equipment and safe processing technologies [19]. P2P can fill the divide between formal and informal recyclers and make the process safer and more efficient.

8.4 Consumer awareness and change of behavior

Consumers play a very important role in the chain of e-waste. Responsible disposal can be encouraged by awareness, product return incentives, and educational programs. The participation rates can also be increased by digital platforms informing about collection centers or buy-back programs that can be found nearby.

8.5 International cooperation and unification of standards

With the transboundary nature of e-waste, it is necessary to have harmonized international regulations. Dumping in vulnerable areas can be decreased by strengthening the Basel Convention and pushing the governments toward more stringent measures in the process of illegal waste shipments [24]. Joint studies and information dissemination systems can be used to trace e-waste flows across the world and hot spots.

8.6 Future policy recommendation

- Global extended producer responsibility framework: A framework to guarantee the application of EPR internationally.
- Innovation funding: Investment in the research and development of sustainable recycling methods.
- Waste cutting aims: Establishing legally binding e-waste generation limits at the individual level.

Green procurement: Green procurement should be encouraged by governments and institutions to buy electronics that have proven to be recyclable and have low impacts on the environment.

8.7 Long-term vision

Green design, sophisticated recycling, involvement of consumers, and alignment with global policies can bring sustainable e-waste in the future. With the adoption of the principles of the circular economy and by utilizing innovation, society will be able to transform the existing linear approach of the take-make-dispose model to a regenerative system in which the future development of technology will no longer be at the cost of the health of the planet.

9 Conclusion and policy recommendations

9.1 Conclusion

The transformation of innovation into contamination by spreading electronic waste is one of the most urgent environmental issues of the twenty-first century. Although technological progress has enhanced the level of connectivity, productivity, and quality of life, it has also produced an unparalleled amount of e-waste, including potentially toxic elements (lead, mercury, cadmium, and brominated flame retardants) and toxicants (Balde et al., 2020).

The effects of e-waste have a very long distance beyond the landfill sites. Poor disposal leads to soil and water pollution, air pollution, and bio-accumulation of toxins in the food chain. Informal recyclers are exposed to occupational risks, such as respiratory infection, neurological and skin damage [4]. In addition, the extraction of raw materials used in the production of electronics only reduces natural resources and carbon emissions in the world.

Nevertheless, e-waste is not only a waste issue but also a wastage of an economy. There are precious resources like gold, silver, copper, and rare earth elements that are still underutilized because the recycling mechanisms are ineffective. The challenge, hence, is to turn this liability into a stream of resources using sustainable management methods.

9.2 Policy recommendations

1. **Strengthen extended producer responsibility (EPR) enforcement**
   Governments should mandate manufacturers to finance and manage the collection, recycling, and safe disposal of their products. EPR programs must be transparent, measurable, and globally standardized [8].
2. **Promote eco-design standards**
   International regulations should incentivize designs that enhance durability, modularity, and repairability while phasing out toxic components [21].
3. **Formalize and train the informal sector**
   Rather than displacing informal workers, governments should integrate them into formal recycling systems, providing safe technologies, protective gear, and fair compensation [23].
4. **Encourage consumer participation**
   Public awareness campaigns, deposit-refund schemes, and take-back programs can significantly increase the volume of electronics returned for proper disposal (Saphores et al., 2012).
5. **Foster technological innovation in recycling**
   Investments should be directed toward hydrometallurgical, bioleaching, and automated sorting technologies to improve recovery rates and reduce environmental impact [22].
6. **Harmonize international regulations**
   Given the global nature of e-waste trade, countries should align their laws with the Basel Convention and establish shared tracking systems to prevent illegal dumping [24].
7. **Implement circular economy policies**
   Shift from a linear “take-make-dispose” model to a circular approach where resources are reused, remanufactured, and recycled, minimizing waste generation at the source [25].

9.3 Final outlook

E-waste management is a multidimensional challenge that demands cooperation among policymakers, industry leaders, scientists, and the public. Without decisive action, the growing volume of discarded electronics will continue to threaten ecosystems and public health. Conversely, through strategic policies, innovation, and global collaboration, the world can shift toward a sustainable technological future where innovation no longer breeds contamination but fosters regeneration.

References

[1] Forti V, Baldé CP, Kuehr R, Bel G, The Global E-Waste Monitor 2020. United Nations University (UNU), International Telecommunication Union (ITU) & International Solid Waste Association (ISWA), Bonn/Geneva/Rotterdam, 2020, 120.
[2] Baldé CP, Forti V, Gray V, Kuehr R, Stegmann P, The Global E-Waste Monitor. United Nations University (UNU), International Telecommunication Union (ITU) & International Solid Waste Association (ISWA), Bonn/Geneva/Vienna, 2017, 1–109.
[3] Robinson BH, E-waste: An assessment of global production and environmental impacts. Sci Total Environ, 2009, 408, 183–191.
[4] Grant K, Goldizen FC, Sly PD, Brune MN, Neira M, Health consequences of exposure to e-waste: A systematic review. Lancet Glob Health, 2013, 1, e350–e361.
[5] Sepúlveda A, Schluep M, Renaud FG, Streicher M, Kuehr R, A review of hazardous substances in e-waste and their environmental fate. Environ Impact Assess Rev, 2010, 30, 28–41.
[6] Lepawsky J, The changing geography of global e-waste. Geogr J, 2015, 181, 147–159.
[7] Baldé CP, Forti V, Gray V, Kuehr R, Stegmann P, The Global E-Waste Monitor. United Nations University (UNU), International Telecommunication Union (ITU) & International Solid Waste Association (ISWA), Bonn/Geneva/Vienna, 2024, 1–109.
[8] Organisation for Economic Co-operation and Development (OECD), Extended Producer Responsibility: A Guidance Manual. OECD Publishing, 2016.
[9] Ellen MacArthur Foundation, Towards the Circular Economy. Ellen MacArthur Foundation, 2013.
[10] Turner A, Filella M, Hazardous metallic additives in waste plastics: Toxicity and risks. Environ Int, 2017, 94, 315–325.
[11] Geyer R, Blass VD, The economics of extended producer responsibility. Environ Plann A, 2010, 42, 675–693.
[12] Cucchiella F, D'Adamo I, Koh SL, Rosa P, Recycling of WEEE: Environmental and economic analysis. Renew Sustain Energy Rev, 2015, 51, 263–272.
[13] Khaliq A, Rhamdhani MA, Brooks G, Masood S, Metal extraction processes from e-waste: A review. Resources, 2014, 3, 152–179.
[14] Cui J, Zhang L, Metallurgical recovery of metals from e-waste. J Hazard Mater, 2008, 158, 228–256.
[15] Park J, Meng F, Lin C, Biological recovery of metals from e-waste via bioleaching. Chemosphere, 2017, 168, 720–731.
[16] Harper G, Sommerville R, Kendrick E, Driscoll L, Slater P, Recycling lithium-ion batteries: Opportunities, challenges, and prospects. Nature, 2019, 575, 75–86.
[17] Gaines L, Lithium-ion battery recycling and recovery of materials. Sustainable Mater Technol, 2018, 17, e00057.
[18] International Renewable Energy Agency (IRENA), International Energy Agency Photovoltaic Power Systems Programme (IEA-PVPS), End-of-Life Management of Solar Photovoltaic Panels. International Renewable Energy Agency (IRENA), 2016.
[19] Schluep M, Programme des Nations Unies pour l'environnement, Recycling, From E-Waste to Resources. UNEP DTIE; StEP Initiative, 2009.

[20] Lundgren K, The Global Impact of E-Waste: Addressing the Challenge. International Labour Organization, 2012.
[21] Geissdoerfer M, Savaget P, Bocken NMP, Hultink EJ, The circular economy paradigm. J Clean Prod, 2017, 143, 757–768.
[22] Ilyas S, Rashid M, Bioleaching of precious metals from e-waste. Waste Manage, 2020, 106, 17–28.
[23] Chi X, Streicher-Porte M, Wang MYL, Reuter MA, Informal e-waste recycling in China. Waste Manage, 2011, 31, 731–742.
[24] Secretariat BC, Basel Convention on the Control of Transboundary Movements of Hazardous Wastes and Their Disposal. United Nations Environment Programme (UNEP), 2019.
[25] Bocken NMP, De Pauw I, Bakker C, Product design and business models for the circular economy. J Ind Prod Eng, 2016, 33, 308–320.
[26] World Health Organization (WHO), Children and Digital Dumpsites: E-Waste Exposure and Child Health. World Health Organization (WHO), 2021.
[27] Eggen T, Moeder M, Arukwe A, Ecotoxicological effects of pollutants from e-waste. Environ Sci Technol, 2010, 44, 8094–8099.
[28] Xu X, Yang H, Chen A, Zhou Y, Wu K, Birth outcomes related to informal e-waste recycling. Reprod Toxicol, 2012, 33, 94–98.
[29] Awasthi AK, Zeng X, Li J, Environmental pollution of e-waste recycling in India: A critical review. Environ Int, 2016, 89–90, 220–242.
[30] European Commission, Directive 2011/65/EU on the restriction of the use of certain hazardous substances in electrical and electronic equipment (RoHS). Off J Eur Union, 2011, 174, 88–110.
[31] Zeng X, Duan H, Wang F, Li J, E-waste recycling impacts in China: A meta-analysis. J Environ Prot, 2016, 7, 311–321.
[32] European Commission, Directive 2012/19/EU on waste electrical and electronic equipment (WEEE). Off J Eur Union, 2012, 197, 38–71.
[33] Ongondo FO, Williams ID, Cherrett TJ, How are WEEE managed around the world?. Waste Manage, 2011, 31, 714–730.
[34] Secretariat of the Stockholm Convention, Stockholm Convention on Persistent Organic Pollutants. United Nations Environment Programme (UNEP), 2019.
[35] Brigden K, Labunska I, Santillo D, Allsopp M, Chemical Contamination at E-Waste Recycling Sites in China and India. Greenpeace International, 2005.
[36] Frazzoli C, Orisakwe OE, Dragone R, Mantovani A, Diagnostic health risk assessment of electronic waste on the general population. Environ Impact Assess Rev, 2010, 30, 388–399.
[37] Zheng G, Xu X, Li B, Huo X, Association between lung function and heavy metals in children at an e-waste site. J Expo Sci Environ Epidemiol, 2013, 23, 67–72.
[38] Manomaivibool P, Extended producer responsibility in Asia: The Indian perspective. Resour Conserv Recycl, 2009, 53, 655–662.
[39] Frazzoli C, Mantovani A, Dragone R, Health risks from electronic waste in developing countries. Environ Impact Assess Rev, 2010, 30, 388–399.
[40] Leung AOW, Duzgoren-Aydin NS, Cheung KC, Wong MH, Heavy metals in human scalp hair at an e-waste site. Environ Sci Technol, 2008, 42, 2674–2680.
[41] Li Y, Xu X, Liu J, Wu K, Guo Y, Hazardous substances and health risks from e-waste exposure. Waste Manage, 2019, 100, 719–729.
[42] Grant R, Oteng-Ababio M, Ayele JM, E-waste flows in Africa: Mapping and socioeconomic implications. Geo J, 2019, 84, 299–311.
[43] Liu J, Xu X, Wu K, Huang J, Guo Y, Lead exposure and temperament in preschool children from Guiyu. Neurotoxicology, 2011, 32, 479–486.
[44] United Nations Environment Programme (UNEP), A New Circular Vision for Electronics: Time for a Global Reboot. United Nations Environment Programme (UNEP), 2019.

[45] Akormedi M, Asampong E, Fobil JN, Working conditions and environmental exposures at Agbogbloshie e-waste recycling site. Int J Occup Environ Health, 2013, 19, 278–286.
[46] Wang DG, Yang M, Qi H, Polybrominated diphenyl ethers in sediments from e-waste regions. Chemosphere, 2011, 84, 70–78.
[47] Nnorom IC, Osibanjo O, Overview of electronic waste management in developing countries. Resour Conserv Recycl, 2008, 52, 843–858.
[48] Wong MH, Wu SC, Deng WJ, Yu XZ, Luo Q, Export of toxic chemicals from e-waste recycling. Chemosphere, 2007, 67, 31–40.
[49] Leung AOW, Cai ZW, Wong MH, Polybrominated dibenzo-p-dioxins/furans at e-waste sites: Distribution and exposure. Environ Sci Technol, 2007, 41, 2721–2727.
[50] Zheng L, Wu K, Li Y, Qi Z, Han D, Blood lead and cadmium levels in children from an e-waste recycling town. Environ Res, 2008, 108, 15–20.
[51] Babayemi JO, Nnorom IC, Osibanjo O, Weber R, POPs and hazardous additives in African e-waste plastics. Environ Sci Pollut Res, 2017, 24, 18131–18145.
[52] Heacock M, Kelly CB, Suk WA, E-waste and harm to vulnerable populations. Environ Health Perspect, 2016, 124, 550–555.
[53] Lepawsky J, Reassembling Rubbish: Worlding Electronic Waste. MIT Press, 2018.
[54] Mandal P, Impact of heavy metals on public health and ecosystem. Indian J Med Res, 2019, 149, 783–785.
[55] Song Q, Li J, Environmental effects and human health impact of e-waste recycling. Environ Pollut, 2014, 187, 201–210.
[56] Luo C, Liu C, Wang Y, Heavy metal contamination and health risk in soils around e-waste sites. J Hazard Mater, 2011, 186, 481–490.
[57] Feldt T, Fobil JN, Wittsiepe J, Wilhelm M, Polycyclic aromatic hydrocarbon exposure among e-waste recycling workers. Sci Total Environ, 2014, 466, 369–374.
[58] Li J, Zeng X, Stevels A, Site-specific risks from heavy metals at e-waste recycling areas. Environ Pollut, 2015, 207, 178–189.
[59] Basel Convention Secretariat, Basel Convention: Texts and Annexes. United Nations Environment Programme (UNEP), 2011.
[60] Puckett J, Smith T, Exporting Harm: The High-Tech Trashing of Asia. Basel Action Network (BAN), 2002.
[61] Widmer, R., Oswald-Krapf, H., Sinha-Khetriwal, D., Schnellmann, M., Böni, H., Global perspectives on e-waste. Environ Impact Assess Rev, 2005, 25, 436–458.
[62] Parajuly, K., Kuehr, R., Awasthi, A. K., Fitzpatrick, C., Lepawsky, J., Smith, E., Widmer, R., Zeng, X., Future e-waste scenarios. StEP Initiative, UNU ViE-SCYCLE, UNEP IETC, Bonn/Osaka, 2019.
[63] Hagelüken, C., Improving metal returns and eco-efficiency in electronics recycling – a holistic approach for interface optimisation between pre-processing and integrated metals smelting and refining. Proc. Eur. Metall. Conf. (EMC), Düsseldorf, Germany, 2006.
[64] Baldé, C. P., Forti, V., Gray, V., Kuehr, R., Stegmann, P., *The Global E-waste Monitor 2020: Quantities, flows and the circular economy potential*. UNU/ITU/ISWA, 2020.

John Tsado Mathew*, Abel Inobeme, Elijah Yanda Shaba, Jonathan Hussaini, Yakubu Azeh, Musah Monday, Tanko Ezekiel, Etsuyankpa Bini Muhammad, Adam Ibn Aliyu, Muhammad Aishetu Ibrahim, Ismaila Aliyu Oga, Mohammed Chado Isah, Aliyu Mohammed Sakpe, Jibrin Noah Akoji, Charles Oluwaseun Adetunji, Simon Imoisi, Amos Mamman

Electronic waste pollution a global challenge: effects on public health and environmental safety

Abstract: Electronic waste (e-waste) pollution has emerged as a critical global problem, driven by rapid technological change, short product cycles, and inadequate recycling regimes. E-waste contains toxic metals, persistent organic pollutants, and hazardous chemicals that have devastating effects on environmental security and public health. Improper disposal and informal recycling dump toxins into the soil, water, and air, causing bioaccumulation, ecosystem degradation, and increased incidence of respiratory, neurological, and reproductive diseases among exposed communities. Resolving the crisis requires sustainable management practices, cleaner recycling technologies, strict regulatory measures, and global cooperation to avert environmental damage and safeguard human health.

Keywords: Electronic waste, bioaccumulation, public health, pollution, remediation

***Corresponding author: John Tsado Mathew**, Department of Chemistry, Ibrahim Badamasi Babangida University, Lapai, Niger State, Nigeria, e-mail: johntsadom@gmail.com
Abel Inobeme, Department of Chemistry, Edo State University, Uzairue, PMB 04, Auchi, Edo State, Nigeria
Elijah Yanda Shaba, Tanko Ezekiel, Adam Ibn Aliyu Department of Chemistry, Federal University of Technology Minna, Nigeria
Jonathan Hussaini, Yakubu Azeh, Musah Monday, Muhammad Aishetu Ibrahim, Amos Mamman, Department of Chemistry, Ibrahim Badamasi Babangida University, Lapai, Niger State Nigeria
Etsuyankpa Bini Muhammad, Ismaila Aliyu Oga, Department of Chemistry, Federal University of Lafia, Nasarawa State
Mohammed Chado Isah, Department of Biology, Ibrahim Badamasi Babangida University, Lapai, Niger State, Nigeria
Aliyu Mohammed Sakpe, Department of Basic Sciences, Niger State College of Agriculture, Mokwa, Nigeria
Jibrin Noah Akoji, Department of Chemistry, Baze University, Abuja, Nigeria
Charles Oluwaseun Adetunji, Applied Microbiology, Biotechnology and Nanotechnology Laboratory, Department of Microbiology, Edo State University Uzairue, PMB 04, Auchi, Edo State, Nigeria
Simon Imoisi, Edo State University, Iyamho, Auchi, Nigeria

 | https://doi.org/10.1515/9783112248072-003

1 Introduction

Electronic waste, often abbreviated to e-waste, has emerged as one of the fastest-growing waste streams worldwide. With the spread of technology and consumer electronics becoming more and more ubiquitous, electronic equipment is being disposed of at a quicker and quicker pace than ever before. Millions of tons of e-waste are produced annually across the globe, and a very small portion of the e-waste is taken to be collected and recycled in an environmentally friendly way. The trend is highly problematic to environmental sustainability and human well-being in that most of the electronic gadgets include lead, mercury, cadmium, and flame-retardant brominated materials, which are toxic substances. This knowledge of e-waste pollution origins, routes, and effects is, thus, essential to the creation of effective prevention and management approaches [1]. The steep rise in the amount of e-waste indicates several overlapping forces: high-technological development, compression of product cycles, and the development of a large consumer base throughout the world. The gadgets, which could take 10 years to wear out, are now outdated or even damaged in a few years. Besides, with the increase in technological advancement of regions, more electronics are placed in households, which eventually results in waste production increasing by a margin. To make matters worse, recycling facilities, particularly in low-/middle-income nations, tend to be far behind the pace at which e-waste is generated, which leads to solutions like unsafe disposal methods and informal recycling [2]. This is not just a waste volume problem because this increasing mountain of discarded electronics poses a severe danger to both environmental safety and the health of the people. Electronic equipment is a complicated composition of materials, such as precious metals, plastics, flame retardants, and extremely toxic ingredients. In case of improper treatment of e-waste, during informal recycling, open dumping, or uncontrolled burning, these toxic substances may leak into the environment, polluting soil, water, and air [3].

The health effect of such environmental pollution is far reaching. The recycling of e-waste is usually practiced in informal systems, especially in developing countries and is done in unsafe conditions. Employees (women, children) could take apart devices manually or burn parts to extract the precious metals, which resulted in exposure to neurotoxic substances. Such exposures may cause severe negative health consequences such as brain hazard, respiratory disease, and developmental issues in children. Moreover, e-waste toxic metals can flow into groundwater due to incorrect disposal and release hazardous materials into drinking water and food security systems of the neighboring populations [4]. In terms of resources, the inability to recycle e-waste in an adequate way is therefore a wasted opportunity. The precious metals in the discarded electronics could be reused and mined, thus minimizing the mining of virgin resources. Yet, the failure of the systems to recycle and the low formal collection rates can lead to the loss of such materials or the reuse in an inefficient manner, most of the time at significant environmental cost. Economic potential of recovered e-waste, which is managed appropriately, may play an important role in terms of material sustainability and saving of resources [5]. The second disturbing pattern is the growing disparity between the production and recycling of e-waste. E-waste has been on the increase, yet there are very low formal recycling rates being recorded. It is estimated that e-waste all over the globe may assume unparalleled dimensions within the years, although there is a very low chance that a significant percentage of this waste will be properly recycled in the future, as this tendency persists. The inadequate recycling system and the absence of enforcement tools escalate the level of environmental pollution and health hazards, especially in the territories where the informal sectors are predominant in the e-waste processing [2].

E-waste is a major socioeconomic and governing problem too. Most areas do not have effective regulatory systems, infrastructure, and technical know-how to handle e-waste in a safe way. Informal recycling industries are more successful under this environment and in many cases have no labor precautions or environmental regulations. Moreover, e-waste also makes the problem complex. E-waste transactions between countries with high and low incomes, involving large-scale sales, both legal and illegal, generate environmental injustice and the exploitation of economically disadvantaged populations [6]. The resolution of the crisis of e-waste thus needs to be a complex, interdisciplinary issue. However, building better infrastructure in the field of collection and recycling is not sufficient; the real causes of e-waste such as consumer behavior, product design, and consumption behavior should also be addressed. The amount of e-waste produced can be minimized by encouraging the product life cycle, repair and reuse, as well as by designing electronics in a way that can be recycled. Simultaneously, it is important to develop efficient and safe formal recycling infrastructure in nations that have ineffective infrastructure. The end-of-life management of the products should be held accountable by the manufacturers through the strengthening of policy mechanisms, e.g., the establishment of extended producer responsibility (EPR) [7].

The growing need to address the problem of e-waste can also be explained by the fact that this approach is in line with universal sustainability. During the production and consumption of goods, waste generation should be reduced and the safe and sustainable practice of recycling must be encouraged. The increasing stream of e-waste is posing a threat to the progress made in the workplace regarding environmental protection, human health, and conservation of resources without effective intervention [2]. E-waste pollution is an urgent worldwide problem that is associated with complicated ecological and health, economic, and governance issues. The magnitude of the issue is such that tens of millions of tonnes are produced every year but the problem is not always properly managed, especially in the prone areas. E-waste has a toxic legacy that endangers the ecosystems, human health, and the security of the resources [8]. It is precisely a multi-stakeholder approach, involving product design, regulation, and infrastructure, as well as behavior change, that can only start to tackle the risks and gain the opportunities presented by a circular electronic economy. This chapter presents various sources, routes, ecological effects, and health impacts of e-waste on the community, providing the basic framework of the concept of sustainable management strategy [9].

2 Sources and types of electronic waste

E-waste has a wide scope of discarded electrical and electronic gadgets that have their origins in various sources in households, industries, and institutions. With the improvement of technology and the availability of devices at lower prices and in general, the production of e-waste has increased manifold. E-waste has a diverse range of sources and types, as they document the many available electronic gadgets in society today. These sources need to be understood to come up with strategies of good management and alleviation of the environmental and health consequences of e-waste disposal [6].

Households and consumers are the major producers of e-waste. Household e-waste is made up of small-scale electronics, large appliances, and entertainment electronics. Mobile phones, tablets, laptops, digital cameras, and MP3 players are examples of small devices that are often substituted since they become technologically obsolete, damaged, or upgraded too quickly. Such devices usually include dangerous elements like the use of lithium-ion batteries which on improper disposal, may release poisonous chemicals. Big-box home appliances (refrigerators, washing machines, microwaves, and air conditioners) are other major contributors to e-waste [6]. Most of these gadgets have refrigerants, heavy metals, and plastics, which are difficult to handle without risks. Also, electronics including televisions, DVD players, audio and game consoles are usually disposed and they frequently find their way to informal disposal or recycling. These substances being piled up in homes and their eventual disposal are not only indicative of the rise in the consumption of electronic products but also the speed of disposing of these products due to the fast-paced modern lifestyles [4].

The other important category of e-waste origin is industrial and institutional sources. Workplaces and offices produce a large amount of e-waste, which is in the form of computers, printers, photocopiers, fax machines, and servers. These devices, which are commonly scrapped off in bulk during technological upgrades, include circuit boards, heavy metals, and plastic parts, which are a risk to the environment when thrown away. Likewise, laboratory and medical facilities help in generating e-waste in the form of diagnostic machines, laboratory equipment, and hospital monitors. Most of these machines have delicate materials, poisonous metals, and other dangerous materials that must be handled well. When factory control systems, sensors, and power tools expire in the course of their working life, industrial machinery, as well, falls under the category of e-waste stream. In contrast to household electronics, industrial e-waste is frequently composed of more concentrated amounts of harmful materials, which put the household in more hazardous danger of polluting the environment in case of improper management [8]. In addition to the e-waste source, it is necessary to understand what materials and components it is composed of, as they will define what risks the disposal and recycling process will have. E-waste normally contains metals, plastic, and hazardous chemicals. Metals may be categorized as base metals, precious metals, and rare earth elements. The use of base metals like copper and aluminum is highly prevalent in the circuit boards, casing, and wiring. Items like gold, silver, and palladium which are precious metals occur in lesser amounts in connectors, circuit boards, and semiconductor devices; hence, they are economically beneficial to recover. Rare earth elements are very important in modern electronics as magnets and batteries, and thousands of high-tech devices rely on them, but only with sophisticated technologies can they be obtained from waste materials [8].

The e-waste consists of hazardous components (toxic metals and chemical substances). Lead, which is present in cathode ray tube (CRT) monitors and in soldered parts, is very toxic and leads to neurological and developmental disorders. Mercury, which is a potent neurotoxin, is used in switches and fluorescent lamps and can be bioaccumulated in water bodies. Cadmium, found in rechargeable batteries and some semiconductors, is cancer causing and may affect the kidneys and skeletal system. Brominated flame retardants (BFRs) that have been frequently added to plastics are persistent organic pollutants, which are resistant to degradation and may affect endocrine functions. These dangerous elements render e-waste disposal a high-priority issue related to the environment and health in particular within the territories where informal recycling or open dumping is a widespread practice [10]. E-waste is also composed of a large percentage of plastics and polymers; in most cases, e-waste consists of more than half the weight of electronics that are discarded. The plastics that are often impregnated with flame retardants are hard to destroy and release toxic elements when burnt or released into soil and water. The mixture of plastics, metals, and the dangerous chemicals that are present in electronic devices is another feature that highlights the complexity of e-waste and contributes to the difficulties in dealing with it [11].

E-waste occurs in various sources such as households, industries, and institutions, which is an indication of the wide application of electronic gadgets within our contemporary society. E-waste at a household level consists of small devices, large appliances, and entertainment electronics, and the industrial and institutional sources include office electronics, laboratory equipment, and industrial machinery. E-waste is also highly varied and consists of metals, plastics, and even hazardous chemicals (lead, mercury, cadmium, and flame retardants containing brominated phospholes). The origin and characteristics of e-waste are essential in ensuring that the environmental and health issues associated with e-waste are addressed. The proper management of e-waste implementation measures should consequently be receptive to the source and structure of e-waste, striving to diminish environmental pollution, recycle usable assets, and secure human health [6].

3 Pathways of e-waste pollution

E-waste pollution can be defined as the release of hazardous substances by discarded electrical and electronic devices into the environment through different channels. E-waste can cause harm to soil, water, and air due to the complexity of modern electronics that contain metals, plastics, flame retardants, and other chemicals and cause massive impacts on the ecosystem and human health. It is crucial to understand the routes by which e-waste contaminants find their way to the environment in order to develop sustainable management techniques, reduce the negative effects, and safeguard the well-being of the population [4]. Landfill disposal, as well as open dumping, is one of the main sources of e-waste pollution. E-waste is often disposed of in unregulated landfills or open dumps within several countries and mostly in the developing regions. When the electronics are disposed of without proper containment, the toxic contents will seep into the soil and groundwater. Some of these metals include lead, cadmium, and mercury, which are very mobile and can survive in the environment, polluting water streams and depositing in plants and animals. Plastics, which comprise a major percentage of electronic waste, degrade slowly and, in many instances, contain flame-retardant substances that leech toxic substances into the soil. In particular, open dumping exposes communities to the above-mentioned hazards: the fact is that in this way, pollution can leak into adjacent water bodies or agricultural lands, with the risk of contaminating food and the health of humans [2].

Informal recycling practices are another key route and this is typical of the low and middle-income countries. On the other hand, informal recyclers are craftsmen dismantling electronics by hand in order to extract precious metals and without protective suits or available safe working conditions. This is done through a process of breaking, cutting, or melting parts to obtain copper, gold, or aluminum. In the course of such operations, toxic substances like lead, cadmium, and flame-retardant brominated materials are being emitted into the environment. Also, as a result of the burning of electronic components to release metals, toxic fumes such as dioxins and furans are produced, which contaminate the air and also expose the workers and the surrounding populations to respiratory hazards. The informal recycling routes are especially alarming to be taken as combining several contamination routes such as soil, water, and air the e-waste effects on health and the environment are disproportionately high [3]. Another important pathway of concern is the airborne emissions due to the incineration or uncontrolled burning of e-waste. At most informal recycling plants, they do their separation of metals and plastic casings with the help of open-air burning. The result of this combustion is the release of air pollutants in the form of fine particulate matter, heavy metals, and persistent organic pollutants. These emissions in the air can travel far away, and they arrive on the soil, plants, and water surfaces miles away from the location of burning. Excessive inhalation of these pollutants has been linked to respiratory diseases, cardiovascular and neurological defects among populations that fall victim to it. Toxic substances dispersed via air also raise the chance of contamination in urban and peri-urban regions, which impact masses of people and generate compound environmental hazards [6].

Another pathway through which e-waste pollution finds its way is through leaching into water systems. The e-waste dumps or unregulated landfills may trigger rainwater and leach the cumbersome metals and other contaminants into the groundwater and surface water. The downstream water bodies near dumping sites tend to have high levels of lead, mercury, and cadmium, which can be concentrated in fish and other water animals. This pollution is not only dangerous to the aquatic ecosystems but also to the human population who will be using this water to drink, cook, and irrigate their lands. Moreover, BFRs and other long-lasting chemicals not only find their way to sediments but also impact aquatic biodiversity and may bioaccumulate their way into the human food chain [12, 13].

Another important route of e-waste pollution is soil contamination. Discarded electronics contain metals and other toxic compounds that bind to particles in the soil and have long-lasting periods of existence and also alter the chemical and physical characteristics of the soil. Polluted soil influences the existence of the microorganisms, decreases the soil fertility, and may adversely influence the agricultural production. In other areas, food crops planted on soils that have been contaminated with e-waste have been reported to accumulate heavy metals, posing a direct threat to human health once ingested in the food. Soil pollution may also serve as a secondary source of pollution since the metals may be released into the water eventually or as dust into the air [14]. Another indirect route that causes pollution of the environment is the transboundary movement of e-waste. A considerable percentage of e-waste produced in high-income countries finds its way into the low and middle-income countries, where there may be no or poor enforcement of regulatory legislation. This is the waste that is exported and usually ends up in informal recycling centers and here the above-discussed routes of pathways; airborne emissions, soil contamination, and water pollution happen in large quantities. This unlawful or inadequately controlled process of e-waste transfer, therefore, is facilitating the expansion of the environmental footprint of pollution internationally, posing environmental and health hazards [15].

There are also cumulative and synergistic pathways since e-waste contaminant usually interacts with other systems of the environment. Indicatively, metals and persistent organic pollutants may be deposited in the sediments and vegetation, resulting in the bioaccumulation of animals and human beings. Soil and water contamination may also be a cause of airborne contamination, which subsequently impacts agricultural products and local food systems. All these interconnections increase the dangers of e-waste and make the clean-up process more difficult, which means that it is necessary to have combined management measures that tackle the various environmental compartments at the same time [16]. Pathways in cities and rural regions are of particular interest in densely populated regions. Minor cases of insensitive disposal of electronic products, including the dumping of mobile phones, batteries, and other electronics in the city's garbage pits, cause pollution in the urban garbage. When these wastes are collected and filled in the land without sorting, the possibilities of leaching and unchecked emission further increase. Likewise, informal recycling on the streets, which in most instances is practiced in and around residential neighborhoods, exposes the local residents, including children to pollution in the air and the soil [6].

Moreover, e-waste pollution propagates in various interrelated channels, such as landfill, informal recycling, uncontrolled burning, leaching into waterways, soil pollution, and transboundary movement. Through these routes, dangerous chemicals, including heavy metals, BFRs, and other toxic chemicals, find their way to environmental compartments and the human food chain. Contamination of air and soil is usually related to contamination of water, which produces the compounding and lasting environmental and health consequences [2]. Knowledge of these (Table 1) pathways is critical in the creation of effective management measures, better regulatory systems, and safeguarding the health of the population. The mitigation strategies need to focus on all the pathways together, waste should be safely disposed of, formal recycling implemented, monitoring of the pollution, and awareness campaigns of the population to minimize the impact of e-waste on the environment and protect the ecosystem and human health [17].

Table 1: Pathways of e-waste pollution.

Pollution pathway	Description	Key environmental/health impacts	References
Soil contamination	Leaching of heavy metals (Pb, Cd, and Hg), flame retardants, and chemicals from dumped or dismantled e-waste into soil layers.	Reduced soil fertility, uptake of toxins by crops, bioaccumulation in food chain, neurological and developmental health risks.	[18, 19]
Water pollution	Runoff from dump sites and improper recycling leads to contamination of surface and groundwater with toxic substances.	Drinking water contamination, aquatic toxicity, kidney damage, endocrine disruption.	[20]
Air pollution	Burning of wires, plastics, and circuit boards releases toxic fumes such as dioxins, furans, and particulate matter.	Respiratory problems, cancer risks, climate impact, long-range atmospheric transport of persistent pollutants.	[21]

Table 1 (continued)

Pollution pathway	Description	Key environmental/health impacts	References
Food chain contamination	Toxins from soil and water enter crops, livestock, and fish consumed by humans.	Chronic exposure leading to organ toxicity, reproductive damage, and bio-magnification of heavy metals.	[22]
Occupational exposure	Informal recyclers inhale fumes or directly contact hazardous substances during dismantling.	Skin diseases, lung disorders, heavy-metal poisoning, increased risk of congenital disabilities.	[4]
Household exposure	Improper storage or dismantling inside homes releases toxins indoors.	Indoor air contamination, risks to children (developmental and neurological issues).	[23]

4 Environmental impact of e-waste

E-waste constitutes a big environmental risk as it is not easily decomposed, given that it is composed of metals, plastics, flame retardants, among other threats. These materials when mishandled can have negative impacts on the soil, water, air, and ecosystems, which in the long run could cause environmental degradation. E-waste has a complex environmental impact, which includes pollution, interference with the natural processes, and endangers biodiversity (Table 2) [3]. One of the most widespread and immediate effects of e-waste is soil contamination. The dumped electronic equipment is usually deposited in landfills or in open dumps and the heavy metals and other harmful elements trickle into the soil over time. The metals that are quite worrying include lead, cadmium, and mercury due to their long-term existence and their capacity to adhere to soil particles. The chemicals and biological compounds change the growth and chemical characteristics of the soil, which influence nutrient cycling and the microorganisms' activity necessary to sustain soil fertility. The agricultural areas might have e-waste pollutants in the soil that may decrease the crop yield and result in similar effects in accumulating the heavy metals in the edible plants, which is a threat in itself to human and animal health. Moreover, the gradual decay of the plastics in e-waste adds to the microplastic contamination in soils, again interfering with the soil structure and microbial communities [8].

Another serious impact of the inappropriate e-waste disposal is water pollution. The release of landfills/dumpsite of metals and hazardous chemicals to the groundwater and water bodies pollutes drinking water sources and aquatic life. Poisonous metals like lead, cadmium, and mercury, and organic toxicants like BFR may be deposited along sediments and bio-accumulated in aquatic organisms. The bioaccumulation is hazardous to humans who eat fish or use the polluted water to drink and irrigate their farms. E-waste contamination of water also affects the aquatic ecosystems, wherein there is disruption to the reproductive processes, growth, and survival of fish, amphibians, and other aquatic organisms. Excessive exposure to these pollutants may cause environmental imbalance in the freshwater and sealands in the long run [24].

The major source of air pollution also occurs through the informal or uncontrolled burning of electronic components to help retrieve valuable metals. When it burns, a complex combination of toxic gases, such as heavy metals, particulate matter, dioxins, and furans, is released. These aerosol pollutants may cover long distances, settling in soil and water surfaces, and hence, increase the return of the environmental footprint of e-waste pollution. Exposure of humans and animals to these harmful emissions is deadly to the health of people and other living creatures, culminating in respiratory illnesses, heart-related issues, and neurological complications. Although the effect may only occur in large cities, even small-scale burning may increase the levels of air pollution at local levels, which may impact community health and quality of life [25]. The e-waste effect is not limited to the soil, water, and air but also includes the biodiversity as well as the performance of the ecosystem. The e-waste contains toxic substances that may accumulate in plants and animals, subjecting them to physiological and reproductive stress. As an example, soil or water contaminated with heavy metals may interfere with the enzymatic and metabolic activities of the plants and animals, resulting in slow growth, a lowering of the reproduction rates, and a higher rate of mortality. Polluted environments can also promote different species in preference over others, changing the composition of species and also decreasing the levels of biodiversity. The ability of the ecosystem to fulfill vital roles like nutrient cycling, pollination, and predator-prey interactions may be affected and cause ecological imbalances in the long run. The surrounding areas of e-waste dumping and recycling can also have considerably less species diversity than natural ecosystems, which points to the harmful impact of e-waste on the natural ecosystem [26, 27]. E-waste is also a serious source of plastic and polymer waste, which contributes to environmental effects. Plastics are also very difficult to dismantle and with the addition of flame retards, other additives plastics can produce some toxic compounds over time. Microplastics formed as a result of the degradation of e-waste plastics may be absorbed by soils, rivers, and oceans, impacting soil fertility, the quality of water, and aquatic organisms. The materials will not disappear in the environment, which is why the effects of e-waste pollution are long-term and can usually be irreversible unless something is done [28].

Poor management of e-waste is also associated with climate-related effects. Electronics open burning and incineration produce greenhouse gases, which include carbon dioxide and methane, that add to global warming. What is more, the decay of organic materials in the landfills produces more greenhouse emissions. The contributions are not as large as those by other contributors but, nevertheless, they demonstrate the interdependence of e-waste pollution with other global environmental factors, such as climate change [2]. The environmental effect of e-waste is increased in the areas where informal recycling activities prevail. In most developing nations, economic interests are motivated, causing people and small-scale operators to extract metals from e-waste without proper safety or environmental regulations. It will result in the concurrent pollution of soil, water, and air, which presents a compounded effect that can be worse than the e-waste problem in places where formal management structures exist. The informal recycling facilities are also the sites of environmental and human exposure, in which the local populations suffer the aggregate impact of a mixture of contaminants [8].

E-waste has a far-reaching and complex environmental effect. Pollution of soil, water, air, ecosystem destruction, loss of biodiversity and the addition to greenhouse gases all help to demonstrate how serious this issue is. The major offenders of environmental pollution are the heavy metals, plastics, and the persistent organic pollutants and their durability is the guarantee that the impacts of the e-wastes are sustainable. Starting with formal recycling, safe disposal, keeping track of the pollution, and creating awareness among the general population, effective mitigation is needed to ensure the consequences of e-waste are minimized, and both the ecosystem and human population are protected [29].

Table 2: Environmental impact of e-wastes.

Environmental medium	Impact/effect	Description/examples	References
Soil	Contamination	Heavy metals like lead, cadmium, and mercury leach into soil from dumped electronics, affecting soil fertility and microbial activity.	[30]
	Reduced agricultural productivity	Toxic metals and chemicals disrupt plant growth, reduce crop yields, and accumulate in edible plants.	[22, 31]
Water	Chemical pollution	Leaching of toxic metals and flame retardants contaminates groundwater, rivers, and lakes, threatening aquatic ecosystems.	[32]
	Bioaccumulation	Metals and persistent organic pollutants accumulate in fish and other aquatic organisms, entering the food chain.	[33]
Air	Airborne toxins	Burning of e-waste releases particulate matter, dioxins, furans, and heavy metals into the atmosphere, affecting air quality.	[34]
Biodiversity/ ecosystem	Habitat degradation	Soil and water pollution disrupt ecosystems, affecting flora and fauna populations.	[35]
	Ecological imbalance	Toxic substances interfere with food chains and reduce species diversity, altering ecosystem functions.	[35]
Climate/global impact	Greenhouse gas emissions	Improper e-waste disposal, particularly incineration, contributes to greenhouse gas emissions and climate change.	[3]

5 Public health implications

E-waste is slowly becoming a major health issue in society because of the dangerous elements in the e-waste devices, which find their way to disposal sites. The increasing rate at which consumer electronics are being utilized and the poor practice of waste management have become avenues through which toxic metals, organic pollutants, and other dangerous materials have been exposed to. Direct and indirect effects of such exposures can be experienced by human health, including both acute injuries and chronic illnesses. Knowing how e-waste impacts human health is crucial to the design of effective prevention methods, legal frameworks, and interventions to target the vulnerable population [3]. One of the key health hazards linked to e-waste is the health risk of being exposed to heavy metals, which are found in most electronic parts. Lead is a neurotoxin that destroys the brain and nervous system of the body; it is a potent neurotoxin that is commonly used in CRTs, batteries, and solder. Continued lead exposure, especially that associated with children, may cause cognitive impairment, developmental retardation, behavioral disorder, and low intelligence quotient (IQ). Cadmium, which is present in batteries and semiconductors, is very toxic to the lungs, kidneys, and skeletal system. Switches, fluorescent lamps, and other constituents contain mercury that may result in neurological and renal dysfunction. Groups of people residing close to the e-waste recycling premises or dumps are also most susceptible to soil, water, air, and polluted food. The persistent intake or inhalation of these metals may have enduring health implications that are differently prolonged to the succeeding generations [36].

Another concern that is extremely significant is organic pollutants like BFRs and polycyclic aromatic hydrocarbons (PAHs). Chemicals are incorporated into plastics and a circuit board to make them less flammable, yet they are not biodegradable in the environment. The e-waste is improperly handled or burned, releasing such compounds into the atmosphere, soil, and water thus forming avenues through which humans get exposed to them. BFRs disrupt the hormone functions, reproduction, and neurological development, whereas PAHs are carcinogenic and mutagenic. Participants of an informal e-waste recycling process, particularly in uncontrolled environments, tend to be highly exposed, which puts them at a higher risk of acquiring chronic illnesses, cancer, and children's developmental disorders [6]. Work hazards in terms of occupational health are very high, especially in informal e-waste recycling. People, both women and children, in most of the developing countries process the electronics manually to extract the valuable metals without protective gear and inadequate ventilation. Some of the activities that expose workers to heavy metals, feasible dusts, and fumes include cutting of wires, circuit board dismantling, and burning of plastic casings. Such job exposures have been linked with respiratory diseases, skin disorders, eye irritation, nervous system problems, and problems with reproduction. In addition, frequent contact with e-waste toxins may interfere with the immune system, leaving the workers prone to infections and other illnesses. Absence of occupational health standards in the informal recycling is also a contribution to these risks, and the outcomes are high morbidity rates of workers and their families [37, 38].

The other public health issue is the exposure at the community level. The e-waste processing and disposal locations are familiar with residential places and as such, the contaminants find their way into the environment in various ways. The pollutants produced during the burning of e-waste can travel long distances, hence exposing the populations living close to the burning sites to toxic substances. Polluted water and soil may influence agricultural products, livestock, and drinking water, providing indirect exposure sources for whole communities. Children around e-waste locations are particularly prone due to the fact that the organ systems are immature, as well as hand-to-mouth habits and elevated levels of absorbing the toxins. The chronicity of these communities can make them highly exposed to neurological and developmental conditions and chronic diseases [4]. The groups that are most vulnerable such as infants, pregnant women, and the elderly are greatly impacted by exposure to e-waste. Heavy metals and organic pollutants during pregnancy may result in birth defects, low birth weight, preterm birth, and poor cognitive growth. Young children and infants who come into contact with dirty soil, water, or air may experience some stunted physical growth, learning problems, and behavioral issues. Older people whose bodies are already physically weaker are more prone to long-term diseases like heart conditions, kidney disorders, and breathing problems in case they are exposed to the toxic substances of the e-wastes. The extensive environmental dispersion of e-waste contaminants fosters the position of these vulnerable groups in danger despite the fact that they might not be directly engaged in e-waste recycling processes [39]. Less recognized are the mental health effects of exposure to e-waste. The communities living around massive casual recycling sites will be stressed, under anxiety, as well as socially disturbed because of the health risks, pollution, and job hazards. The threat of illness, the inability to use essential resources, which is caused by unsafe recycling jobs, and the deterioration of the environment can adversely impact the mental state. Also, children who are exposed to neurotoxic substances may acquire behavior disorders or learning disabilities and these can have long-term psychological and social impacts. The physical and mental health risks are a complex representation of the multi-factor nature of the effects of e-waste on human health [4, 40].

The other pathway, which is also of great concern, in relation to e-waste and human health, concerns food chain contamination. Myriad organic pollutants and heavy metals are dumped in soil and water systems that find their way into the food system, animals, and fish consumed by human consumers. Repeatedly using contaminated food can cause bioaccumulation of toxic substances in the human body, which has long-term implications on the person's health. An example of such cases is the accumulation of cadmium in the kidney, lead in bones, and mercury in the nervous system, which are all linked to dietary exposure. This contamination is a severe problem in the environmental health of the general population because it is prevalent in areas of elevated informal processing of e-waste or inadequate waste disposal facilities [41]. The solution to the problem of e-waste regarding its impact on public health involves very complex interventions. To begin with, formal e-waste collection and recycling systems can be enhanced, thereby minimizing exposure to hazardous substances. Workers' safe handling, worker protective equipment, and controlled processing facilities are important factors in reducing the risks of the occupation and communities. Second, the international standards and regulations, such as the prohibition of informal dumping, the international conventions concerning hazardous wastes, and the program of EPR, can force all manufacturers and consumers to deal with safe disposal. Third, communities can be educated on the risks of informal recycling and safe disposal, and encouraged to take part in recycling programs that would minimize human exposure. Lastly, it is necessary to monitor/research to know the levels of exposure, encouraging high-risk groups of people, and instituting evidence-based policies to make sure it reduces health effects [2].

E-waste is a looming mass health problem that has far-reaching impacts. Having exposure to heavy metals, persistent organic pollutants, and other toxic substances may influence the neurologic development, reproductive conditions, renal functioning, respiratory and cardiovascular conditions, and overall health. Key exposure routes that consequently impact vulnerable groups (children, pregnant women, and the elderly), particularly are informal recycling processes, location near e-waste disposal sites, e-waste consumption, and consumption of contaminated food or water [42]. Those risks can be managed through joint efforts comprising formal waste management, regulation enforcement, community education, and further research. These measures can decrease the exposure of the human population, safeguard the health of populations, and remove several harmful impacts of e-waste on the populations worldwide [8].

6 Analytical approaches for e-waste monitoring

Monitoring e-waste and its associated contaminants is essential for understanding the environmental and public health risks posed by discarded electronics. Analytical approaches provide quantitative and qualitative information about the composition of e-waste, the concentration of hazardous substances, and the pathways of contamination. By applying advanced analytical techniques, researchers, policymakers, and waste management professionals can assess pollution levels, design effective remediation strategies, and ensure compliance with regulatory standards. This section highlights the main analytical approaches used in e-waste monitoring, emphasizing their methodology, applications, and significance [43, 44]. Chemical characterization is a foundational approach for understanding the composition of e-waste. Electronic devices contain metals, plastics, flame retardants, and other materials in varying concentrations. To quantify these components, samples of e-waste or environmental media (soil, water, air, or sediment) are collected and analyzed. Atomic absorption spectroscopy (AAS) and inductively coupled plasma optical emission spectroscopy (ICP-OES) are widely used to determine metal concentrations, including lead, cadmium, mercury, and copper. AAS measures the absorption of light by metal ions in a sample, providing sensitive and accurate quantification. ICP-OES uses plasma to excite atoms, which emit light at specific wavelengths that can be detected and quantified. These techniques are particularly effective for assessing heavy metal contamination in soils, sediments, and water bodies near e-waste disposal or recycling sites [45].

Inductively coupled plasma mass spectrometry (ICP-MS) is another powerful tool for e-waste monitoring, offering high sensitivity and the ability to detect trace levels of metals and rare earth elements. ICP-MS is capable of analyzing multiple elements simultaneously, making it suitable for complex samples derived from electronics. This technique is particularly valuable for assessing low-concentration pollutants that may still pose significant environmental and health risks. It can detect metals in parts per billion, allowing researchers to evaluate subtle contamination trends and bioaccumulation in environmental media or food sources [46, 47].

X-ray fluorescence (XRF) spectroscopy is a nondestructive analytical technique frequently employed for rapid screening of e-waste. XRF identifies the elemental composition of electronic components by detecting the characteristic X-rays emitted from a sample when it is exposed to a primary X-ray source. This method is particularly useful for in-situ monitoring of e-waste, allowing operators to identify hazardous metals without extensive sample preparation. Portable XRF analyzers have been widely used in field assessments, enabling rapid decision-making regarding the classification, segregation, and treatment of e-waste. XRF is also effective for identifying valuable metals such as gold, silver, and palladium, supporting recovery efforts while monitoring environmental risks [48, 49]. Chromatographic techniques, including gas chromatography (GC) and high-performance liquid chromatography (HPLC), are essential for analyzing organic pollutants in e-waste. BFRs, PAHs, and phthalates are examples of persistent organic compounds released from electronics. GC and HPLC separate and quantify these compounds based on their chemical properties. When coupled with mass spectrometry (GC-MS or LC-MS), these techniques provide highly sensitive and accurate detection of trace organic contaminants. Chromatographic analysis is particularly important for assessing environmental and health risks associated with burning, leaching, or degradation of plastics in e-waste [50, 51].

Spectroscopic techniques, such as Fourier-transform infrared spectroscopy (FTIR) and Raman spectroscopy, are also used to characterize polymers and chemical additives in electronic devices. FTIR identifies functional groups and chemical bonds in plastics, flame retardants, and other organic components, providing information about their composition and potential toxicity. Raman spectroscopy complements FTIR by offering detailed molecular information and identifying specific compounds even in complex matrices. These techniques are crucial for understanding the behavior of e-waste components in the environment, predicting their persistence, and evaluating their potential for harmful effects [52]. Microscopy and imaging techniques provide additional insights into the physical and structural characteristics of e-waste and contaminated media. Scanning electron microscopy (SEM), often coupled with energy-dispersive X-ray spectroscopy (EDS), allows detailed visualization of surface structures while providing elemental composition analysis. SEM-EDS is particularly valuable for analyzing fine particles, dust, and residues generated during e-waste recycling or disposal. Transmission electron microscopy (TEM) can provide nanoscale characterization, especially when assessing the release of nanoparticles from e-waste components, which may pose emerging environmental and health concerns [53].

Biological monitoring complements chemical and physical analyses by assessing the bioavailability and toxicity of e-waste contaminants. Bioassays using microorganisms, plants, or model organisms such as fish or invertebrates can evaluate the ecological impact of e-waste pollutants. For example, the inhibition of microbial growth or plant germination in contaminated soil provides an indirect measure of heavy metal toxicity. Bioaccumulation studies, where metals and persistent organic pollutants are measured in tissues of fish or crops, offer critical information about the transfer of contaminants through food chains and human exposure risks [54, 55].

Remote sensing and geographic information systems (GIS) are increasingly integrated into e-waste monitoring programs to map contamination and identify high-risk areas. Satellite imagery, aerial surveys, and GIS mapping help locate informal e-waste recycling hubs, assess the extent of landfill contamination, and monitor spatial patterns of pollution over time. Coupled with analytical data, GIS enables policymakers to prioritize interventions, allocate resources effectively, and track changes in environmental quality resulting from management initiatives [56, 57]. Quality control and standardization are essential aspects of analytical monitoring. Reliable data require standardized sampling protocols, consistent laboratory procedures, and inter-laboratory validation. International guidelines, such as those provided by environmental protection agencies and international waste conventions, guide sample collection, handling, and analysis to ensure accuracy, reproducibility, and comparability of results across studies. This standardization is critical for assessing global trends in e-waste pollution, informing regulatory compliance, and evaluating the effectiveness of mitigation strategies [58].

Analytical approaches are central to monitoring and managing e-waste pollution. Techniques such as AAS, ICP-OES, ICP-MS, XRF, chromatographic and spectroscopic methods, microscopy, biological monitoring, and GIS provide comprehensive information on the composition, concentration, and pathways of hazardous substances. These methods enable the identification of contamination hotspots, the assessment of environmental and health risks, and the evaluation of mitigation strategies [59, 60]. By combining chemical, physical, and biological analyses with geospatial tools and standardized protocols, e-waste monitoring can support sustainable management, regulatory compliance, and protection of both ecosystems and human health. Effective monitoring not only informs decision-making but also promotes responsible practices in collection, recycling, and disposal, ultimately contributing to the reduction of the global e-waste burden [6].

7 Remediation and management strategies

The rapid growth of e-waste has necessitated the development of effective remediation and management strategies to mitigate environmental and public health risks. E-waste contains hazardous substances such as heavy metals, persistent organic pollutants, and plastics, which, if not managed properly, can contaminate soil, water, and air. Remediation and management strategies aim to reduce the environmental burden of e-waste, recover valuable resources, and ensure safe disposal or recycling. These strategies span physical, chemical, biological, regulatory, and societal approaches, each playing a complementary role in sustainable e-waste management [6, 61].

Physical methods are among the most common strategies for e-waste management. These include manual dismantling, shredding, sorting, and mechanical separation of electronic components. Through physical separation, metals, plastics, and other materials can be recovered for reuse or recycling. Manual dismantling is particularly effective for high-value components, allowing the recovery of precious metals such as gold, silver, and palladium. Mechanical shredding and separation reduce the overall volume of waste and facilitate subsequent treatment processes. Physical methods are relatively straightforward, require moderate investment, and serve as the foundation for safe recycling programs [5]. Chemical methods complement physical approaches by extracting metals and other valuable substances from e-waste. Chemical leaching, solvent extraction, and acid or base treatments are commonly used to recover metals from printed circuit boards, connectors, and other electronic components. These techniques can efficiently isolate precious metals and remove hazardous elements, thereby reducing the environmental toxicity of the remaining waste. Although chemical methods are effective, they require careful handling to prevent secondary contamination and often necessitate neutralization of chemical residues before disposal. When applied properly, these methods significantly enhance resource recovery and reduce the environmental footprint of e-waste [63, 64]. Biological or bioremediation techniques represent a more environmentally friendly approach to managing e-waste. These strategies employ microorganisms, fungi, or plants to degrade, absorb, or immobilize hazardous compounds in contaminated soils, sediments, or water. Certain bacteria can bioleach metals from e-waste, while plants used in phytoremediation absorb heavy metals from soils, reducing toxicity and restoring ecosystem function. Bioremediation is a low-energy, sustainable approach that complements physical and chemical methods, offering long-term benefits for environmental restoration and reducing the risks associated with conventional waste management practices [65].

Thermal treatment methods, including controlled incineration, pyrolysis, and high-temperature processing, are used to recover metals and degrade organic contaminants. Pyrolysis, for example, decomposes organic materials such as plastics and flame retardants at high temperatures in the absence of oxygen, preventing the release of dioxins and furans. Controlled incineration can reduce waste volume and recover energy in the form of heat or electricity. Thermal methods, when implemented under regulated conditions, can safely handle hazardous materials while supporting energy recovery. However, improper thermal treatment can release toxic emissions, highlighting the importance of regulation and monitoring [66].

EPR policies are an essential management strategy for sustainable e-waste handling. EPR requires manufacturers to take responsibility for the collection, recycling, and disposal of their products at the end of life. This approach encourages eco-friendly product design, reduces the environmental burden on consumers and municipalities, and incentivizes producers to minimize the generation of hazardous materials. EPR frameworks have been successfully implemented in several countries, improving recycling rates, reducing informal e-waste processing, and promoting corporate accountability [67]. Formal recycling programs are critical for effective e-waste management. These programs, managed by government agencies or certified private companies, provide structured collection, processing, and recycling systems. Formal recycling ensures that hazardous materials are handled safely, valuable resources are efficiently recovered, and environmental contamination is minimized. Unlike informal recycling, formal programs operate under standardized protocols, employ protective equipment for workers, and adhere to environmental regulations, thereby reducing occupational and community exposure to toxic substances [6, 7].

Public awareness and education are complementary strategies that enhance the effectiveness of e-waste management programs. Consumer knowledge about proper disposal, recycling options, and the hazards of informal processing can significantly reduce the volume of e-waste entering landfills or being processed unsafely. Educational campaigns can target schools, communities, and industrial sectors to promote responsible disposal practices and encourage participation in recycling initiatives. Public engagement also fosters a culture of sustainability, reinforcing the social dimension of e-waste management [69]. Policy and regulatory measures provide the legal and institutional framework necessary for effective e-waste management. Governments can enact laws, set standards for safe disposal, regulate informal recycling activities, and establish monitoring and enforcement mechanisms. International conventions, such as the Basel Convention, regulate the transboundary movement of hazardous waste, preventing the export of e-waste to countries with weak environmental governance. Effective policies ensure compliance, reduce environmental contamination, and facilitate coordinated efforts among stakeholders, including manufacturers, recyclers, and consumers [70, 71].

Remediation and management of e-waste require a comprehensive, multi-pronged approach that combines physical, chemical, biological, thermal, regulatory, and societal strategies. Physical dismantling and separation, chemical extraction, bioremediation, and thermal processing provide practical and technical solutions for reducing environmental toxicity and recovering valuable materials [72]. Policy frameworks, EPR, formal recycling programs, and public education create systemic support to ensure safe handling and sustainable management of e-waste. By integrating these strategies, countries can mitigate the environmental and public health risks of e-waste, promote resource recovery, and foster sustainable electronic consumption and disposal practices (Table 3) [73].

Table 3: E-waste remediation and management strategies.

Strategy/ approach	Description/method	Key benefits/notes	References
Physical methods	Manual dismantling, shredding, sorting, and mechanical separation of electronic components.	Reduce volume of waste, facilitate recycling, allow recovery of valuable metals, and lower environmental contamination.	[2]
Chemical methods	Use of chemical leaching, solvent extraction, or acid/base treatments to recover metals and hazardous substances.	Efficient recovery of precious metals like gold, silver, and palladium; help remove toxic materials from waste streams.	[74]
Biological/ bioremediation	Utilization of microorganisms or plants to degrade, absorb, or immobilize hazardous compounds.	Eco-friendly, low energy requirement, reduce heavy metal concentrations, and support soil and water restoration.	[75, 76]
Thermal/ pyrolysis	Controlled incineration, pyrolysis, or high-temperature treatment to recover metals and degrade organics.	Reduce volume of organic waste, recover energy, and can destroy persistent organic pollutants if done safely.	[77]
Extended producer responsibility (EPR)	Policies requiring manufacturers to take back and manage end-of-life products.	Encourage sustainable product design, reduce environmental burden, and shift responsibility to producers.	[67]
Formal recycling programs	Government or private sector-managed collection, processing, and recycling of e-waste.	Ensure safe handling of hazardous materials, reduce informal recycling risks, and recover valuable resources efficiently.	[6]
Public awareness and education	Campaigns to educate consumers and communities about proper disposal and recycling.	Promote responsible disposal, reduce informal dumping, and support participation in recycling programs.	[78, 79]
Policy and regulatory measures	Implementation of laws, regulations, and standards for e-waste management.	Provide legal framework for safe disposal, recycling, and environmental protection.	[80]

8 Global case studies and lessons learned

The global proliferation of e-waste has prompted countries to develop diverse strategies for its management, reflecting differences in regulatory frameworks, economic resources, technological capacity, and public awareness. Examining global case studies provides critical insights into best practices, challenges, and lessons that can inform sustainable e-waste management strategies worldwide. These cases highlight the importance of regulatory enforcement, formal recycling infrastructure, stakeholder engagement, and innovation in mitigating the environmental and public health impacts of e-waste [2].

Sweden is often cited as a model for effective e-waste management. The country has established a robust system of EPR, where manufacturers are responsible for the collection, recycling, and safe disposal of their products. Municipalities provide collection centers for households and small businesses, while certified recycling companies process e-waste to recover metals, plastics, and other valuable components. Sweden also integrates public awareness campaigns, ensuring that citizens understand proper disposal methods. As a result, more than 90% of e-waste is collected and recycled in the country, minimizing environmental contamination and conserving resources. The Swedish model demonstrates that a combination of strong policy enforcement, accessible collection infrastructure, and public engagement is critical to achieving high recycling rates and reducing informal disposal practices [67]. Japan provides another instructive example, emphasizing technological innovation and regulation. The Home Appliance Recycling Law mandates that consumers pay a recycling fee when disposing of televisions, refrigerators, air conditioners, and washing machines. Manufacturers are required to take back discarded products and ensure environmentally sound recycling. Advanced mechanical and chemical recycling technologies are used to recover rare metals and other valuable materials, while emissions and effluents are strictly monitored. Japan's approach illustrates how legislative frameworks can be combined with technological investment to recover critical resources efficiently, prevent pollution, and encourage sustainable product design [81]. Ghana, particularly the Agbogbloshie area in Accra, represents a contrasting case where informal e-waste processing has led to severe environmental and public health challenges. Agbogbloshie is one of the largest e-waste recycling hubs in West Africa, where discarded electronics are manually dismantled, and valuable metals are extracted through burning and acid leaching. These informal practices expose workers and nearby communities to high concentrations of heavy metals, dioxins, and other toxic substances. Soil, water, and air contamination are widespread, and children often participate in recycling activities, increasing their risk of neurodevelopmental disorders. The Ghana case highlights the dangers of informal recycling, including occupational hazards, environmental degradation, and community health risks. It underscores the need for formal recycling programs, regulatory enforcement, and public awareness campaigns to address uncontrolled e-waste processing [1, 82].

India offers a mixed scenario, with formal e-waste collection systems emerging alongside extensive informal recycling. The E-Waste (Management) Rules, first introduced in 2011 and updated in 2016, require manufacturers to implement EPR and provide take-back services. Large metropolitan areas such as Bengaluru and Delhi have established authorized collection and recycling facilities. However, a significant portion of e-waste still enters informal sectors, where rudimentary techniques, including open burning and acid leaching, are used to extract metals. India's experience demonstrates that regulatory frameworks alone are insufficient without effective enforcement, infrastructure development, and incentives for formal recycling. Collaboration between government agencies, private sector stakeholders, and nongovernmental organizations is essential to transition informal operations into safer, more sustainable systems [83].

European Union (EU) member states collectively provide valuable lessons through the Waste Electrical and Electronic Equipment (WEEE) Directive. The WEEE Directive establishes legal obligations for producers, importers, and distributors of electronic goods, requiring them to finance collection and recycling efforts. EU countries have developed standardized reporting, monitoring, and enforcement mechanisms, ensuring compliance and promoting high recycling rates. Public awareness campaigns and convenient collection points encourage citizen participation, while technological innovation supports efficient recovery of metals and reduction of hazardous waste. The EU example highlights the effectiveness of harmonized regional policies, strict monitoring, and integrated systems for managing e-waste on a large scale [84, 85]. China, as the largest producer and consumer of electronics, illustrates both the challenges and opportunities associated with e-waste management in rapidly industrializing countries. While formal recycling facilities exist in urban areas, informal recycling hubs, such as Guiyu, have historically posed significant environmental and health hazards due to uncontrolled burning, acid leaching, and water contamination. In response, China has introduced the Circular Economy Promotion Law and e-waste collection systems in major cities, emphasizing formalized recycling, technological upgrades, and pollution control. The Chinese experience underscores the importance of transitioning informal sectors to regulated frameworks, investing in technology, and implementing regional policies to manage the high volumes of e-waste generated by industrial growth [86].

Lessons learned from these global case studies reveal several key principles for effective e-waste management. First, regulatory frameworks and EPR programs are critical for ensuring accountability among manufacturers, importers, and consumers. Second, formalized recycling infrastructure, equipped with modern technologies for safe and efficient recovery, is essential to prevent environmental contamination and health hazards. Third, public awareness and participation significantly enhance collection rates and reduce informal disposal practices. Fourth, technology and innovation play a crucial role in improving recovery efficiency, minimizing emissions, and processing complex materials safely. Finally, addressing informal sectors through incentives, training, and regulation is necessary to transition hazardous practices into safer, sustainable operations [3]. Global experiences in e-waste management offer valuable insights for countries seeking to mitigate environmental and health risks associated with discarded electronics. Sweden and Japan demonstrate the benefits of robust regulatory frameworks, public participation, and technological investment, while Ghana, India, and China illustrate the challenges posed by informal recycling, high volumes of waste, and weak enforcement. Lessons learned emphasize the importance of integrating policy, infrastructure, technology, and education to achieve sustainable e-waste management. By applying these lessons, countries can reduce environmental contamination, recover valuable resources, protect public health, and develop resilient systems for the growing challenge of e-waste worldwide [8, 87].

9 Conclusion

E-waste has emerged as a pressing global challenge due to its rapid generation, complex composition, and potential hazards to both human health and the environment. Improper disposal and informal recycling practices have led to widespread contamination of soil, water, and air, while exposing workers and surrounding communities to toxic metals, persistent organic pollutants, and other harmful substances. The environmental and public health implications of e-waste underscore the urgency of implementing effective monitoring, management, and remediation strategies. Analytical approaches, including chemical, spectroscopic, and biological techniques, play a crucial role in assessing contamination levels and guiding interventions. Global case studies reveal that robust regulatory frameworks, EPR, formal recycling infrastructure, technological innovation, and public awareness are essential for sustainable e-waste management. While countries like Sweden and Japan demonstrate successful strategies through policy enforcement and advanced recycling, regions with informal recycling, such as Ghana and parts of India, highlight the risks associated with uncontrolled e-waste handling. Effective e-waste management requires an integrated approach combining policy, technology, education, and community engagement. Addressing these challenges will not only reduce environmental degradation and health risks but also promote resource recovery and sustainability in the rapidly expanding digital world.

References

[1] Njoku A, Agbalenyo M, Laude J, Ajibola TF, Attah MA, Sarko SB, Environmental injustice and electronic waste in Ghana: Challenges and recommendations. Int J Environ Res Public Health, 2023, 21, 25.

[2] Quinto S, Law N, Fletcher C, Le J, Antony Jose S, Menezes PL, Exploring the e-waste crisis: Strategies for sustainable recycling and circular economy integration. Recycling, 2025, 10, 72.

[3] Ghulam ST, Abushammala H, Challenges and opportunities in the management of electronic waste and its impact on human health and environment. Sustainability, 2023, 15, 1837.
[4] Parvez SM, Jahan F, Brune MN, Gorman JF, Rahman MJ, Carpenter D, Islam Z, Rahman M, Aich N, Knibbs LD, Sly PD, Health consequences of exposure to e-waste: An updated systematic review. Lancet Planet Health, 2021, 5, e905–e920.
[5] Mohamed D, Fayad A, Mohamed AMO, Al Nahyan MT, The role of e-waste in sustainable mineral resource management. Waste, 2025, 3, 27.
[6] Srivastava AN, Sikarwar VS, Bisen D, Fathi J, Maslani A, Lopez Nino BN, Barmavatu P, Kaviti AK, Pohořelý M, Buryi M, E-waste unplugged: Reviewing impacts, valorization strategies and regulatory frontiers for efficient e-waste management. Processes, 2025, 13, 2014.
[7] Nisha B, Shajil S, Dutta R, Jain T, Consumer awareness and perceptions about e-waste management in semi-urban area of northern Tamil Nadu: A mixed-method approach. J Fam Community Med, 2022, 29, 132–137.
[8] Andeobu L, Wibowo S, Grandhi S, Environmental and health consequences of e-waste dumping and recycling carried out by selected countries in Asia and Latin America. Sustainability, 2023, 15, 10405.
[9] Abujder Ochoa WA, Torrico Arce AG, Iarozinski Neto A, Munaro MR, Calabokis OP, Ballesteros-Ballesteros VA, Interlinking urban sustainability, circular economy and complexity: A systematic literature review. Sustainability, 2025, 17, 7118.
[10] Dockrell M, Purchase D, Price RG, E-waste and metal contamination in the environment: Health effects. IntechOpen, 2023, doi: 10.5772/intechopen.1001826.
[11] Prata JC, The environmental impact of e-waste microplastics: A systematic review and analysis based on the DPSIR framework. Environments, 2024, 11, 30.
[12] Parvin F, Tareq SM, Impact of landfill leachate contamination on surface and groundwater of Bangladesh: A systematic review and possible public health risks assessment. Appl Water Sci, 2021, 11, 100.
[13] Mathew JT, Inobeme A, Shaba EY, Musah M, Azeh Y, Abubakar H, Adam IB, AI M, HA M, Ismail H, Mt U, MS A, SP Y, AO I, MB E, ST M, Mamman A, Adsorptive removal of Cu^{2+}, Pb^{2+} and Cr^{6+} from pharmaceutical wastewater using graphene/rutile (TiO_2) nanocomposites. Sci World J, 2025, 20, 1263–1272.
[14] Correia AAS, Rasteiro MG, A review of persistent soil contaminants: Assessment and remediation strategies. Environ, 2025, 12, 229.
[15] Maes T, Preston-Whyte F, E-waste it wisely: Lessons from Africa. SN Appl Sci, 2022, 4, 72.
[16] Petruzzelli G, Pezzarossa B, Pedron F, The fate of chemical contaminants in soil with a view to potential risk to human health: A review. Environments, 2025, 12, 183.
[17] Fayshal MA, Current practices of plastic waste management, environmental impacts, and potential alternatives for reducing pollution and improving management. Heliyon, 2024, 10, e40838.
[18] Bouida L, Rafatullah M, Kerrouche A, Qutob M, Alosaimi AM, Alorfi HS, Hussein MA, A review on cadmium and lead contamination: Sources, fate, mechanism, health effects and remediation methods. Water, 2022, 14, 3432.
[19] Mathew JT, Shaba EY, Musah M, Azeh Y, Muhammad EB, Tanko E, Tanko MS, Muhammad AI, Hussaini J, Mamman A. Aerogel nanocomposites: Properties and applications, In: Inobeme A, Mathew J, Adetunji C, Eds, Advances, Applications, and Future Perspectives of Aerogels. IGI Global, 2025, 25–62.
[20] Babuji P, Thirumalaisamy S, Duraisamy K, Periyasamy G, Human health risks due to exposure to water pollution: A review. Water, 2023, 15, 2532.
[21] De Guzman R, Schiller J, Air pollution and its impact on cancer incidence, cancer care and cancer outcomes. BMJ Oncol, 2025, 4, e000535.
[22] Alengebawy A, Abdelkhalek ST, Qureshi SR, Wang MQ, Heavy metals and pesticides toxicity in agricultural soil and plants: Ecological risks and human health implications. Toxics, 2021, 9, 42.

[23] Holden KA, Lee AR, Hawcutt DB, Sinha IP, The impact of poor housing and indoor air quality on respiratory health in children. Breathe, 2023, 19, 230058.
[24] Siddiqua A, Hahladakis JN, Al-Attiya WAKA, An overview of the environmental pollution and health effects associated with waste landfilling and open dumping. Environ Sci Pollut Res, 2022, 29, 58514–58536.
[25] Jakhar R, Samek L, Styszko K, A comprehensive study of the impact of waste fires on the environment and health. Sustainability, 2023, 15, 14241.
[26] Alabi OA, Adeoluwa YM, Huo X, Xu X, Bakare AA, Environmental contamination and public health effects of electronic waste: An overview. J Environ Health Sci Eng, 2021, 19, 1209–1227.
[27] Mathew JT, Hussaini J, Inobeme A, Adetunji CO, Tanko MS. Nanocarriers for delivering pesticides, In: Chen J, Ed, Nanocarriers in Plant Science and Agriculture. IGI Global. 2025, 273–296.
[28] Abahussain AAM, Nasr FA, Jumah AB, Saravanan P, Siva Kumar N, Al-Zharani M, Guganathan L, Sasikumar G, Alsalamah SA, Qurtam AA, Senthilkumar J, Tamizhdurai P, Toxic threats from plastic waste: Human health impacts, challenges, and policy solutions. RSC Adv, 2025, 15, 40761–40788.
[29] Owusu-Sekyere K, Aladago DA, Leverenz D, Oteng-Ababio M, Kranert M, Environmental impacts on soil and groundwater of informal e-waste recycling processes in Ghana. Sustainability, 2024, 16, 4347.
[30] Tang H, Xiang G, Xiao W, Yang Z, Zhao B, Microbial mediated remediation of heavy metals toxicity: Mechanisms and future prospects. Front Plant Sci, 2024, 15, 1420408.
[31] Mathew JT, Azeh Y, Adetunji CO, Monday M, Hussaini J, Tanko MS. Catalytic application of aerogels, In: Inobeme A, Mathew J, Adetunji C, Eds, Advances, Applications, and Future Perspectives of Aerogels. IGI Global. 2025, 147–178.
[32] Zhang P, Yang M, Lan J, Huang Y, Zhang J, Huang S, Yang Y, Ru J, Water quality degradation due to heavy metal contamination: Health impacts and eco-friendly approaches for heavy metal remediation. Toxics, 2023, 11, 828.
[33] Jamil Emon F, Rohani MF, Sumaiya N, Tuj Jannat MF, Akter Y, Shahjahan M, Abdul Kari Z, Tahiluddin AB, Goh KW, Bioaccumulation and bioremediation of heavy metals in fishes: A review. Toxics, 2023, 11, 510.
[34] Matei E, Râpă M, Mateș IM, Popescu AF, Bădiceanu A, Balint AI, Covaliu-Mierlă CI, Heavy metals in particulate matter: Trends and impacts on environment. Molecules, 2025, 30, 1455.
[35] Bashir I, Lone FA, Bhat RA, Mir SA, Dar ZA, Dar SA, Concerns and threats of contamination on aquatic ecosystems. Bioremed Biotechnol, 2020, 1, 1–26.
[36] Desye B, Tesfaye AH, Berihun G, Ademas A, Sewunet B, A systematic review of the health effects of lead exposure from electronic waste in children. Front Public Health, 2023, 11, 1113561.
[37] Decharat S, Kiddee P, Health problems among workers who recycle electronic waste in southern Thailand. Osong Public Health Res Perspect, 2020, 11, 34–43.
[38] Mathew JT, Hussaini J, Azeh Y, Musah M, Shaba EY, Tanko E, Muhammad AI, Etsuyankpa MB, Musa TS, Mamman A. Environmental and health impact assessment on aerogel application, In: Inobeme A, Mathew J, Adetunji C, Eds, Advances, Applications, and Future Perspectives of Aerogels. IGI Global. 2025, 179–202.
[39] Chen C, Ma C, Li Q, Hang JG, Shen J, Nakayama SF, Kido T, Lin Y, Feng H, Jung CR, Sun XL, Lou J, Prenatal exposure to heavy metals and adverse birth outcomes: Evidence from an e-waste area in China. GeoHealth, 2023, 7, e2023GH000897.
[40] Mathew JT, Musah M, Azeh Y, Shaba EY, Muhammad EB, Hussaini J, Tanko MS, Mamman A. Exploring future perspectives in aerogel-based research, In: Inobeme A, Mathew J, Adetunji C, Eds, Advances, Applications, and Future Perspectives of Aerogels. IGI Global. 2025, 203–228.
[41] Laoye B, Olagbemide P, Ogunnusi T, Akpor O, Heavy metal contamination: Sources, health impacts, and sustainable mitigation strategies with insights from Nigerian case studies. F1000Research, 2025, 14, 134.

[42] Mathew JT, Hussaini J, Inobeme A, Azeh Y, Musah M, Shaba YE, Etsuyankpa MB, Tanko E, Ndejiko JM, Adetunji CO, Muhammad AI, Musa ST, Mamman A, X-ray fluorescence analysis of vegetation tissues via chemometric tools, In: Singh VK, Ed, X-Ray Fluorescence Spectroscopy and Chemometrics. Springer, 2025, 209–231.
[43] Eze C, Vinken M, E-waste: Mechanisms of toxicity and safety testing. FEBS Open Bio, 2024, 14, 1420–1440.
[44] Mathew JT, Inobeme A, Azeh Y, Musah M, Shaba EY, Tanko E, Etsuyankpa MB, Musa ST, Ndejiko JM, Hussaini J, Adetunji CO, Muhammad AI, Mamman A, X-ray fluorescence in analyzing agricultural contamination via chemometric tools, In: Singh VK, Ed, X-Ray Fluorescence Spectroscopy and Chemometrics. Springer, 2025, 233–254.
[45] Ferreira-Filipe DA, Hursthouse A, Duarte AC, Rocha-Santos T, Patrício Silva AL, E-waste plastics in the environment: Assessment, characterisation, and bioprocessing. Appl Sci, 2025, 15, 2122.
[46] Klika KD, Han J, Busse MS, Soloshonok VA, Javahershenas R, Vanhaecke F, Makarem A, Inductively coupled plasma-mass spectrometry (ICP-MS): An emerging tool in radiopharmaceutical science. J Am Chem Soc, 2024, 146, 30717–30727.
[47] Mathew JT, Inobeme A, Azeh Y, Musah M, Shaba EY, Tanko T, Etsuyankpa MB, Musa ST, Hussaini J, Adetunji CO, Muhammad AI, Mamman A, X-ray fluorescence spectroscopy for analyzing food products coupled with chemometrics, In: Singh VK, Ed, X-Ray Fluorescence Spectroscopy and Chemometrics. Springer, 2025, 255–273.
[48] Guagliardi I, Ricca N, Cicchella D, Comparative evaluation of inductively coupled plasma mass spectrometry (ICP-MS) and X-ray fluorescence (XRF) analysis techniques for screening potentially toxic elements in soil. Toxics, 2025, 13, 314.
[49] Mathew JT, Inobeme A, Azeh Y, Musah M, Shaba EY, Tanko E, Etsuyankpa MB, Muhammad AI, Musa ST, Ndejiko MJ, Hussaini J, Adetunji CO, Mamman A, Total reflection X-ray fluorescence: Technological developments and expanding applications, In: Singh VK, Ed, X-Ray Fluorescence Spectroscopy and Chemometrics. Springer, 2025, 303–323.
[50] Chibwe L, De Silva AO, Spencer C, Teixera CF, Williamson M, Wang X, Muir DCG, Target and nontarget screening of organic chemicals and metals in recycled plastic materials. Environ Sci Technol, 2023, 57, 3380–3390.
[51] Idris AY, Elele UU, Mathew JT, Preparation and characterization of MoO_3 nanoparticles for the photocatalytic degradation of dyeing wastewater. Sci World J, 2024, 19, 1006–1011.
[52] Huang Z, Hu B, Wang H, Analytical methods for microplastics in the environment: A review. Environ Chem Lett, 2023, 21, 383–401.
[53] Sacco MA, Gualtieri S, Santos A, Mendes B, Raffaele R, Tarallo AP, Verrina MC, Ranno F, Monterossi MD, Ricci P, Aquila I, Scanning electron microscopy techniques in the analysis of gunshot residues: A literature review. Appl Sci, 2025, 15, 2634.
[54] Igbo JK, Chukwu LO, Oyewo EO, Blum JL, Schanzer A, Wirgin I, Meltzer GY, Roy NK, Zelikoff JT, The chemistry and health outcomes of electronic waste (e-waste) leachate: Exposure to e-waste is toxic to Atlantic killifish embryos. Sustainability, 2022, 14, 11304.
[55] Mathew JT, Musah M, Azeh Y, Musa M, Removal of some toxic metals from pharmaceutical wastewater using geopolymer/Fe_3O_4/ZnO nanocomposite: Isotherm, kinetics and thermodynamic studies. Confluence Univ J Sci Technol, 2024, 1, 50–58.
[56] Sakshi NK, Singh R, Diverse applications of remote sensing and geographic information systems in implementing integrated solid waste management: A short review. Eng Proc, 2023, 56, 109.
[57] Etsuyankpa BM, Musa ST, Ambo AI, Sulaiman LA, Mathew JT, Removal of Cu^{2+}, Fe^{3+} and Pb^{2+} from abattoir wastewater using TiO_2/CdS nanocomposite: Isotherm and kinetics studies. Sch Int J Chem Mater Sci, 2025, 8, 300–310.
[58] Gazeau B, Zaman A, Minunno R, Shaikh F, Developing traceability systems for effective circular economy of plastic: A systematic review and meta-analysis. Sustainability, 2024, 16, 9973.

[59] Mathew JT, Inobeme A, Azeh Y, Musah M, Abdulkadir A, Shaba EY, Etsuyankpa MB, Musa TS, Muhammad AI, Ismail H, Kanwa AM, Mamman A, Hussaini J, Carbon capture and storage via electrochemical and bioelectrochemical techniques: A review. Caliphate J Sci Technol, 2024, 2, 141–158.
[60] Patra B, Das MT, Pradhan SN, Dash SR, Bhuyan PP, Pradhan B, Ferrochrome pollution and its consequences on groundwater ecosystems and public health. Limnol Rev, 2025, 25, 23.
[61] Mathew JT, Inobeme A, Abdulkadir A, Etsuyankpa MB, Shaba EY, Musah M, Azeh Y, Adetunji CO, Ibrahim AM, Tanko E, Mamman A, Properties and characteristics of nanocoatings for medicinal applications, In: Sustainable Approach to Protective Nanocoatings. IGI Global, 2024, 203–227.
[62] Mathew JT, Inobeme A, Adetuyi BO, Falana YO, Adetunji CO, Shahnawaz M, Application of microplastics in toiletry products, In: Shahnawaz M, Adetunji CO, Dar MA, Zhu D, Eds, Microplastic Pollution. Springer, 2024, 73–84.
[63] Chicardi E, Lopez-Paneque A, García-Orta VHG, Sepúlveda-Ferrer RE, Gallardo JM, Enrichment methods for metal recovery from waste from electrical and electronic equipment: A brief review. Metals, 2025, 15, 140.
[64] Das K, Masud MAA, Sarker A, Arafa RA, Patel M, Unveiling the sustainable and biological remediation of heavy metals contaminations in soils and water ecosystems through potential microbes: A review. Sustainability, 2025, 17, 7357.
[65] Xu R, Qu X, He Y, Chen F, Zhong Y, Zhang H, Ding J, Dou J, Methodological insights into the occurrence, conversion, and control of polychlorinated dibenzo-p-dioxins/dibenzofurans from waste incineration. Molecules, 2025, 30, 4106.
[66] Hammoud R, Massoud MA, Chalak A, Abiad MG, Exploring the feasibility of extended producer responsibility for efficient waste management in Lebanon. Sci Rep, 2025, 15, 15444.
[67] Etsuyankpa MB, Augustine AU, Musa ST, Mathew JT, Ismail H, Salihu AM, Mamman A, An overview of wastewater characteristics, treatment and disposal: A review. J Appl Sci Environ Manage, 2024, 28, 1553–1572.
[68] Nisha B, Shajil S, Dutta R, Jain T, Consumer awareness and perceptions about e-waste management in semi-urban area of northern Tamil Nadu: A mixed-method approach. J Fam Community Med, 2022, 29, 132–137.
[69] Abalansa S, El Mahrad B, Icely J, Newton A, Electronic waste, an environmental problem exported to developing countries: The good, the bad and the ugly. Sustainability, 2021, 13, 5302.
[70] Adetunji CO, Mathew JT, Singh KRB, Bodunrinde RE, Inobeme A, Olaniyan O, Ogundolie FA, Singh J, Nayak V, Singh RP, Molecular characterization of multidrug-resistant genes in wastewater treatment plants, In: Microbial Community Studies in Industrial Wastewater Treatment. Taylor & Francis eBooks, 2023, 127–141.
[71] Magalhães S, Alves L, Medronho B, Svanedal I, Norgren M, Rasteiro MG, Innovative approaches to mitigating microplastic pollution in effluents and soils. Sustainability, 2025, 17, 9014.
[72] Nandan A, Suresh AC, Saole P, Jeevanasai SA, Chandrasekaran R, Meili L, Wan Azelee NI, Selvasembian R, An integrated approach for electronic waste management: Sources, toxicological effects, assessment, governance and mitigation approaches. Sustainability, 2023, 15, 16946.
[73] Kumari S, Panda R, Prasad R, Alorro RD, Jha MK, Sustainable process to recover metals from waste PCBs using physical pre-treatment and hydrometallurgical techniques. Sustainability, 2024, 16, 418.
[74] Raklami A, Meddich A, Oufdou K, Baslam M, Plants–microorganisms-based bioremediation for heavy metal cleanup: Recent developments, phytoremediation techniques, regulation mechanisms and molecular responses. Int J Mol Sci, 2022, 23, 5031.
[75] Musah M, Mathew JT, Azeh Y, Synthesis, characterization and application of ZnO/GO/Zeolite-A nanocomposite in the sorption of selected heavy metals from pharmaceutical effluent. Sch Int J Chem Mater Sci, 2025, 8, 202–212.

[77] Li Y, Duan Y, Wang Z, Maurice NJ, Claire MJ, Ali N, Giwa AS, Leveraging municipal solid waste management with plasma pyrolysis and IoT: Strategies for energy byproducts and resource recovery. Processes, 2025, 13, 321.
[78] Abubakar IR, Maniruzzaman KM, Dano UL, AlShihri FS, AlShammari MS, Ahmed SMS, Al-Gehlani WAG, Alrawaf TI, Environmental sustainability impacts of solid waste management practices in the global south. Int J Environ Res Public Health, 2022, 19, 12717.
[79] Musa AV, Musah M, Mathew JT, Production and characterization of Zeolite-A nanoparticles for the treatment of pharmaceutical wastewater. Sci World J, 2024, 19, 987–993.
[80] Odeyingbo OA, Deubzer OK, Ogunmokun OA, Assessment of the impact of the revised national e-waste framework on the informal e-waste sector of Nigeria. Recycling, 2025, 10, 117.
[81] Portugaise MK, Jóhannsdóttir L, Murakami S, Extended producer responsibility's effect on producers' electronic waste management practices in Japan and Canada. Discov Sustain, 2023, 4, 8.
[82] Inobeme A, Natarajan A, Safalmani P, Adetunji CO, Ariola D, Chetan C, Mathew JT, Inobeme J, Kaushik P, Kshitij RBS, Shyam SP, Probing heavy metals in tropospheric particulate matter: An approach towards sustainability. J Mol Liq, 2024, 412, 125822.
[83] Borthakur A, Design, adoption and implementation of electronic waste policies in India. Environ Sci Pollut Res, 2023, 30, 8672–8681.
[84] Andersen T, A comparative study of national variations of the European WEEE directive: Manufacturer's view. Environ Sci Pollut Res, 2022, 29(14), 19920–19939.
[85] Abdulkadir A, Musah M, Lakan II, Mathew JT, Production of MgO/ZnO nanocomposite for the removal of selected toxic metals from tannery wastewater. Sch Int J Chem Mater Sci, 2025, 8, 250–262.
[86] Li B, Liu D, Zhang L, Wu Y, Ding X, Zeng X, Challenges of e-waste dismantling in China. Toxics, 2024, 12, 867.
[87] Inobeme A, Mathew JT, Devolli A, Adetunji CO, Sharma N, Maliki N, Hussaini J, Metal components in industrial wastes and methods for metal ions recovery, In: Metal Value Recovery from Industrial Waste Using Advanced Physicochemical Treatment Technologies. Elsevier, 2025, 1–15.

R. Pavithra, S. Sanjupriya, M. Poonkothai*

E-waste management: mitigating risks of persistent organic pollutants and advancing sustainable solutions

Abstract: The rapid rise in electronic waste presents complex global challenges, containing both inorganic and organic components that threaten the environment and human health. A significant concern is the presence of organic pollutants such as plastics, flame retardants, plasticizers, polycyclic aromatic hydrocarbons, dioxins, and furans, many of which persist in nature and can bioaccumulate in living organisms. Improper disposal and informal recycling practices facilitate the release of these hazardous substances into soil, groundwater, and the atmosphere through leaching and open burning, allowing entry into food chains and widespread human exposure. Organic pollutants in e-waste, especially persistent organic pollutants, pose significant toxic risks to soil microbiota, aquatic systems, and human populations, leading to endocrine disruption, carcinogenicity, and occupational health hazards for workers handling electronics without proper safeguards. Addressing contamination requires effective detection and monitoring using advanced analytical and biomonitoring techniques. Remediation strategies include physical adsorption, chemical degradation such as advanced oxidation, photocatalysis, and biological approaches utilizing microbes and enzymes with innovative technologies like nanotechnology and green catalysis. International conventions and national regulations exist, but gaps in policy, high technology costs, and inadequate infrastructure in many regions impede progress. Future efforts must leverage artificial intelligence, digital tools, and collaborative stakeholder engagement to advance sustainable solutions for e-waste management and reduce global risks linked to organic pollutants.

Keywords: E-waste, persistent organic pollutants, bioaccumulation, environmental persistence, remediation

***Corresponding author: M. Poonkothai,** Department of Zoology, Avinashilingam Institute for Home Science and Higher Education for Women, Coimbatore 641 043, Tamil Nadu, India, e-mail: poonkothaii_zoo@avinuty.ac.in

R. Pavithra, S. Sanjupriya, Department of Zoology, Avinashilingam Institute for Home Science and Higher Education for Women, Coimbatore 641 043, Tamil Nadu, India

 | https://doi.org/10.1515/9783112248072-004

1 Introduction

The exponential increase in the generation of electronic waste (e-waste) is one of the most pressing global environmental challenges of the twenty-first century. Rapid technological advancement, short product life cycles, and consumer-driven upgrades have accelerated the accumulation of discarded electrical and electronic devices. E-waste comprises a complex mixture of valuable materials such as metals and plastics, alongside hazardous substances including heavy metals, flame retardants, and persistent organic pollutants (POPs). Improper disposal and informal recycling practices, particularly in developing countries, result in severe contamination of soil, water, and air, posing significant risks to environmental and human health [1].

Among the toxic constituents of e-waste, organic pollutants such as polybrominated diphenyl ethers (PBDEs), polychlorinated biphenyls (PCBs), dioxins, and furans are of major concern due to their persistence, bioaccumulation, and toxicity. These compounds are commonly used as flame retardants and plasticizers in electronic components and are released during dismantling, open burning, or incineration. Once in the environment, these pollutants resist degradation and can be transported over long distances, accumulating in biota and entering food chains. Their exposure is associated with endocrine disruption, carcinogenicity, neurotoxicity, and reproductive effects, particularly among vulnerable populations involved in informal e-waste recycling [2].

Addressing the growing threat of organic pollutants in e-waste necessitates the development of efficient detection, monitoring, and remediation technologies. Analytical methods such as gas chromatography-mass spectrometry and liquid chromatography are widely used to identify and quantify pollutants, while biomonitoring frameworks provide insights into human and ecological exposure. Remediation approaches encompass physical, chemical, and biological methods, including adsorption, advanced oxidation processes, photocatalysis, and microbial degradation. The integration of nanotechnology, green catalysis, and bio-based adsorbents offers new possibilities for sustainable pollutant removal and resource recovery, minimizing environmental impact [3].

The transition toward sustainable e-waste management requires a holistic approach that combines technological innovation with effective policy implementation. International conventions such as the Basel and Stockholm Conventions, along with national regulations like India's E-Waste Management Rules, promote safe recycling and extended producer responsibility. However, challenges remain in enforcement, infrastructure, and public awareness. Emphasizing circular economy principles, reuse, repair, recycling, and resource recovery, can transform e-waste from a liability into a valuable resource stream. Integrating artificial intelligence, digital traceability, and eco-design into waste management systems will further enhance sustainability and reduce the risks posed by POPs, contributing to global environmental protection and public health [4].

2 Sources of organic pollutants

Electronic waste contains a diverse range of plastics and polymers that often have chemical additives such as PBDEs and PCBs, which serve as flame retardants to reduce fire risk in electronic components. Studies show these substances can be released during dismantling and recycling processes, especially in poorly regulated environments [5]. PBDEs and PCBs persist in the environment and pose significant health risks due to their toxicity and bioaccumulative nature. Research conducted in e-waste recycling workshops in southern China revealed elevated PBDE and PCB concentrations, often exceeding levels found in developed countries, indicating significant occupational exposure risks [6].

Plasticizers, including phthalates, are another class of additives commonly found in e-waste plastics. These compounds can leach out during improper disposal or open burning, contributing to environmental contamination and endocrine disruption in humans and wildlife. Polycyclic aromatic hydrocarbons (PAHs), dioxins, and furans are toxic byproducts formed during the thermal degradation or incineration of e-waste plastics and electronic circuit boards. These POPs result from incomplete combustion and are associated with carcinogenic and mutagenic effects. Advanced spectrometric analyses have identified these pollutants in airborne particles and soils near informal recycling sites (Figure 1). Proper management and regulation of e-waste processing are essential to minimizing these environmental and health hazards [7].

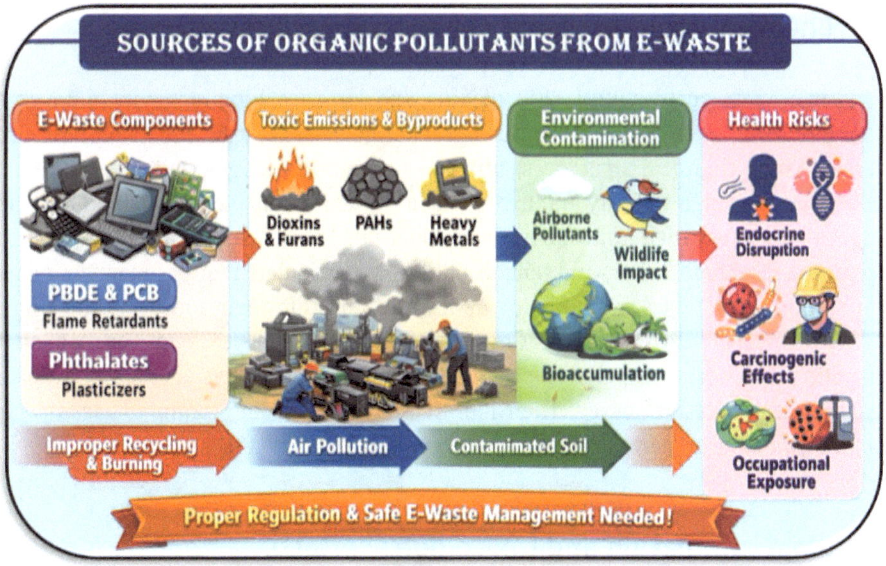

Figure 1: Sources of organic pollutants from e-waste.

3 Types and characteristics of organic pollutants

Organic pollutants from e-waste arise mainly from plastics, flame retardants, and other synthetic materials used in electronic components. These compounds are persistent in the environment, bioaccumulate in living organisms, and pose serious risks to human health and ecosystems. Understanding their types and characteristics is crucial for developing effective management and remediation strategies.

3.1 Persistent organic pollutants

POPs in e-waste pose significant environmental and health risks due to their persistence, bioaccumulation, and toxicity. Key POPs identified in e-waste recycling sites include PBDEs, PCBs, dioxins, and furans. These chemicals are primarily used as flame retardants and are released during dismantling, recycling, and open burning of electronic components. Studies from e-waste sites in China demonstrate elevated levels of these POP in air and soil, leading to increased human exposure through inhalation and ingestion, especially among workers and local residents. The pollutants disrupt endocrine functions and pose carcinogenic risks. Monitoring and regulation of POPs in electronic waste is crucial for safeguarding public health, and advanced treatment methods are needed to mitigate their release during waste processing [8].

3.2 Chemical properties and environmental persistence

POPs in e-waste, such as PBDE, PCB, dioxins, and furans, exhibit longevity in the environment and resist degradation through chemical, biological, and photolytic processes. Their chemical structures, often comprised of halogenated aromatic compounds, confer high stability and hydrophobicity, facilitating bioaccumulation in fatty tissues and long-range atmospheric transport [9]. These compounds are lipophilic, sparingly soluble in water, and tend to adhere to particulate matter, enhancing environmental persistence and allowing deposition far from their source. Studies have demonstrated that POP concentrations are markedly higher in regions near e-waste recycling facilities, with pollutants remaining in soils and air for years even after recycling activities have ceased. Their half-lives can vary from months to decades, depending on environmental conditions and chemical nature [10].

3.3 Bioaccumulation and toxicological effects

Bioaccumulation of organic pollutants in e-waste primarily involves POP such as PCB, PBDE, and dioxins, which are resistant to degradation and accumulate in living tissues. These POPs bioaccumulate through soil, water, and food chains, affecting various organisms and magnifying in higher trophic levels [11]. Toxicological effects include endocrine disruption, immunotoxicity, neurotoxicity, reproductive impairment, and carcinogenicity. For example, PBDE disrupts thyroid hormone function, and PCBs accumulate in human fatty tissues, posing cancer risks. Such toxicity impacts biodiversity and ecosystem function, highlighting the need for sustainable e-waste management to mitigate these risks and protect human and environmental health [12].

4 Pathways of environmental contamination

Environmental contamination from electronic waste arises when toxic substances from discarded devices enter soil, water, and air. These pollutants disperse through improper disposal, informal recycling, and leaching processes, leading to lasting impacts on ecosystems and human health.

4.1 Leaching into soil and groundwater

Leaching of toxic substances from electronic waste into soil and groundwater is a critical environmental concern. Heavy metals such as lead, mercury, and cadmium, prevalent in e-waste, can seep through porous soil layers, eventually contaminating groundwater resources. This process is exacerbated by improper disposal methods like illegal dumping and uncontrolled landfilling, especially in areas lacking effective legislation [13]. Leachates from e-waste contain harmful metals and organic pollutants, including PBDEs, which persist in the environment and degrade soil quality by harming microorganisms and plants. This results in bioaccumulation through the food chain, posing significant health and ecological risks. Remediation methods are essential to mitigate these impacts and restore soil and water quality [14].

4.2 Toxic emissions from e-waste burning

Open burning and informal recycling of e-waste release toxic emissions, including heavy metals (e.g., lead, copper, and zinc) and halogenated organic compounds such as polychlorinated dibenzo-*p*-dioxins and dibenzofurans. These pollutants are emitted as airborne particulate matter and gases, undergoing atmospheric transport and subsequent deposition onto soil and water surfaces [15]. In places like Guiyu, China, air concentrations of PBDE during open burning reached levels hundreds of times higher than those in surrounding areas, posing severe inhalation exposure risks. Deposited pollutants enter food chains via soil and water contamination, bioaccumulating in plants and animals, ultimately leading to human exposure through diet and air. Workers in informal recycling sectors face the highest risks due to inadequate protective measures and poor ventilation. These pathways contribute significantly to ecological and human health hazards, emphasizing the urgent need for regulated e-waste management and emission controls to mitigate POP risks [16].

5 Ecological and health impacts

Electronic waste poses substantial ecological and health risks through the release of hazardous substances such as heavy metals and POPs into the environment. These contaminants degrade soil and water quality, disrupt ecosystems, and threaten biodiversity [17]. Epidemiological studies reveal that exposure to e-waste toxins can cause serious health problems, including respiratory diseases, neurological damage, reproductive issues, and increased cancer risk. Vulnerable groups like children and pregnant women bear the greatest burden of adverse effects. Informal recycling practices in developing countries amplify exposure risks through inhalation, dermal contact, and dietary intake of contaminated food [18]. Research highlights severe contamination in areas with poor e-waste management, where high levels of lead, cadmium, mercury, and brominated flame retardants have been detected in human tissues and environmental samples. To mitigate these impacts, sustainable e-waste management solutions involving stricter regulations, formal recycling infrastructure, and international cooperation are urgently needed. Emphasis on safe recycling technologies and minimizing hazardous chemical use in electronics manufacturing is critical for protecting both environmental and public health globally [19].

6 Detection and monitoring of organic pollutants

Detection and monitoring of organic pollutants from e-waste are essential to assess environmental contamination and human health risks. Persistent organic compounds released during improper disposal accumulate in soil and water systems. Continuous monitoring enables effective risk mitigation and supports sustainable e-waste management strategies.

6.1 Sampling techniques for soil, water, and air

Sampling techniques for organic pollutants in e-waste environments have been extensively studied to ensure accurate detection and monitoring. Soil sampling often targets topsoil (0–10 cm depth), collected systematically across sites using soil augers and trowels, with samples composited for representativeness. Dust sampling involves fiber brushes and dustpans to collect floor and roadside dust, avoiding contamination [20]. Air sampling utilizes passive diffusion samplers or active air samplers to capture volatile organic compounds across different time intervals. Samples are carefully labeled, stored away from sunlight, and kept cool to maintain integrity before analysis. These methods have been applied in informal recycling hotspots, revealing elevated pollutant levels, especially in dismantling and burning sites, highlighting the need for standardized, rigorous sampling protocols for risk assessment and remediation planning [21].

6.2 Analytical methods

Analytical methods for detecting organic pollutants from e-waste predominantly rely on advanced chromatographic and spectrometric techniques. Gas chromatography-mass spectrometry (GC-MS) is widely applied for identifying and quantifying volatile and semivolatile organic compounds, including flame retardants, PAHs, and phthalates [22]. Techniques such as liquid chromatography (LC) coupled with time-of-flight mass spectrometry (TOF-MS) enable nontargeted screening of hundreds of hazardous organic compounds in electronic waste polymers. Sample preparation typically involves solid-phase extraction or liquid-liquid extraction to concentrate analyses before analysis. Emerging methods include chemiluminescence flow injection for sensitive pesticide detection and micro-solid phase extraction. These approaches provide both qualitative and quantitative data crucial for risk assessment and remediation planning in e-waste contexts [23].

6.3 Biomonitoring and risk assessment frameworks

Biomonitoring and risk assessment frameworks are critical for evaluating human and ecological exposure to POPs from e-waste. Human biomonitoring (HBM) assesses internal chemical exposure by measuring biomarkers in biological samples, offering direct insight into health risks. Integrating HBM data with physiologically based pharmacokinetic models enhances exposure and health risk assessments, especially for populations near e-waste sites [24]. Risk assessment frameworks combine exposure data, toxicological information, and health effect modeling to characterize adverse outcomes and inform management strategies. Challenges include harmonizing methodologies, addressing vulnerable populations, and translating data into actionable policies. Advanced frameworks advocate integrating biomonitoring with environmental and epidemiological monitoring for comprehensive risk evaluation and sustainable e-waste management [25].

7 Remediation strategies

Effective remediation strategies are essential for mitigating the environmental and health impacts of escalating e-waste generation. These approaches focus on reducing hazardous pollutants, restoring contaminated sites, and promoting safe material recovery. Integrating technological, biological, and policy-driven methods ensures sustainable and responsible e-waste management.

7.1 Physical methods

Physical methods are among the primary approaches used in e-waste management to separate, recover, and process valuable materials without altering their chemical composition. These methods rely on mechanical and physical properties such as size, density, magnetic susceptibility, and conductivity to segregate different components [26]. Common techniques include shredding, crushing, magnetic separation, eddy current separation, and gravity-based sorting. By efficiently breaking down and sorting e-waste, these methods enable the recovery of metals, plastics, and other reusable materials. Physical treatment serves as an essential preliminary step before chemical or biological processes, reducing the volume of hazardous waste and enhancing resource recovery efficiency [27].

7.1.1 Adsorption using activated carbon and biochar

Adsorption is an effective physical method employed in e-waste management to remove toxic metals and organic contaminants from leachates and effluents generated during waste processing. Activated carbon and biochar are widely used adsorbents due to their large surface areas, high porosities, and excellent adsorption capacities. Activated carbon, derived from carbonaceous materials, effectively captures heavy metals such as lead, cadmium, and chromium through surface complexation and electrostatic interactions [28]. Biochar, a sustainable and cost-effective alternative produced via pyrolysis of biomass, offers similar adsorption efficiency with added environmental benefits. Its surface functional groups facilitate the binding of contaminants, thereby reducing their mobility and toxicity. Incorporating adsorption techniques using these materials not only enhances pollutant removal but also promotes circular economy practices by converting agricultural or organic waste into valuable adsorbents for e-waste treatment [29].

7.1.2 Thermal desorption and incineration

Thermal desorption and incineration are key physical methods used for treating and reducing the volume of hazardous components in e-waste. Thermal desorption involves heating contaminated materials at moderate temperatures to volatilize organic pollutants, which are then captured or destroyed in a secondary treatment system. This method effectively removes volatile organic compounds (VOCs), hydrocarbons, and other thermally degradable contaminants. Incineration, on the other hand, subjects e-waste to high temperatures, leading to complete combustion of organic matter and reduction of waste volume by up to 90% [30]. It also facilitates the recovery of energy in the form of heat or electricity. However, careful control of operating conditions is essential to prevent the release of toxic emissions such as dioxins and heavy metal vapors. When properly managed, these thermal processes provide an efficient and reliable approach for safe e-waste disposal and resource recovery [31].

7.2 Chemical methods

Chemical methods are widely applied in e-waste management to transform, dissolve, or stabilize hazardous substances through controlled chemical reactions. These processes aim to detoxify pollutants, recover valuable metals, and minimize the environmental risks associated with improper disposal. Unlike physical techniques, chemical treatments can alter the molecular structure of contaminants, converting them into less toxic or more manageable forms. Common approaches include leaching, precipitation, oxidation-reduction, and chemical stabilization [32]. Advanced oxidation processes (AOPs), photocatalytic degradation, and dehalogenation are among the most effective modern techniques for treating e-waste-derived pollutants. These methods not only ensure efficient removal of persistent organic compounds and heavy metals but also support sustainable recycling and recovery. By integrating chemical treatments with physical and biological approaches, e-waste management can achieve cleaner production and environmental protection goals [33].

7.2.1 Advanced oxidation processes

AOPs are powerful chemical treatment techniques designed to degrade complex organic pollutants in e-waste leachates and effluents. They rely on the *in-situ* generation of highly reactive hydroxyl radicals, which possess strong oxidative potential to mineralize contaminants into harmless end products such as water and carbon dioxide. Common AOPs include Fenton and photo-Fenton reactions, ozonation, and UV/H_2O_2 systems [34]. These processes effectively break down persistent compounds like brominated flame retardants, PCBs, and heavy organic residues commonly found in e-waste recycling operations. AOP can be integrated with physical pretreatment methods for enhanced efficiency and reduced chemical usage. Despite their effectiveness, factors such as pH, catalyst concentration, and energy demand influence their overall performance. The versatility and eco-friendly nature of AOP make them a promising technology for sustainable e-waste treatment and environmental remediation [35].

7.2.2 Photocatalytic degradation

Photocatalytic degradation is an emerging green technology that utilizes semiconductor nanoparticles such as titanium dioxide (TiO_2) and zinc oxide (ZnO) to degrade organic pollutants under light irradiation. When exposed to UV or visible light, these nanoparticles generate electron-hole pairs that initiate redox reactions, leading to the breakdown of hazardous organic compounds into less toxic substances. In e-waste management, photocatalysis has shown remarkable efficiency in decomposing dyes, heavy organic molecules, and flame retardants present in leachates and residues [36]. TiO_2 is particularly favored for its chemical stability, nontoxicity, and strong photocatalytic activity, while ZnO offers high efficiency under UV exposure and is cost-effective. The process can be further enhanced through doping or nanocomposite formation to extend light absorption into the visible range. Thus, photocatalytic degradation offers a sustainable, low-energy solution for detoxifying e-waste contaminants and advancing eco-friendly waste treatment technologies [37].

7.2.3 Chemical stabilization and dehalogenation

Chemical stabilization and dehalogenation are crucial chemical methods for reducing the toxicity and mobility of hazardous components in e-waste. Chemical stabilization involves adding reagents that immobilize heavy metals or toxic elements by converting them into stable, insoluble compounds, thus preventing leaching into the environment. Common stabilizing agents include phosphates, silicates, and lime-based materials. Dehalogenation, on the other hand, targets halogenated organic compounds such as PCBs and brominated flame retardants present in circuit boards and plastics [38]. Through reductive or catalytic reactions, halogen atoms are removed, transforming these persistent pollutants into less harmful hydrocarbons. Technitreatments such as base-catalyzed decomposition and subcritical water treatment are often used to enhance dehalogenation efficiency. Together, these processes play a pivotal role in mitigating environmental hazards, ensuring safer disposal, and facilitating the recycling of valuable materials from e-waste streams [4].

7.3 Biological methods

Biological methods offer eco-friendly and sustainable alternatives for managing e-waste by utilizing microorganisms, fungi, and enzymes to degrade or transform hazardous pollutants into nontoxic forms. These techniques harness the natural metabolic capabilities of living organisms to detoxify POP, heavy metals, and other contaminants found in e-waste. Unlike chemical or physical methods, biological treatments are cost-effective, energy-efficient, and generate minimal secondary pollution [39]. Processes such as microbial degradation, fungal bioremediation, and enzyme-mediated reactions are gaining attention for their effectiveness in degrading complex organic compounds and immobilizing toxic elements. The integration of biological approaches with chemical and physical treatments enhances overall remediation efficiency and promotes green waste management. Thus, biological methods play a vital role in advancing environmentally sustainable e-waste management strategies [40].

7.3.1 Microbial degradation of persistent organic pollutants

Microbial degradation is a promising biological approach for the remediation of POPs present in e-waste, such as PCBs, dioxins, and brominated flame retardants. Certain bacteria possess specialized enzymes capable of breaking down these toxic compounds through oxidative and reductive pathways. Aerobic bacteria such as *Pseudomonas*, *Bacillus*, and *Sphingomonas* utilize oxygen-dependent enzymes like oxygenases and peroxidases to initiate the degradation of POP, converting them into simpler organic molecules [41]. Under anaerobic conditions, reductive dechlorination occurs, where halogen atoms are replaced by hydrogen, thereby detoxifying the compound. Factors like temperature, pH, and nutrient availability significantly influence microbial activity and degradation efficiency. Genetic engineering and microbial consortia have further enhanced the ability of microbes to degrade complex mixtures of pollutants. Therefore, microbial degradation provides an environmentally sound solution for managing POP-contaminated e-waste and restoring ecological balance [42].

7.3.2 Fungal bioremediation

Fungal bioremediation, or mycoremediation, utilizes the enzymatic and absorptive capabilities of fungi to degrade and detoxify hazardous compounds found in e-waste. Fungi, especially white-rot species such as *Phanerochaete chrysosporium* and *Trametes versicolor*, produce extracellular enzymes like laccases, lignin peroxidases, and manganese peroxidases that can oxidize complex organic pollutants [43]. These enzymes enable fungi to break down persistent compounds such as PAHs, brominated flame retardants, and dyes into simpler, less toxic molecules. Additionally, fungal mycelia have strong metal-binding properties, allowing them to adsorb and immobilize heavy metals from contaminated e-waste leachates. Fungal bioremediation is advantageous due to its low cost, scalability, and ability to function under varied environmental conditions. Its integration with microbial and chemical processes further enhances the overall detoxification efficiency, making it a sustainable approach for e-waste treatment and environmental restoration [44].

7.3.3 Enzyme-based degradation mechanisms

Enzyme-based degradation represents a cutting-edge biological approach in e-waste management, focusing on the direct catalytic breakdown of pollutants by purified or immobilized enzymes. These enzymes, derived from bacteria and fungi, act on specific chemical bonds within contaminants, transforming them into less toxic or biodegradable forms. Key enzymes such as laccases, peroxidases, dehalogenases, and hydrolases play crucial roles in degrading POPs, halogenated compounds, and dyes found in e-waste residues [45]. Laccases and peroxidases oxidize aromatic and halogenated compounds, while dehalogenases specifically target chlorinated or brominated molecules, facilitating detoxification. The use of immobilized enzyme systems enhances stability, reusability, and efficiency under diverse environmental conditions. Enzyme-based remediation is highly specific, rapid, and environmentally benign, making it suitable for large-scale application. By integrating enzyme technology with nanomaterials or biosorbents, the degradation of e-waste-derived pollutants can be significantly accelerated, contributing to sustainable waste management practices [46].

7.4 Emerging technologies

Emerging technologies in e-waste management focus on innovative, sustainable, and high-efficiency methods for pollutant removal and resource recovery. These advanced techniques aim to overcome the limitations of conventional physical, chemical, and biological treatments by incorporating nanotechnology, electrochemical processes, and green catalytic systems. Such technologies enhance degradation rates, minimize environmental impact, and promote circular economy principles. Their integration into existing treatment systems holds great promise for achieving sustainable e-waste remediation and resource conservation.

7.4.1 Nanotechnology-based remediation

Nanotechnology-based remediation represents a revolutionary approach in e-waste management, utilizing nanomaterials with exceptional surface reactivity, adsorption capacity, and catalytic efficiency. Nanoparticles such as zero-valent iron, titanium dioxide, and carbon nanotubes are employed to remove or degrade toxic metals and organic pollutants from e-waste leachates. These nanomaterials can adsorb heavy metals like lead, cadmium, and mercury and catalytically break down persistent organic compounds [47]. Their small particle size and large surface area allow faster reaction kinetics and enhanced contaminant removal. Additionally, functionalized nanoparticles can selectively target specific pollutants, improving remediation precision. Nanotechnology also enables recovery of valuable metals through nanosorbents and nanofiltration membranes, supporting resource circularity. However, the potential environmental risks of nanoparticle release must be carefully evaluated. Hence, nanotechnology-based remediation provides a highly efficient, adaptable, and sustainable solution for the complex challenges of e-waste management [48].

7.4.2 Electrochemical oxidation

Electrochemical oxidation is an advanced, eco-friendly technology used to degrade organic pollutants and recover valuable metals from e-waste effluents. The process employs an electric current passed through electrodes immersed in contaminated solutions, generating powerful oxidizing agents such as hydroxyl radicals and ozone. These species effectively mineralize hazardous compounds like brominated flame retardants, PCBs and dyes into nontoxic products such as CO_2 and water. Materials like boron-doped diamond, graphite, and titanium-based electrodes enhance oxidation efficiency and durability [49]. Electrochemical oxidation is advantageous due to its minimal chemical use, easy controllability, and potential for integration with renewable energy sources. Moreover, it can be coupled with electrocoagulation for simultaneous pollutant removal and metal recovery. This method represents a sustainable and scalable solution for treating e-waste leachates, reducing environmental toxicity, and advancing clean technological applications in waste management [50].

7.4.3 Green catalysis and sustainable approaches

Green catalysis and sustainable approaches are emerging as key strategies for environmentally responsible e-waste management. These techniques focus on employing nontoxic, recyclable, and energy-efficient catalysts to transform hazardous pollutants into harmless or reusable products. Catalysts such as bio-based materials, metal-organic frameworks, and ionic liquids are used to degrade POPs and recover valuable metals under mild reaction conditions. Green catalysis minimizes secondary pollution, reduces energy consumption, and promotes the circular use of materials [51]. Additionally, combining catalytic degradation with renewable energy-driven systems such as solar photocatalysis and bio-catalytic oxidation enhances treatment efficiency while maintaining environmental safety. Sustainable approaches emphasize the integration of eco-design, resource recovery, and waste-to-value conversion within the e-waste management cycle. Together, these innovations align with global sustainability goals by advancing cleaner production, reducing ecological footprints, and fostering a greener, more resilient technological future [52].

8 Policy and regulatory framework

Effective policy and regulatory frameworks are crucial for ensuring the safe handling, recycling, and disposal of e-waste, minimizing its environmental and health impacts. International conventions such as the Basel Convention (1989) and the Stockholm Convention (2001) play a pivotal role in this regard. The Basel Convention regulates the transboundary movement of hazardous waste, promoting waste minimization and cleaner technologies, while the Stockholm Convention targets the elimination of POPs found in e-waste, including brominated flame retardants and PCBs. Together, they provide a global foundation for responsible e-waste management, though challenges in enforcement and compliance persist, especially in developing nations. At the national level, India's E-Waste (Management) Rules (2011, 2016, and 2022) establish a strong legal framework emphasizing Extended Producer Responsibility (EPR), formal recycling, and traceability through a digital EPR portal. Despite these advancements, weak enforcement, limited awareness, and the dominance of the informal sector undermine effectiveness. Many producers fail to meet EPR targets, and inadequate infrastructure and coordination among agencies further impede progress. Bridging these gaps requires stronger institutional mechanisms, harmonized global cooperation, enhanced public participation, and incentives for green technologies to build a transparent, efficient, and sustainable e-waste management system [53].

9 Challenges and limitations

Despite technological and policy advancements, e-waste management faces persistent challenges that limit effective implementation. The highly heterogeneous composition of e-waste, comprising metals, plastics, glass, ceramics, and hazardous substances like lead, mercury, and brominated flame retardants, makes dismantling, segregation, and material recovery complex. The coexistence of valuable and toxic components complicates recycling, often resulting in secondary pollution and resource loss. Rapid innovation and product miniaturization further hinder standardization in recycling processes [54]. Additionally, the high cost of advanced treatment technologies such as hydrometallurgy, pyrometallurgy, and electrochemical oxidation poses economic barriers, particularly in developing nations. These methods demand significant investment in infrastructure, energy, and expertise, rendering formal recycling less profitable compared to informal, low-cost operations. The absence of efficient collection systems and unstable markets for recovered materials further reduces financial viability. Compounding these issues, developing countries suffer from inadequate infrastructure, limited formal recycling units, and unsafe informal practices like open burning and acid leaching, leading to severe environmental and health risks [55]. Weak enforcement, poor public awareness, and insufficient technical and financial support exacerbate these limitations. Overcoming these challenges requires technological innovation, affordable and energy-efficient treatment methods, capacity building, and international cooperation. Strengthening institutional frameworks, expanding formal recycling infrastructure, and promoting public-private partnerships are essential for achieving safe, economically viable, and sustainable e-waste management systems globally [56].

10 Future directions

The future of e-waste management depends on adopting sustainable, technology-driven, and circular strategies that enhance resource recovery and environmental protection. Integrating circular economy principles shifts the focus from the linear "take–make–dispose" model to one emphasizing reuse, repair, remanufacturing, and recycling. This approach encourages modular product designs, buy-back programs, and product-as-a-service models that extend device lifecycles and reduce raw material extraction. Complementary policies like EPR and eco-design standards can further drive innovation and sustainable manufacturing. Another crucial direction is green material design, which aims to replace hazardous substances such as brominated flame retardants and PCBs with biodegradable, recyclable, or low-toxicity alternatives. Using green chemistry and life cycle assessment (LCA), manufacturers can develop halogen-free flame retardants, bio-based polymers, and recyclable substrates that minimize environmental pollution and simplify recycling processes [57]. Examining the effectiveness of environmental policies and governance frameworks in addressing environmental challenges in e-waste management and promoting sustainability in Accra, Ghana. Additionally, Artificial Intelligence (AI) and digital monitoring offer transformative potential by enabling automation, real-time data tracking, and predictive analytics. AI-powered image recognition can optimize waste sorting and recovery, while blockchain and IoT-based systems enhance transparency in e-waste flows and EPR compliance. Machine learning can forecast waste trends and improve collection logistics, making recycling operations more efficient and traceable. Together, these forward-looking strategies, circular economy integration, green material innovation, and AI-driven digitalization, can create a regenerative, closed-loop e-waste management system that promotes economic efficiency, technological advancement, and long-term environmental sustainability [58].

11 Conclusion

Electronic waste poses significant environmental and health risks due to the presence of hazardous substances such as POPs, heavy metals, and flame retardants. Improper disposal and informal recycling practices lead to the release of these pollutants into soil, water, and air, causing ecological damage and human health issues. Exposure to e-waste toxins can cause respiratory diseases, neurological damage, reproductive problems, and increased cancer risk. Detection and monitoring of organic pollutants in e-waste involve advanced analytical techniques like chromatography and spectrometry. Remediation strategies include physical methods, chemical approaches, and biological techniques. Emerging technologies such as nanotechnology, electrochemical oxidation, and green catalysis offer promising solutions for sustainable e-waste management. However, challenges persist, including the complex composition of e-waste, high treatment costs, and inadequate infrastructure in developing countries. Effective policy frameworks, international cooperation, and the adoption of circular economy principles are crucial for addressing the e-waste crisis and promoting sustainable solutions.

References

[1] Vishwakarma A, Kanaujia K, Hait S, Global scenario of E-waste generation: Trends and future predictions, In: Global E-Waste Management Strategies and Future Implications. Elsevier, 2023, 13–30.

[2] Zhao X, Li W, Wang W, Liu J, Yu Y, Li Y, Liu Y, Legacies and health risks of heavy metals, polybrominated diphenyl ethers, and polychlorinated dibenzo-dioxins/furans at e-waste recycling sites in South China. Front Environ Sci Eng, 2023, 17(7), 79.

[3] Goyal S, Gupta SA, Comprehensive review of current techniques, issues, and technological advancements in sustainable e-waste management. E-Prime Adv Electr Eng Electron Energy, 2024, 9, 100702.

[4] Gaur TS, Yadav V, Prakash S, Panwar A, Integration of Industry 4.0 and circular economy for sustainable e-waste management. Manag Environ Qual Int J, 2025, 36(5), 1304–1325.

[5] Ike-Eze ICE, Ucheji ON, Ubi PA, Oji EO, Aigbodion VS, Omah AD, Ogbuefi UC, An overview of flame retardants in printed circuit boards for LEDs and other electronic devices. J Mater Environ Sci, 2023, 14(4), 410–420.

[6] Lin C, Zeng Z, Xu R, Liang W, Guo Y, Huo X, Risk assessment of PBDEs and PCBs in dust from an e-waste recycling area of China. Sci Total Environ, 2022, 803, 150016.

[7] Ge X, Hashmi MZ, Lin S, Qi Z, Yu Y, An T, Emission characteristics of (halogenated) polycyclic aromatic hydrocarbons during printed circuit board combustion and estimated emission intensity of a typical e-waste dismantling site in South China. Environ Pollut, 2023, 334, 122162.
[8] Huang W, Huang Y, Chen Y, Tan W, Wu K, A comprehensive review of the human body burden of persistent organic pollutants and associated health effects in an e-waste recycling area in China. Discov Environ, 2023, 1(1), 13.
[9] Mateescu C, Lungulescu EM, Nicula NO, Effectiveness of biological approaches for removing persistent organic pollutants from wastewater: A mini-review. Microorganisms, 2024, 12(8), 1632.
[10] Purchase D, Abbasi G, Bisschop L, Chatterjee D, Ekberg C, Ermolin M, Wong MH, Global occurrence, chemical properties, and ecological impacts of e-wastes (IUPAC Technical Report). Pure Appl Chem, 2020, 92(11), 1733–1767.
[11] Saha L, Kumar V, Tiwari J, Rawat S, Singh J, Bauddh K, Electronic waste and their leachates impact on human health and environment: Global ecological threat and management. Environ Technol Innov, 2021, 24, 102049.
[12] Renzelli V, Gallo M, Morviducci L, Marino G, Ragni A, Tuveri E, Silvestris N, Polybrominated diphenyl ethers and human health: Effects on metabolism, diabetes and cancer. Cancers, 2023, 15(17), 4237.
[13] Khan J, Kumar A, Giri A, Db P, Tripathi A, Giri DD . Impact of electronic waste pollutants on underground water, In: Groundwater Geochemistry: Pollution and Remediation Methods. John Wiley & Sons Ltd., 2021, 265–281.
[14] Owusu-Sekyere K, Aladago DA, Leverenz D, Oteng-Ababio M, Kranert M, Environmental impacts on soil and groundwater of informal e-waste recycling processes in Ghana. Sustainability, 2024, 16(11), 4347.
[15] Fujimori T, Itai T, Goto A, Asante KA, Otsuka M, Takahashi S, Tanabe S, Interplay of metals and bromine with dioxin-related compounds concentrated in e-waste open burning soil from Agbogbloshie in Accra, Ghana. Environ Pollut, 2016, 209, 155–163.
[16] Angon PB, Islam MS, Das A, Anjum N, Poudel A, Suchi SA, Sources, effects and present perspectives of heavy metals contamination: Soil, plants and human food chain. Heliyon, 2024, 10(7), e28357.
[17] Alabi OA, Adeoluwa YM, Huo X, Xu X, Bakare AA, Environmental contamination and public health effects of electronic waste: An overview. J Environ Health Sci Eng, 2021, 19(1), 1209–1227.
[18] Parvez SM, Jahan F, Brune MN, Gorman JF, Rahman MJ, Carpenter D, Sly PD, Health consequences of exposure to e-waste: An updated systematic review. Lancet Planet Health, 2021, 5(12), e905–e920.
[19] Murthy V, Ramakrishna S, A review on global e-waste management: Urban mining towards a sustainable future and circular economy. Sustainability, 2022, 14(2), 647.
[20] Lawrence PG, Roper W, Morris TF, Guillard K, Guiding soil sampling strategies using classical and spatial statistics: A review. Agron J, 2020, 112(1), 493–510.
[21] Kalina J, Scheringer M, Borůvková J, Kukučka P, Přibylová P, Bohlin-Nizzetto P, Klánová J, Passive air samplers as a tool for assessing long-term trends in atmospheric concentrations of semivolatile organic compounds. Environ Sci Technol, 2017, 51(12), 7047–7054.
[22] Bika SH, Adeniji AO, Okoh AO, Okoh OO, Spatiotemporal distribution and analysis of organophosphate flame retardants in the environmental systems: A review. Molecules, 2022, 27(2), 573.
[23] Shyamalagowri S, Shanthi N, Manjunathan J, Kamaraj M, Manikandan A, Aravind J, Techniques for the detection and quantification of emerging contaminants. Phys Sci Rev, 2023, 8(9), 2191–2218.
[24] Wang JR, Kuang HX, Liu Y, Li XY, Chen TH, Zhu XH, Yu YJ, Associations between volatile organic compounds exposure and multiple oxidative damage biomarkers: Method development, human exposure, and application for e-waste pollution prediction. Sci Total Environ, 2024, 956, 177402.
[25] Nandan A, Suresh AC, Saole P, Jeevanasai SA, Chandrasekaran R, Meili L, Selvasembian R, An integrated approach for electronic waste management – overview of sources of generation,

toxicological effects, assessment, governance, and mitigation approaches. Sustainability, 2023, 15(24), 16946.
[26] Agbim A, Schumacher KA, Sharp N, Paul R, Corzo R, Elemental characterization of electronic waste: A review of research methodologies and applicability to the practice of e-waste recycling. Waste Manag, 2024, 187, 91–100.
[27] Sarkhoshkalat MM, Afkham A, Bonyadi Manesh M, Sarkhosh M, Circular economy and the recycling of e-waste, In: New Technologies for Energy Transition Based on Sustainable Development Goals. Springer Nature Singapore, 2024, 319–354.
[28] Al-Gailani A, Bongosia JG, Hornsby K, Taylor MJ, Maximising e-waste leachate adsorption: Multi-component isotherm models and mechanisms with scalable mesoporous sugarcane bagasse-derived biochar. J Water Process Eng, 2025, 75, 107979.
[29] Rezvani B, Innovative approaches in adsorbent production through pyrolysis of waste biomass: Effective parameters, modifications, and techno-economic analysis. Can J Chem Eng, 2025, 103(11), 5296.
[30] Gadipelli P, Eco strategies for e-waste collection and responsible disposal, In: E-Waste: From Generation to Management. Springer Nature Singapore, 2025, 141–164.
[31] Walke S, Mandake M, Naniwadekar M, Review study of e-waste management and resource recovery system for controlling environmental pollution. Int J Environ Waste Manag, 2025, 37(2), 215–243.
[32] Bhattacharjee J, De A, Kamila B, Mandal A, Recent scenario of e-waste recycling: Chemical engineering. Int J Chem React Eng, 2025, 23(6), 675–700.
[33] Chen Z, Wei W, Chen H, Ni BJ, Recent advances in waste-derived functional materials for wastewater remediation. Eco-Environ Health, 2022, 1(2), 86.
[34] Jin Q, Zhan Y, Tao D, Wang T, Khim JS, He Y, Removing emerging e-waste pollutant DTFPB by synchronized oxidation-adsorption Fenton technology. J Hazard Mater, 2023, 445, 130587.
[35] Farag S, Farag Y, Attia M, Recycling technologies for brominated flame-retardant plastics in e-waste. Cleaner Waste Syst, 2025, 12, 100418.
[36] Erkmen AN, Ulber R, Jüstel T, Altendorfner M, Towards sustainable recycling of critical metals from e-waste: Bioleaching and phytomining. Resour Conserv Recycl, 2025, 215, 108057.
[37] Yang BC, Lee CH, Suryawan IWK, Resilient socio-technical systems for adaptive consumer e-waste management. Sustain Cities Soc, 2025, 118, 106026.
[38] Kavil YN, Alshemmari H, Alkasbi MM, Alelyani SS, Orif MI, Al-Farawati RK, Electronic waste and environmental diplomacy: How GCC e-waste management interfaces with the Stockholm Convention. J Hazard Mater Adv, 2025, 17, 100610.
[39] Bajpai M, Nigam M, Singh LK, Awasthi G, Electronic waste: An emerging pollutant of human health concern and biological techniques for their treatment, In: Environmental Toxicants and Lifestyle Diseases. Springer Nature Singapore, 2025, 141–156.
[40] Chakraborty A, Amrit R, Wagh MS, Ray MK, Panda BP, Tchouya GRF, From lab to landfill: Exploring nanobiotechnological approaches for e-waste remediation and environmental sustainability, In: Nano-microbiology for Sustainable Development. Springer Nature Switzerland, 2025, 243–267.
[41] Dutta B, Chatterjee D, Guha A, Ray RR, Green treatments for polyaromatic hydrocarbons in e-wastes. Biodegradation, 2025, 36(3), 48.
[42] Souza MCO, Domingo JL, Dioxins and the One Health paradigm: An interdisciplinary challenge in environmental toxicology. Toxics, 2025, 13, 964.
[43] Tiwari S, Aachhera S, Verma P, Gahan CS, Bioremediation of e-waste: A sustainable approach, In: Biotechnology for Environmental Sustainability. Springer Nature Singapore, 2025, 709–724.
[44] Ferreira-Filipe DA, Duarte AC, Hursthouse AS, Rocha-Santos T, Silva ALP, Biobased strategies for e-waste metal recovery: A critical overview of recent advances. Environments, 2025, 12(1), 26.
[45] Quinto S, Law N, Fletcher C, Le J, Jose SA, Menezes PL, Exploring the e-waste crisis: Strategies for sustainable recycling and circular economy integration. Recycling, 2025, 10(2), 72.

[46] Selvakumar P, Dahiya R, Manjunath TC, Nathsharma SK, Chemical and biological processes for ewaste recycling, In: Integrated Approaches for Sustainable E-Waste Management. IGI Global Scientific publishing, New York, London, Beijing, 2025, 151–176.
[47] Mukherjee T, Mitra S, Herat S, Kaparaju P, Nanotechnology as a cleaner and greener approach toward e-waste recycling and valorization, In: Development in E-waste Management. CRC Press, Boca Raton, 2023, 87–103.
[48] Dodamani S, Kurangi B, Teradal N, Kurjogi M, Nanobiotechnology for e-waste management, In: Nanobiotechnology. Elsevier, United States, 2021, 271–281.
[49] Saadaoui M, Mohamed AT, Anwer AH, Al-Qaradawi SY, Khaled M, Benamor A, Revitalizing e-waste: Eco-friendly electrochemical sensor for Hg (II) detection enhanced by oxygen vacancy in metal oxide nanostructures based on recycled LCD. Mater Today Sustain, 2025, 29, 101015.
[50] Zhang H, Dai X, Chen P, Song S, Jia F, Efficient and eco-friendly gold extraction from electronic waste via thiosulfate electrochemical oxidation. Sep Purif Technol, 2025, 361, 131375.
[51] Zheng S, Yang S, Chen J, Wu D, Qi B, Zhang C, Wan L, New progresses in efficient, selective, and environmentally friendly recovery of valuable metal from e-waste and industrial catalysts. Adv Sustain Syst, 2024, 8(6), 2300512.
[52] Hossain R, Sahajwalla V, Current recycling innovations to utilize e-waste in sustainable green metal manufacturing. Philos Trans A, 2024, 382(2284), 20230239.
[53] Bimpong FAK, Asibey MO, Inkoom DKB, Ghana's recently introduced e-waste regulatory policy: A hope for a better e-waste sector?. Waste Manag Res, 2024, 42(11), 1031–1041.
[54] Wirtu YD, Godana UA, Tucho GT, A review of public perception and awareness of e-waste management in developing countries: Challenges and opportunities for sustainable solutions. Environ Rev, 2025, 33, 1–12.
[55] Doppalapudi PK, Iqbal SS, Dagwar PP, Dutta D, E-waste management: Balancing environmental challenges and economic opportunities, In: A Vision for Environmental Sustainability. Springer Nature Switzerland, 2025, 25–53.
[56] Rauf AU, Electronic waste problem in developing nations: Mismanagement, health implications, and circular economy opportunities. J Environ Health, 2024, 16(1), 18–31.
[57] Shanti Q, Law N, Fletcher C, Le J, Subin AJ, Menezes PL, Exploring the e-waste crisis: Strategies for sustainable recycling and circular economy integration. Recycling, 2025, 10(2), 72.
[58] Singh V, Bhattacharya S, Chugh P, Global perspectives on e-waste management: Challenges, innovations, and sustainable solutions, In: E-Waste: From Generation to Management. Springer Nature Singapore, 2025, 23–39.

John Tsado Mathew*, Abel Inobeme, Jonathan Hussaini, Elijah Yanda Shaba, Yakubu Azeh, Musah Monday, Tanko Ezekiel, Adam Ibn Aliyu, Etsuyankpa Bini Muhammad, Ismaila Aliyu Oga, Muhammad Aishetu Ibrahim, Mohammed Chado Isah, Jibrin Noah Akoji, Charles Oluwaseun Adetunji, Aliyu Mohammed Sakpe, Amos Mamman

Organic pollutants from electronic wastes and their remediation strategies

Abstract: Organic pollutants from electronic waste (e-waste), such as flame retardants, plasticizers, and persistent organic compounds, pose great risks to environmental and human health. The toxic pollutants leach into soil, water, and air during uncontrolled disposal or informal recycling and result in bioaccumulation and long-term environmental impairment. Exposure to them has been traced to endocrine disruption, carcinogenicity, and developmental disorders in humans. Proper remediation measures are necessary, from advanced oxidation processes, bioremediation through microbes and enzymes, and adsorption methods to green chemical recycling. Adoption of these measures, together with sustainable waste management policies, is mandatory for risk mitigation and environmental safeguarding.

Keywords: Electronic wastes, organic pollutants, remediation, sustainable management, Exposure

***Corresponding author: John Tsado Mathew,** Department of Chemistry, Ibrahim BadamasiBabangida University, Lapai, Niger State, Nigeria, e-mail: johntsadom@gmail.com
Abel Inobeme, Department of Chemistry, Edo State University, Uzairue, PMB 04, Auchi, Edo State, Nigeria
Jonathan Hussaini, Yakubu Azeh, Musah Monday, Muhammad Aishetu Ibrahim, Amos Mamman, Department of Chemistry, Ibrahim BadamasiBabangida University, Lapai, Niger State, Nigeria
Elijah Yanda Shaba, Tanko Ezekiel, Adam Ibn Aliyu, Department of Chemistry, Federal University of Technology, Minna, Nigeria
Etsuyankpa Bini Muhammad, Ismaila Aliyu Oga, Department of Chemistry, Federal University of Lafia, Nasarawa State
Mohammed Chado Isah, Department of Biology, Ibrahim Badamasi Babangida University, Lapai, Niger State, Nigeria
Jibrin Noah Akoji, Department of Chemistry, Baze University, Abuja, Nigeria
Charles Oluwaseun Adetunji, Applied Microbiology, Biotechnology and Nanotechnology Laboratory, Department of Microbiology, Edo State University, Uzairue, PMB 04, Auchi, Edo State, Nigeria
Aliyu Mohammed Sakpe, Department of Basic Sciences, Niger State College of Agriculture, Mokwa, Nigeria

 | https://doi.org/10.1515/9783112248072-005

1 Introduction

Electronic waste (e-waste) has emerged as one of the most rapidly growing solid waste streams worldwide due to the increasing consumption of electrical and electronic equipment and the reduced lifespan of modern devices. Technological advancement, urbanization, population growth, and consumer demand for sophisticated electronics have contributed significantly to the global surge in e-waste generation. According to recent global assessments, more than 62 million tons of e-waste were generated in 2022, and the figure is projected to exceed 75 million tons by 2030 as digitalization expands across developing and developed nations. E-waste consists of discarded electronic appliances such as computers, mobile phones, televisions, printers, batteries, refrigerators, and other digital devices. While many discussions on e-waste emphasize heavy metals such as lead, cadmium, and mercury, substantial attention is now being directed toward organic pollutants embedded in or released from e-waste because of their toxicity, persistence, and potential to bioaccumulate in the environment [1].

Organic pollutants from e-waste include polybrominated diphenyl ethers (PBDEs), polychlorinated biphenyls (PCBs), polycyclic aromatic hydrocarbons (PAHs), phthalates, dioxins, and furans, among others. These chemicals are used extensively as flame retardants, plasticizers, insulating fluids, and additives in electronic components. During informal recycling processes – such as open burning of cables, dismantling of plastic casings, crude extraction of printed circuit boards, or acid leaching – these toxic organic compounds are released into soil, water, and air. In developing nations, particularly across Africa and parts of Asia, informal e-waste recycling sites have become hotspots for organic pollutants, exposing workers, nearby residents, and the environment to significant health hazards [2]. Research from countries such as China, Ghana, and Nigeria has shown elevated concentrations of PBDEs, PAHs, and dioxins in soils, dust, sediments, and biological samples around e-waste dumpsites. These contaminants are known to persist for many years, undergo long-range atmospheric transport, and enter food chains, posing risks at local and global levels [3]. The toxicity of organic pollutants from e-waste is well-documented. PBDEs, for instance, interfere with thyroid hormone regulation and neurodevelopment; PCBs are classified as carcinogens; PAHs can induce DNA damage; and dioxins are among the most toxic chemicals known, with effects ranging from endocrine disruption to immunotoxicity. Human exposure occurs through inhalation of contaminated air, ingestion of polluted food or water, and dermal contact with contaminated soil or dust. Vulnerable groups such as children are particularly at risk due to their frequent hand-to-mouth activities and developing physiological systems. The environmental impact is equally severe, as these pollutants reduce soil fertility, impair aquatic ecosystems, decrease biodiversity, and alter microbial communities essential for nutrient cycling. Thus, the increasing recognition of organic pollutants as critical constituents of e-waste has positioned them as emerging contaminants of global concern [4].

Managing the risks associated with organic pollutants in e-waste requires effective and sustainable remediation strategies. Conventional remediation approaches include physical, chemical, and biological methods. Physical methods such as soil washing, thermal desorption, and adsorption using activated carbon are commonly applied but may be costly and energy-intensive. Chemical remediation strategies – including oxidation, reduction, and advanced oxidation processes – can degrade contaminants but sometimes generate secondary pollutants [5]. Biological methods, or bioremediation, have gained significant attention due to their cost-effectiveness, environmental friendliness, and ability to degrade complex organic compounds. Microbial degradation, phytoremediation using plants, and enzymatic treatments are increasingly explored for removing organic pollutants from contaminated sites. Recent advancements have also introduced emerging strategies such as nanoremediation, biochar-assisted remediation, and the use of genetically engineered microorganisms to enhance pollutant degradation. These modern approaches aim to improve efficiency, reduce costs, and minimize environmental impacts while supporting a circular and sustainable waste management system [6]. Despite the growing body of research, many developing countries still lack proper infrastructure, regulatory frameworks, and technological capacity to manage e-waste safely. The informal sector dominates e-waste handling in regions such as Nigeria, where crude recycling techniques release large quantities of toxic organic pollutants into the environment. Weak enforcement of environmental regulations, limited public awareness, inadequate waste collection systems, and poor recycling technologies continue to exacerbate the problem. As a result, the burden of organic pollution persists, disproportionately affecting low-income communities situated near dumpsites or informal recycling centers. Understanding the nature, sources, environmental pathways, and health implications of organic pollutants from e-waste is therefore crucial for designing effective mitigation and remediation strategies [7].

2 Generation and global trends of e-waste

E-waste, commonly referred to as e-waste, has become one of the fastest-growing waste streams worldwide. Its rapid increase is attributed to constant technological innovation, rising consumer demand for electronics, and the reduced lifespan of modern devices. E-waste includes discarded electrical and electronic equipment such as mobile phones, laptops, televisions, refrigerators, printers, and batteries. As digital technologies continue to penetrate all aspects of human life – including communication, healthcare, education, banking, transportation, and industry – the volume of discarded electronic equipment rises at an alarming rate. Recent global assessments report that the world produced approximately 62 million tons of e-waste in 2022, and this figure is projected to reach between 75 and 82 million tons by 2030 if current trends persist. This unprecedented increase has positioned e-waste as a critical global environmental challenge [1]. The distribution of e-waste generation varies greatly across regions, influenced by economic development, technological advancement, and population size. High-income regions such as Europe, North America, and Oceania generate the highest per-capita e-waste due to strong purchasing power and rapid turnover of electronic products. Europe produces about 17.6 kg of e-waste per person annually, while Oceania and North America generate approximately 16.1 kg and 13.5 kg per person, respectively. In contrast, Africa generates the lowest amount – around 2.5 kg per capita – yet suffers disproportionately from the harmful effects of e-waste due to significant imports of used electronics from developed nations. Asia, being the most populous continent, generates the highest total volume at over 24 million tons per year, largely driven by rapid industrialization, large populations, and growing middle-class demand for electronic devices [8].

Several key factors contribute to the global rise in e-waste. Rapid technological advancement results in constant upgrades and a steady stream of new products, making older devices obsolete quickly. Consumerism and changing lifestyles also play major roles, as individuals increasingly use multiple electronic devices such as smartphones, tablets, smartwatches, and laptops. Urbanization and industrialization in developing nations further increase the demand for digital systems in commerce, governance, and education. Another significant contributor is planned obsolescence, where manufacturers design products with shorter lifespans or limited repairability, encouraging frequent replacement rather than repair. Additionally, global population growth naturally increases the number of consumers purchasing electronics, thereby raising e-waste generation [9]. A major global concern is the transboundary movement of e-waste. A substantial portion of discarded electronics from developed countries is transported to developing nations under the label of "used electronics" for reuse. Despite international regulations such as the Basel Convention, illegal shipments of near-end-of-life devices continue. Countries such as Nigeria, Ghana, India, and Vietnam have become major destinations for imported e-waste. Weak regulations, low labor costs, and thriving informal recycling sectors make these regions vulnerable to unsafe recycling practices. Informal recyclers often dismantle electronic devices manually and use crude methods such as open burning and acid leaching to recover valuable metals. These practices release toxic organic pollutants – including dioxins, furans, PAHs, and brominated flame retardants (BFRs) – into the air, soil, and water, posing significant environmental and health risks [1].

Global management and recycling practices reveal substantial disparities between regions. While the world generates millions of tons of e-waste annually, only about 17.4% is formally collected and recycled. Europe records the highest formal recycling rate of over 42%, thanks to strong policies, advanced recycling technologies, and established collection systems. In stark contrast, Africa's formal recycling rate is below 1%, and most e-waste is handled through informal channels. Formal recycling typically involves advanced technological processes such as automated sorting, mechanical shredding, hydrometallurgy, and controlled smelting [10, 11]. These methods maximize resource recovery while minimizing environmental pollution. However, they require significant capital investment, technical expertise, and strong policy support, which many developing countries lack. Consequently, informal recycling remains dominant in low-income regions, leading to widespread contamination and exposure to hazardous pollutants [12]. Recent global trends indicate a growing shift toward more sustainable e-waste management. Many countries are adopting Extended Producer Responsibility (EPR) policies, which require manufacturers to take back and manage their products at end-of-life. Such approaches encourage eco-friendly designs, repairability, and recycling-friendly materials. Circular economy initiatives are also gaining momentum, emphasizing product repair, refurbishment, reuse, and material recovery to minimize waste generation. Innovations such as biodegradable electronic components, modular device designs, and digital tracking systems using blockchain and IoT are being explored to improve transparency and reduce illegal dumping. Public awareness campaigns worldwide are educating consumers on the importance of proper disposal and responsible consumption to reduce the environmental impact of e-waste [13]. Furthermore, global e-waste generation is increasing at an unprecedented rate due to technological advancement, lifestyle changes, and shorter product lifespans. While high-income countries generate the most e-waste per person, developing countries shoulder the greatest environmental burden due to informal recycling and illegal imports. Without strengthened policies, improved recycling technologies, and increased public awareness, e-waste will continue to pose significant environmental and public health challenges. Understanding global generation and trends is therefore vital for developing sustainable strategies that reduce the impacts of e-waste and support a safer, circular, and environmentally responsible future [1].

3 Composition of e-waste

E-waste is a highly complex and diverse waste stream composed of discarded electrical and electronic devices that contain a mixture of valuable, hazardous, and nonhazardous materials. Its composition varies depending on the type of equipment, manufacturing technology, age of the device, and the purpose for which it was produced. Generally, e-waste is made up of metals, plastics, glass, ceramics, and a wide range of chemical components such as flame retardants, organic pollutants, and battery electrolytes. Understanding this composition is essential because it determines the environmental impacts of e-waste, the risks associated with improper handling, and the most suitable recycling or remediation approaches [1]. Metals constitute a significant portion of e-waste, both in terms of weight and value. Common metals include iron, copper, aluminum, and nickel, which are used in structural components, wiring, coils, and electronic casings. Precious metals such as gold, silver, palladium, and platinum are present in printed circuit boards, semiconductors, connectors, and microprocessors. These precious metals make ICT waste – such as computers and mobile phones – highly valuable for recycling. However, the concentration of these metals varies by device type and age, with older electronics often containing higher amounts of valuable metals. Hazardous heavy metals such as lead, mercury, cadmium, and chromium are also common. Lead is widely used in solder and cathode ray tubes (CRTs), mercury exists in fluorescent lamps and LCD screens, while cadmium is found in rechargeable batteries and some semiconductors. These heavy metals pose serious environmental and health risks when released into soil, water, or air through informal recycling and burning [11]. Plastics make up another major portion of e-waste, often accounting for 15–25% of an electronic device's weight. They are used in casings, circuit board substrates, cable insulation, switches, and connectors. Many plastics in e-waste contain toxic additives such as BFRs. These compounds are used to reduce flammability but become hazardous pollutants when e-waste is burned, releasing persistent organic pollutants (POPs) such as dioxins and furans. Polyvinyl chloride (PVC), a common material in wire insulation and casings, also generates toxic chlorine-based compounds when combusted [14].

Glass is another important component, particularly in older television screens and computer monitors. CRTs contain leaded glass, which is both difficult to recycle and hazardous. Newer devices using LCD and LED technologies contain specialized glass with mercury, indium, and rare earth elements. These components require careful handling to avoid environmental contamination [15]. Batteries present in many electronic devices add further complexity to e-waste composition. Lithium-ion, nickel–cadmium, nickel–metal hydride, and lead–acid batteries contain metals such as lithium, cobalt, cadmium, and lead, as well as electrolytes that can leak into the environment. These batteries contribute significantly to the toxicological profile of e-waste if not properly managed [16]. Other parts of e-waste include circuit boards, which contain resins, fiberglass, metals, and chemicals; wiring and cables made from copper or aluminum encased in PVC; and various electronic components such as resistors, capacitors, and sensors containing ceramics and polymers. Solar panels, medical devices, and industrial electronics add even more specialized materials such as silicon wafers, silver pastes, and thick-film coatings [17].

4 Organic pollutants present in e-waste

E-waste contains several hazardous organic pollutants that pose serious risks to environmental and human health. These pollutants originate from plastics, circuit boards, cables, insulation materials, flame-retardant additives, and various synthetic components used in electronic equipment. During disposal, especially through informal recycling practices such as open burning, crude dismantling, and acid leaching, these organic contaminants are released into the surrounding air, soil, water, and food chains [18]. One major class of organic pollutants in e-waste is BFRs, including PBDEs, tetrabromobisphenol-A (TBBPA), and hexabromocyclododecane (HBCD). They are used to reduce flammability in plastics, printed circuit boards, televisions, and computer casings. These compounds are persistent in the environment, bioaccumulative, and linked to endocrine disruption, reproductive abnormalities, and neurodevelopmental disorders. When electronics are burned, BFRs degrade into even more toxic compounds such as polybrominated dioxins and furans, which are similar in toxicity to chlorinated dioxins [19].

Phthalates, commonly found in cable coatings, PVC plastics, and flexible electronic components, are another group of organic contaminants. These chemicals can leach into soil and water and act as endocrine disruptors, affecting hormone regulation and fetal development. Similarly, bisphenol A (BPA) – used in polycarbonate plastics and epoxy resins in electronics – is known for its estrogenic activity and potential to affect the reproductive system [20]. E-waste also contains PAHs, which are produced during the open burning of wires, plastics, and circuit boards. PAHs are carcinogenic and mutagenic, and their presence in contaminated sites is often associated with high cancer risks among workers and nearby residents. Additionally, organostannic compounds, such as tributyltin (TBT) and triphenyltin (TPT), used as stabilizers in PVC materials, are potent immunotoxins and endocrine disruptors [2].

4.1 Polybrominated diphenyl ethers

PBDEs are a class of brominated flame-retardant chemicals widely used in plastics, textiles, and electronic equipment to inhibit fire. Introduced in the 1970s, the most common commercial mixture, penta-PBDE, was employed for furniture and carpet fibers; octa- and deca-PBDE formulations replaced it in the 1990s and 2000s for more robust fire protection. PBDEs exhibit low acute toxicity but persist in the environment due to their lipophilicity, high thermal stability, and strong resistance to biodegradation. They bioaccumulate in fatty tissues of wildlife and humans, and are frequently detected in house dust, indoor air, and human breast milk at concentrations ranging from 0.1 to 10 µg/g [30].

Human exposure occurs mainly through ingestion of contaminated dust and consumption of animal products, with infants and toddlers at greatest risk. Epidemiological studies link PBDE body burdens to neurodevelopmental delays, endocrine disruption, and impaired thyroid function. In vitro and animal experiments have demonstrated that PBDEs interfere with hormone receptors, alter lipid metabolism, and induce oxidative stress. The International Agency for Research on Cancer (IARC) classifies certain PBDE congeners as possibly carcinogenic to humans (Group 2B) [31]. Regulatory responses have intensified worldwide. The United States banned the production of pen-

Table 1: Organic pollutants present in e-waste.

Organic pollutant	Sources in e-waste	Environmental/health concerns	References
Polybrominated diphenyl ethers (PBDEs)	Flame-retardant plastics, circuit boards, cables, and casings	Persistent; endocrine disruption; neurodevelopmental toxicity; bioaccumulation	[21]
Polychlorinated biphenyls (PCBs)	Transformers, capacitors, and old electronic components	Carcinogenic; immunotoxicity; liver damage; long-term environmental persistence	[22]
Polycyclic aromatic hydrocarbons (PAHs)	Produced during burning of plastics, wires, and casings	Mutagenic; carcinogenic; respiratory issues; soil and water contamination	[23]
Phthalates (DEHP, DBP, etc.)	Plasticizers in cables, coatings, and PVC components	Endocrine disruption; reproductive toxicity; developmental effects	[20]
Dioxins and furans	Emitted from informal burning of halogenated plastics	Highly toxic; carcinogenic; immune and reproductive system damage	[24]
Bisphenol A (BPA)	Polycarbonate plastics and epoxy resins in electronics	Hormonal imbalance; fertility issues; developmental defects	[25]
Organotin compounds (TBT, TPT)	Heat stabilizers in PVC cables and flexible plastics	Immunotoxicity; endocrine disruption; aquatic toxicity	[26]
Nonylphenols	Circuit boards, plastic additives, and surfactants	Hormone disruption; aquatic ecosystem toxicity	[27]
Polyvinyl chloride (PVC) by-products	Cable insulation and electronic housings	Releases toxic hydrochloric acid and organics when burned	[28]
Short-chain chlorinated paraffins (SCCPs)	Lubricants, plastic additives, and flame retardants	Persistent; bioaccumulative; suspected carcinogens	[29]

ta-PBDE in 2004 and octa-PBDE in 2008, and the European Union phased out all commercial PBDEs by 2006 under the POPs Convention. Current risk management focuses on the phase-out of remaining products, substitution with alternative flame-retardants, and remediation of contaminated indoor environments. Ongoing research aims to better quantify exposure pathways, refine toxicokinetic models, and develop cost-effective decontamination strategies. As global awareness grows, PBDEs exemplify the challenges of balancing fire safety with long-term environmental and public health. Future investigations will assess the long-term health outcomes of reduced PBDE exposure and evaluate the effectiveness of emerging flame-retardant technologies that promise lower ecological footprints while maintaining fire safety standards (Table 1) [30].

4.2 Polychlorinated biphenyls

PCBs are a series of artificial organic chemicals, which were broadly utilized in industrial and business use throughout the 1920s and then prohibited in the late 1970s. Their major use was due to the fact that they were chemically stable, low flammable, and had good dielectric characteristics, hence they were suitable in transformers capacitors and hydraulic systems as coolants and insulating fluids. PCBs represent a family of congeners (209 in total) differing in the number and location of chlorine atoms on the biphenyl backbone that gives them a range of physicochemical behaviors. Congeners that have the highest chlorination levels are more lipophilic and bioaccumulative in tissues that are fatty and the less chlorinated ones are more easily degraded [22]. The ecological and human health issues involve environmental persistence and bioaccumulation, which are issues of interest. The PCBs flow to the ecosystems either by accidental releases, improper disposal, or degradation of the contaminated equipment. They survive in soils, sediments, and water, whereby they can be consumed by aquatic life and consequently increase in the food chain. Long-term exposure to man has been associated with endocrine imbalance, immunological release, carcinogenicity, and neurodevelopmental handicap in children. IARC categorizes PCBs as carcinogenic to humans (Group 1) and most agencies at the national level set strict limits on concentrations in food and the environment that should be allowed [22, 32]. There are regulatory frameworks that have obliged the global elimination of all PCB production, including the Stockholm Convention on Persistent Organic Pollutants, which has established regulations for remediation and disposal. Although these bans are in place, legacy PCB pollution is still common at numerous industrial sites and the common ways of remedying these situations are through excavation of soil, burning, or stabilization in situ. Continuous surveillance and risk analysis are important measures to reduce the exposure levels, defend the susceptible people and also, to get the ecosystems back on their feet, following the long-term effects of PCB poisoning [33, 34].

4.3 Polycyclic aromatic hydrocarbons (PAHs)

PAHs exist in oil drains and are implicated in acid rain, among other activities. human PAHsare found in oil drains and are involved in acid rain, among other things [35]. PAHs are a class of organic compounds that consist of two or more benzene rings, which are fused together. The large amount of PAHs is the result of incomplete burning of organic substances, even though this gas is naturally found in volcanic emissions and in wildfires [36]. The major point sources are industrial and residential. Plants generating power through fossil fuel, steel mills, and tire-rolling plants produce high-molecular -weight PAHs, which are benzo[a]pyrene or indeno[1,2,3-,c,d]pyrene. Naphthalene and phenanthrene species are emitted by urban traffic, particularly by diesel engines. Cooking, heating, and the removal of agricultural residues are referred to as biomass burning, which contributes to the atmospheric load, particularly in developing areas. After being released, PAHs adsorb themselves on fine particulate matter (PM 2.5), have a long range, and are deposited on the soil, water, and vegetation [37].

Exposure of humans takes place mainly by breathing polluted air, consumption of contaminated water and food, and contact of the skin with the ground or surfaces. Sniffing of particulate-bound PAHs may introduce carcinogenic compounds directly to the lungs, whereas eating contaminated fish/shellfish/leafy greens may cause the final concentration of PAHs in the liver and the digestive system. Those chronically exposed to it have a higher risk of cancer of the lungs, skin, and bladder, reproductive and developmental problems. PAHs are listed as the eighth most recognized human carcinogen according to the American Cancer Society [38]. The determination of PAHs is dependent on EPA Method 8270 E Air (GC-MS) and EPA Method 8270F Water and soil. Regulations are established with exposure limits: the U.S. EPA, 8-h inhalation reference concentration (RfC) of benzo a pyrene is 2 $\mu g/m^3$, and state regulations, like the Tier 2 air quality regulations, mandate ambient concentrations of some PAHs as low as 35 ng/m^3. The European Union, under the REACH program, internationally categorizes high-molecular-weight PAHs as carcinogenic in human beings (Group 1) [39]. The prevention depends on the control of the emission of sources. The release of PAH can be reduced by up to 70% through cleaner fuels, upgrades in the particulate filters, or clean-burn devices. The indoor PAH levels are reduced by appropriate ventilation, the application of the High Efficiency Particulate Air (HEPA) filter, and not having open-flame cooking in indoor settings. In the case of polluted areas, ecological and human health can be recovered through the means of phytoremediation and soil washing. Further study of the PAH metabolism and more rigorous surveillance will aid in safeguarding the at-risk groups against such chronic pollutants [40].

4.4 Phthalates and other plasticizers

The plasticizers are known to be additives that make otherwise hard polymers flexible, soft, and durable. Best known is their association with PVC, although they also become part of polyurethane foams, elastomers, and other thermoplastic blends and improve their performance. The phthalate esters, which include di-2-ethylhexyl phthalate (DEHP), diisononyl phthalate (DINP), diisodecyl phthalate (DIDP), and di-*n*-butyl phthalate (DBP), are the most widely used types of compounds in the consumer-goods industry. There are other plasticizers, such as citrate esters (e.g., acetyl tributyl citrate), sebacates, adipates, and oxylactone derivatives and they are commonly promoted as being phthalate-free [41, 42]. Phthalates are incorporated between 5% and 30% of the weight of PVC to fill up or make soft films used in medical devices, food packaging, and car interiors. Plasticizers like 2-ethylhexyl terephthalate or trimellitate are used as cell-uniformity-facilitating and brittleness-reducing agents in polyurethane foams. The plasticizers are also used as lubricants during the extrusion process because they reduce the viscosity of mortar and increase the life of the extrusion lines [43].

One of the reasons why they should be scrutinized by regulators is the endocrine-disrupting nature of phthalates. Exposure to human beings is by ingestion (packaging materials of food, toys), by inhalation (vapor loss of vinyl flooring), and by dermal exposure (personal care products containing plasticized PVC). Prenatal exposure to phthalates has been implicated in epidemiologic studies as having effects on the development of sex, reduced fertility, and neurobehavioral alterations. Further, the phthalates are lipophilic and may build up in fatty tissue, increasing the chronic toxicity. Environmental stability is average: phthalates are not very fast in degrading both in soil and in aquatic environments, but they may be metabolized by microbes into more water soluble and potentially more bioactive monoesters [44].

In the European Union, DEHP and DBP are registered in the database named REGistration, Evaluation, Authorization and Restriction of Chemicals (EC) as substances of very high concern (SVHC), and their application in plasticizers to toys, childcare items, and foods is highly limited. The U.S. Consumer Product Safety Improvement Act (CPSIA) also restricts phthalates in the content of products that are sold to children to 0.1% by weight. Continuous risk evaluation is provided by new chemical registries, including the US EPA and the European registries, specifically the ECHA database. Such restrictions are accompanied by voluntary industry actions, like the Phthalate-Free labeling campaign or the use of citrates or sebacates as car interior materials that will satisfy the needs of the consumer, who wants less toxic substances [45]. The migration rates and the dimension of endocrine activity of nonphthalate plasticizers are lower. Citrate esters (e.g., acetyltributyl citrate, diethyl citrate) are finding applications in packaging foods and cosmetics, whereas alkyl sebacates (e.g., ethyl phenylhexyl sebacate) are employed in flexible PVC as medical tubing. Plasticizers of lubricant-type 2-ethylhexyl terephthalate 2-ethylhexyl terephthalate (DEHP derived) can be designed to have the right balance between migration and mechanical performance. Nonetheless, every substitute will have to be considered in terms of toxicokinetics, bioaccumulation, and environmental persistence because in some cases substitution will result in the so-called regrettable outcomes [46]. Phthalates have been central to the development of flex plastics, but the endocrine-disruptive nature of phthalates has triggered a regulatory change toward safer and more sustainable plasticizers. It is quite imperative that further research and cooperation between chemists, toxicologists, and policymakers should lead to materials fulfilling human health and environmental goals [47].

4.5 Dioxins and furans

Dioxins and furans are a family of highly toxic, POPs that belong to the broader class of chlorinated aromatic hydrocarbons. Although naturally occurring in trace amounts through volcanic activity and forest fires, the majority of environmental dioxins (especially the most toxic congener, 2,3,7,8-tetrachlorodibenzo-*p*-dioxin, TCDD) arise from industrial processes such as waste incineration, chemical manufacturing, and the production of certain herbicides and pesticides. Household combustion – particularly the burning of tires, plastics, and other chlorinated materials – also contributes to local atmospheric releases [48]. Chemically, dioxins and furans possess a rigid, planar structure that resists biodegradation, leading to long atmospheric transport and deposition onto soils and water bodies. Their lipophilicity causes bioaccumulation in the food chain, especially in fatty tissues of large mammals and fish, thereby magnifying human exposure through diet. The World Health Organization (WHO) estimates that the global average intake of dioxin-like compounds is around 66 pg TEQ (toxic equivalency) per kilogram of body weight per day, exceeding the provisional tolerable weekly intake (PTWI) in many regions [49].

Health impacts are diverse and largely mediated by activation of the aryl hydrocarbon receptor (AhR). Chronic low-dose exposure is linked to immunosuppression, endocrine disruption, reproductive toxicity, and increased risk of cancers (particularly liver, breast, and prostate). In children, dioxin exposure has been correlated with impaired neurodevelopment and decreased IQ. Animal studies also reveal developmental delays and altered organogenesis at exposure levels far below acute toxicity thresholds [50]. Regulatory frameworks have evolved to curb dioxin emissions. The United Nations' Stockholm Convention (1992) designated dioxins as POPs, mandating measures to reduce industrial releases. In the United States, the Clean Air Act (CAA) and the EPA's National Emission Standards for Hazardous Air Pollutants (NESHAP) regulate incinerator emissions, while the Toxic Substances Control Act (TSCA) governs the production and use of dioxin-forming chemicals. Internationally, the European Union's RoHS directive restricts the use of hazardous substances, including dioxins, in electrical and electronic equipment [51]. Monitoring efforts combine atmospheric sampling, sediment cores, and biomonitoring of sentinel species. Advances in high-resolution mass spectrometry now allow detection of individual dioxin congeners at attomole levels, improving the precision of exposure assessments. Moreover, predictive dispersion models help identify hotspots for remediation.

Mitigation strategies focus on source control and waste management: high-temperature incineration with catalytic or noncatalytic destruction systems, flue-gas treatment (e.g., activated carbon injection), and comprehensive diversion of chlorinated plastics from landfills. In communities near legacy sites, soil remediation techniques such as bioremediation with specialized microbes or phytoremediation using hyperaccumulator plants are being trialed, though their efficacy for dioxins remains limited [52, 53]. Dioxins and furans continue to pose a significant environmental and public-health challenge due to their persistence, bioaccumulation potential, and diverse toxicological endpoints. Addressing this issue requires coordinated regulatory action, robust monitoring, and innovative remediation technologies to safeguard ecosystems and human populations [54].

5 Sources and release mechanisms of organic pollutants from e-waste

Organic pollutants released from e-waste arise from the diverse materials used in the manufacture of modern electronic devices. E-waste contains metals, plastics, ceramics, glass, and a wide range of organic compounds incorporated into casings, printed circuit boards, connectors, coatings, and numerous additive systems. These organic constituents include polymers, plasticizers such as phthalates, stabilizers, surfactants, lubricants, and multiple classes of flame retardants embedded within both plastics and metal finishes. When e-waste is poorly handled – through informal dismantling, open burning, mechanical shredding, or passive disposal in landfills – these organic substances escape into the environment as volatile organic compounds (VOCs), PAHs, PCBs, BFRs, and dioxin-like compounds. Understanding their sources and release pathways is important for effective control and reduction strategies [18]. A major source of organic pollutants is the large volume of polymer-based materials used in electronic devices, including polyethylene, polypropylene, PVC, polycarbonates, and epoxy resins. These polymers contain additives such as phthalates, organotin stabilizers, PBDE flame retardants, and antioxidants, all of which can leach or volatilize when exposed to heat or chemical stress. Printed circuit boards also contribute significantly, as they are made of epoxy or polyester substrates bound with organic adhesives, fluxes, and solder pastes containing organophosphorus compounds. When heated, these materials break down to release phenols, halogenated species, and other organic compounds [55].

Electronic components like integrated circuits, LEDs, and OLEDs contain encapsulating polymers, phosphor compounds, and organometallic materials that release halogenated organics upon degradation. Batteries further add to the organic load through their electrolytes, polymer separators, and flame-retardant additives, which may vaporize or leach after thermal or mechanical damage. Even connectors and switches, which often contain polymer housings and organosilicon coatings, can release organic pollutants when mechanically or chemically disturbed [56]. These pollutants are released through several mechanisms during both informal and formal recycling practices. Thermal degradation, especially open burning at informal sites, breaks down polymer matrices at temperatures between 700 °C and 1,200 °C, generating PAHs, VOCs, dioxins, and chlorinated or brominated aromatics. Even controlled pyrolysis used in formal recycling can produce volatile organic gases unless properly captured. Mechanical processes such as manual dismantling and wire cutting generate fine plastic dust and flakes that contain embedded additives capable of leaching into surrounding soils [57].

Chemical leaching using acids to recover metals exposes polymeric materials to corrosive conditions, resulting in the release of dissolved organic compounds, including aromatic amines and phenols. In landfills, microbial breakdown of biodegradable organics produces greenhouse gases while mobilizing halogenated by-products into groundwater. Outdoor storage of discarded electronics also promotes slow release of VOCs through polymer aging, while airborne dust settles on vegetation and soils, creating long-term exposure pathways [58]. These emissions have measurable impacts on environmental quality and human health. Open-burning operations have been shown to release significant quantities of PAHs and dioxins into the air, with VOC concentrations such as benzene and xylene reaching hazardous levels near burning sites. Soils at e-waste dumps often accumulate high levels of PBDEs and PCBs, while groundwater beneath informal sites may contain elevated concentrations of chlorinated VOCs. These persistent pollutants bioaccumulate in aquatic and terrestrial organisms, concentrating especially in fatty tissues of livestock and fish, and eventually entering the human food chain [59, 60]. Effective mitigation requires integrated thermal, chemical, and mechanical management strategies. High-temperature incinerators equipped with advanced flue-gas cleaning systems can greatly reduce emissions of dioxins and organic vapors. Chemical controls, including acid neutralization and activated-carbon filtration, help remove dissolved organics from recycling effluents. Preprocessing steps such as shredding and magnetic separation reduce unnecessary heating of polymer-rich materials, thereby lowering the release of organics. Improving practices in the informal sector through training, provision of protective equipment, and discouragement of open burning is essential. Postrecycling treatment of plastic streams using solvent extraction or additive recovery technologies can further reduce residual toxicity while reclaiming valuable chemicals [61]. Looking ahead, the organic composition of e-waste will continue to evolve as electronics incorporate new materials such as graphene, perovskites, and liquid-metal components. Newer flame retardants and polymer additives may introduce different environmental risks that require updated monitoring and safer material design. Future research should focus on detailed profiling of organic emissions at each recycling stage, life-cycle assessments that track indirect release pathways, the development of safer biodegradable additives, and design approaches that minimize polymer use and facilitate safer disassembly. A comprehensive understanding of how organic pollutants originate and move through e-waste recycling systems is crucial for building evidence-based policies that protect ecosystems and human health (Table 2) [11, 62].

Table 2: Sources and release mechanisms of organic pollutants from e-waste.

Source of e-waste	Organic pollutants released	Mechanism of release	References
Circuit boards (computers, printers, and mobile phones)	PBDEs, PCBs, and PAHs	Open burning, acid leaching, thermal processing, and weathering	[63]
Plastic casings and housings (televisions, appliances, and toys)	Phthalates, PBDEs, and BPA	Incineration, fragmentation, and leaching into soil and water	[64]
Cables and wiring (PVC-insulated wires and power cords)	PAHs, phthalates, dioxins, and furans	Open burning, shredding, and chemical leaching	[65]
Batteries (lithium-ion, lead-acid, and Ni-Cd)	Organophosphates and PAHs	Acid extraction, improper disposal, and chemical degradation	[66]
Fluorescent and LED lighting	Mercury compounds and brominated dioxins	Crushing, burning, and landfill leachate	[67]
CRT monitors and old TVs	PCBs, PBDEs, and dioxins	Mechanical breakage, smelting, and open burning	[68]
Electrical tools and industrial equipment	PBDEs, PAHs, and phthalates	Dismantling, combustion, and chemical treatment	[69]
Solar panels and photovoltaic cells	BPA, PAHs, and organophosphates	Crushing, thermal degradation, and weathering	[70]
Household appliances (refrigerators and air conditioners)	PBDEs, PAHs, and organotin compounds	Dismantling, incineration, and improper landfill disposal	[2]
Medical electronic devices	PCBs, phthalates, and BPA	Breakage, thermal processing, and leaching into soil or water	[71]

6 Environmental and health impacts of organic pollutants

Organic pollutants are carbon-based chemical compounds that persist in the environment and pose significant threats to ecological systems and human health. Unlike inorganic contaminants that often leave visible residues, organic pollutants tend to infiltrate soils, water bodies, and the atmosphere in subtle ways, accumulating gradually in living organisms and magnifying their toxic effects through the food chain [72].

These pollutants belong to several major families, including PCBs commonly used in electrical equipment, polyaromatic hydrocarbons generated from fossil fuel combustion, organohalogen pesticides applied in agriculture, volatile industrial solvents released from paints and fuels, per- and polyfluoroalkyl substances found in firefighting foams and nonstick cookware, and plastic additives such as BPA and phthalates present in various consumer products. Many of these chemicals interact synergistically in the environment, increasing their hazard potential [73].

Once released, many organic pollutants exhibit high persistence and bioaccumulation. Compounds such as PCBs may remain in soil for decades, entering groundwater or being taken up by plants, while their strong affinity for lipids allows them to concentrate in the fatty tissues of aquatic organisms and terrestrial wildlife. PAHs can attach to suspended particles in water, making them readily available to filter-feeding organisms and leading to biomagnification in higher trophic levels. These pollutants also alter soil chemistry by disrupting microbial communities essential for nutrient cycling, reducing soil fertility, and interfering with key ecological processes such as carbon sequestration in wetland environments. In aquatic ecosystems, solvent contamination can result in developmental abnormalities in fish, affecting locomotion and increasing vulnerability to predators [55].

VOCs contribute to atmospheric pollution by evaporating easily and participating in photochemical reactions that form smog and tropospheric ozone. This oxidizing environment can damage plant tissues, reduce photosynthetic efficiency, and decrease crop yields. Furthermore, long-range atmospheric transport allows pollutants released in industrial regions to travel thousands of kilometers, contaminating remote areas such as the Arctic. Biodiversity suffers significantly from chronic exposure to organic pollutants, with evidence showing declines in amphibian populations due to pesticide exposure and developmental disruptions in fish caused by Perfluoroalkyl and Polyfluoroalkyl Substances (PFAS) contamination, leading to altered sex ratios and population imbalances that ripple through entire ecosystems [74]. Humans are exposed to organic pollutants through inhalation of polluted air, ingestion of contaminated food and water, and dermal contact with consumer products or industrial materials containing hazardous compounds. Because many of these pollutants are lipophilic, they accumulate in body fat and persist long after external exposures have ceased. The health effects associated with these compounds include carcinogenicity, endocrine disruption, neurotoxicity, immune suppression, reproductive harm, and metabolic disorders [75].

For example, benzene and several PAHs are well-documented carcinogens; phthalates and BPA interfere with hormone signaling; organophosphate pesticides impair neurological function; and PFAS exposure has been linked to weakened immune responses in children. Importantly, low-dose exposures can still produce significant health effects, especially among vulnerable groups such as pregnant women, children, and workers in industrial settings [76]. These impacts are unequally distributed, with communities near industrial zones, waste sites, or high-traffic areas experiencing disproportionately high exposure levels. Urban neighborhoods with historical PCB contamination face increased cancer risks and reduced life expectancy, while rural populations who rely on locally grown food may unknowingly consume crops contaminated by PAHs or other pollutants. Such disparities highlight the need for equitable regulatory interventions and community-centered remediation initiatives [77, 78].

Moving forward, mitigating the environmental and health consequences of organic pollutants requires integrated efforts that strengthen monitoring programs, enforce more rigorous regulations, and incorporate affected communities into planning and decision-making. Continued research is essential for developing innovative detection tools, advancing bioremediation technologies, and understanding emerging contaminants such as PFAS. Education remains a critical tool for raising awareness of exposure risks and promoting safer choices at the household, industrial, and policy levels. Through coordinated scientific, regulatory, and societal action, it is possible to minimize the widespread impacts of organic pollutants and protect both ecosystems and public health (Table 3) [79].

Table 3: Environmental and health impacts of organic pollutants from e-waste.

Organic pollutant	Environmental impacts	Health impacts	References
Polybrominated diphenyl ethers (PBDEs)	Persistent in soil and water, bioaccumulative, and disrupts aquatic and terrestrial ecosystems	Endocrine disruption, thyroid hormone imbalance, neurodevelopmental delays, and reproductive toxicity	[80]
Polychlorinated biphenyls (PCBs)	Contaminates soil, sediments, and water; persistent and bioaccumulative	Immunosuppression, liver damage, reproductive disorders, cancer, and neurobehavioral impairments	[22]
Polycyclic aromatic hydrocarbons (PAHs)	Contaminates air, soil, and water; toxic to aquatic organisms; promotes long-term soil pollution	Carcinogenic and mutagenic effects, respiratory problems, liver toxicity, and skin disorders	[35]
Phthalates and other plasticizers	Leaches into soil and water; disrupts ecosystems; persistent in sediments	Endocrine disruption, reproductive abnormalities, developmental defects, and metabolic disorders	[46]
Dioxins and furans	Highly persistent in environment; bioaccumulative; contaminates soil, water, and food chain	Carcinogenic, immunotoxic, endocrine disruption, developmental toxicity, and liver damage	[54]
Bisphenol A (BPA)	Leaches into water and soil; disrupts aquatic species	Hormone disruption, reproductive toxicity, developmental abnormalities, and metabolic disorders	[81, 82]

Table 3 (continued)

Organic pollutant	Environmental impacts	Health impacts	References
Organostannic compounds (e.g., TBT and TPT)	Bioaccumulative in aquatic ecosystems; toxic to aquatic invertebrates	Immunotoxicity, endocrine disruption, developmental toxicity, and liver damage	[83]

7 Analytical methods for detecting organic pollutants in e-waste

Analytical methods for detecting organic pollutants in e-waste have advanced significantly as concerns over environmental toxicity and human exposure continue to grow. E-waste is a complex assemblage of materials – including circuit boards, plastic housings, batteries, and wiring – each containing a variety of hazardous organic compounds. These pollutants range from PAHs and organochlorine pesticides to organophosphorus and BFRs, as well as organosiloxanes used in silicone-based components. Accurate detection and quantification of these contaminants remain central to environmental risk assessment, regulatory enforcement, and the development of sustainable recycling and remediation strategies. Over the past decade, continuous improvements in sensitivity, selectivity, and miniaturization have broadened the analytical toolbox, enhancing the reliability and speed of pollutant monitoring in e-waste matrices [84].

The process begins with careful sampling and preparation, as e-waste is inherently heterogeneous. Representative sampling is therefore crucial, and standardized protocols such as ISO 16786:2019 recommend stratifying materials – separating printed circuit boards, plastics, or batteries – before homogenizing and subsampling. For organic analysis, the sample is typically dried, ground, and sieved to achieve uniformity. Moisture removal prevents interference during extraction, while grinding increases surface area and improves extraction efficiency. Extraction of organic pollutants from the solid matrix is commonly performed using techniques such as Soxhlet extraction, accelerated solvent extraction, or pressurized liquid extraction [85, 86]. These methods rely on solvents like hexane, acetone, or acetonitrile, depending on target analytes. To minimize matrix effects that could distort instrumental responses, the extract is often subjected to cleanup using silica gel, Florisil, or alumina cartridges. Finally, the extract is concentrated, frequently by rotary evaporation or nitrogen blowdown, to achieve the sensitivity required for trace-level analysis. Newer approaches such as solid-phase micro-extraction and stir-bar sorptive extraction now allow preconcentration directly on-site, reducing laboratory workload while maintaining method robustness [87].

Gas chromatography coupled with mass spectrometry remains the leading technique for analyzing volatile and semi-volatile organic pollutants found in e-waste. GC-MS is exceptionally powerful for identifying PAHs, chlorinated solvents, and many flame retardants that possess sufficient volatility or can be derivatized to enhance chromatographic behavior. Derivatization, such as silylation or trifluoroacetylation, is sometimes necessary to improve analyte volatility or stability. Separation is typically achieved using nonpolar capillary columns, while mass spectrometric detection employing electron impact ionization at 70 eV ensures reproducible fragmentation patterns [88]. Modern systems using high-resolution time-of-flight or Orbitrap mass analyzers provide mass accuracies of less than five parts per million, enabling confident identification even in complex matrices. Internal standards, often isotopically labeled analogs, ensure accurate quantification by compensating for extraction inefficiencies and instrumental variability. Detection limits for PAHs using GC-MS/MS can be as low as 0.05 μg/kg, and for PBDEs, they typically range between 0.2 and 1 μg/kg. To overcome challenges such as co-elution and thermal degradation of sensitive compounds, advanced methods, including two-dimensional gas chromatography and softer ionization sources like chemical ionization, may be employed [89].

While GC-MS is indispensable for volatile organics, liquid chromatography coupled with mass spectrometry is equally important for nonvolatile or thermally labile compounds. LC-MS is particularly valuable for organophosphorus flame retardants, chlorinated hydrocarbons that do not volatilize easily, and per- and polyfluoroalkyl substances found in various consumer electronics. Sample extraction for LC-MS commonly employs ultrasonic agitation or QuEChERS protocols, which are favored for their simplicity and high recoveries. Chromatographic separation typically utilizes reverse-phase C18 columns under gradient elution, with acetonitrile-water mixtures containing formic acid to improve ionization [90]. Electrospray ionization is used in either positive or negative mode depending on compound classes, and quantification relies on multiple reaction monitoring transitions for high sensitivity and selectivity. LC-MS/MS can detect flame retardants at levels as low as 0.02 μg/kg, and PFAS compounds often below 0.1 μg/kg. Matrix suppression, a common challenge in LC-MS, is addressed through techniques such as solid-phase extraction cleanup and matrix-matched calibration. For compounds with multiple isomers, chiral chromatography or ion mobility spectrometry may be required to ensure accurate identification [91].

Complementary spectroscopic techniques provide additional insights at various stages of analysis. Fourier-transform infrared spectroscopy is frequently used as a rapid screening tool to identify functional groups associated with organosiloxanes, epoxy resins, or plasticizers. Raman spectroscopy, with its minimal sample preparation requirements, is valuable for mapping the distribution of aromatic compounds on circuit boards. Nuclear magnetic resonance spectroscopy, though less commonly used due to higher cost and sample requirements, offers powerful structural elucidation capabilities. X-ray photoelectron spectroscopy is particularly useful for studying organic contaminants adsorbed on metallic surfaces – conditions common in printed circuit boards and connectors. These spectroscopic tools complement chromatographic techniques and help refine targeted analysis workflows [92, 93]. Emerging trends are transforming how organic pollutants in e-waste are monitored. Miniaturized systems such as micro-GC and micro-LC platforms facilitate on-site screening at recycling facilities, reducing turnaround times and enabling immediate decision-making. Paper-based analytical devices equipped with smartphone readout allow rapid field detection of PAHs and flame retardants. High-throughput screening approaches using automated solid-phase extraction and robotic systems now allow the processing of dozens of samples simultaneously, supporting large-scale monitoring and epidemiological studies. The integration of machine learning for spectral deconvolution is another promising development, helping to resolve overlapping chromatographic peaks and enhancing the detection of low-abundance pollutants. The development of compound libraries specifically tailored to e-waste pollutants also strengthens nontargeted screening approaches. Increasing regulatory focus on mixture toxicity is driving analytical workflows to shift from evaluating single compounds to assessing cumulative risks using toxic equivalency models [94, 95]. A practical example of these analytical methods is demonstrated in recent research examining PBDE concentrations in discarded laptop circuit boards. Using accelerated solvent extraction combined with comprehensive two-dimensional gas chromatography and time-of-flight mass spectrometry, researchers were able to resolve twelve flame retardant congeners, ranging from 0.5 to 35 μg/kg. Lower-brominated congeners were the most abundant, reflecting earlier manufacturing practices. Such data enabled waste managers to prioritize high-risk components for specialized treatments such as vitrification, illustrating how analytical results can directly guide management decisions [96].

Effective analysis requires careful adherence to best practices, such as using matrix-matched calibration to correct for ion suppression, validating methods according to ISO 17025 requirements, and keeping up with evolving regulatory expectations that increasingly demand full traceability of hazardous organics in recycled products. Collaboration among chemists, materials scientists, and toxicologists is essential for developing holistic strategies that link pollutant detection to health and environmental outcomes [97]. The analytical landscape for detecting organic pollutants in e-waste has become highly sophisticated. While GC-MS and LC-MS/MS continue to serve as the gold standards for quantification, spectroscopic tools, and emerging portable technologies are expanding the scope and flexibility of monitoring. Advancements in automation, miniaturization, and data analytics promise faster and more comprehensive assessment methods, which are crucial for ensuring environmental safety and public health in a world increasingly dependent on electronic devices [95].

8 Remediation strategies for organic pollutants

Remediation strategies for organic pollutants encompass a diverse range of approaches designed to remove, contain, or transform hazardous compounds such as volatile hydrocarbons, pesticides, industrial solvents, and PAHs. Because these pollutants persist in the environment, accumulate in living systems, and pose serious risks to ecological and human health, remediation efforts must draw from physical, chemical, biological, and emerging technological methods. The choice of strategy depends on the contaminant profile, site conditions, regulatory requirements, cost considerations, and long-term environmental objectives. Effective remediation often involves combining methods to maximize removal efficiency and ensure sustainable recovery of contaminated ecosystems [98, 99]. Physical remediation remains one of the most direct approaches for managing heavily polluted sites. Methods such as excavation and landfilling remove contaminated soil or sediments and relocate them to engineered disposal facilities, providing rapid containment for high-concentration hotspots. Capping, which involves installing impermeable layers over polluted zones, can effectively prevent contaminant migration with minimal disturbance, although it does not eliminate the pollutant mass [100, 101].

Techniques including cold-sluicing and thermal desorption rely on temperature manipulation – either freezing or heating – to mobilize and extract volatile and semi-VOCs. Additionally, in situ vapor extraction creates pressure gradients to draw out vapor-phase contaminants from saturated soils or aquifers. These physical techniques are valued for their immediacy but often fall short of achieving long-term chemical transformation, making them better suited for initial containment than complete remediation [102]. Chemical remediation offers more aggressive transformation of organic pollutants through oxidation, reduction, or stabilization processes. Advanced oxidation processes, which employ hydrogen peroxide, ozone, ultraviolet light, or Fenton reagents, generate highly reactive radicals capable of breaking down complex organic molecules into simpler, often less harmful products. While these methods can rapidly mineralize persistent contaminants such as PAHs and chlorinated solvents, they require careful management to avoid producing toxic intermediate compounds. Electrochemical oxidation is another promising technique, relying on applied electrical potentials to generate powerful oxidants directly at electrode surfaces, though practical challenges include electrode fouling and energy demands. Redox-stimulated bioremediation introduces electron donors or acceptors to alter subsurface conditions and promote natural degradative pathways [103].

Other chemical approaches, such as solidification and stabilization, incorporate binders like cement or pozzolanic materials to immobilize contaminants within hardened matrices, effectively reducing leachability without physically removing the pollutants. Co-precipitation processes that form insoluble organo-metal complexes offer similar containment benefits but require strict pH management to remain stable [104]. Biological remediation harnesses the metabolic abilities of microorganisms and plants to degrade or immobilize pollutants in a more environmentally harmonious and cost-effective manner. Microbial methods such as bioventing increase oxygen supply to stimulate aerobic degradation of VOCs. Biostimulation introduces nutrients or electron donors to enhance indigenous microbial activity, while bioaugmentation introduces specialized microbial cultures capable of degrading recalcitrant pollutants like PCBs or high-molecular-weight PAHs [99]. Combined methods, such as sequential chemical oxidation followed by biological treatment, capitalize on the strengths of each approach by first breaking down complex contaminants into biodegradable intermediates and then relying on microbial communities to achieve complete mineralization. Phytoremediation offers plant-based solutions that either extract contaminants into plant tissues, enhance microbial degradation through root exudates, or stabilize pollutants within the root zone to prevent erosion and transport. Although biological remediation is typically slower than physical or chemical approaches, it is often the most sustainable long-term option and can restore ecological balance in affected sites [5].

Integrated remediation strategies are increasingly recognized as the most effective means of addressing complex contamination scenarios. For example, initial excavation or thermal treatment may remove the bulk of highly concentrated pollutants, after which bioventing or biostimulation can address residual contamination in surrounding soils and groundwater. Sequential oxidation-biodegradation approaches are especially valuable for degrading persistent organics by increasing the bioavailability of compounds that are otherwise resistant to microbial attack. Phytoremediation can also serve as a final polishing step, offering an aesthetically pleasing and low-impact method for reducing remaining trace contaminants in shallow soils. These combined approaches not only enhance overall efficiency but also provide more durable, long-term remediation outcomes [105]. Recent advancements are expanding the future possibilities of organic pollutant remediation. Bioremediation-based membrane systems integrate microbial biofilms with filtration technologies to capture and degrade organics in flowing water. Microbial electrochemical systems employ bacteria to transfer electrons to electrodes while breaking down pollutants, generating electricity as a by-product. The development of genetically engineered microbes using CRISPR technologies offers the potential for highly specific and efficient degradation of stubborn organic compounds, although regulatory and ecological concerns must be carefully considered. Hydroponic phytoremediation accelerates contaminant uptake by cultivating plants in controlled, nutrient-rich water systems, bypassing the limitations associated with soil heterogeneity. While these technologies remain in various stages of development, they illustrate the growing trend toward innovative, decentralized, and energy-efficient remediation solutions [106].

Selecting an appropriate remediation strategy requires comprehensive site characterization, including detailed analysis of contaminant types, distribution patterns, geology, hydrology, and ecological risks. Regulatory frameworks guide permissible approaches, especially in cases involving genetically modified organisms or the handling of hazardous waste. Economic evaluations help balance capital and operational costs, while stakeholder engagement ensures community concerns – such as noise, health risks, and aesthetics – are addressed. Long-term monitoring remains essential for verifying treatment effectiveness, preventing rebound of residual contaminants, and sustaining microbial activity in biological systems [107]. A practical illustration of integrated remediation can be seen at a former paint manufacturing site contaminated with BTEX compounds. Initial excavation and thermal treatment removed the most heavily contaminated soil layers. Subsequent in situ bioventing and nutrient-based biostimulation significantly reduced dissolved BTEX levels in the surrounding groundwater. Finally, perimeter planting with willow species helped stabilize soils and minimize plume migration. Within eighteen months, contaminant concentrations fell below regulatory thresholds, demonstrating the effectiveness of a staged, multimethod approach [105]. The remediation of organic pollutants demands a nuanced combination of physical removal, chemical transformation, biological degradation, and innovative emerging technologies. Each method contributes uniquely to the remediation process, and when applied thoughtfully and sequentially, they provide robust, sustainable, and cost-effective pathways for restoring contaminated environments and protecting public health [99].

9 Conclusion

E-waste is becoming one of the most pressing environmental issues of our time. Its complex composition releases a cocktail of POPs – BFRs, PCBs, PAHs, dioxins, furans, phthalates, bisphenol A, and others – through improper disposal, informal recycling, and accidental degradation. These chemicals are stubbornly persistent in the environment, accumulate through food webs, and are toxic to both ecosystems and human health. They contaminate air, water, and soil, disrupt ecological processes, and are linked to endocrine disruption, reproductive harm, neurodevelopmental deficits, and cancer in exposed populations. Understanding how e-waste is generated, its global trends, and its chemical makeup is therefore essential for pinpointing pollution sources and assessing their full impact. Modern analytical tools – gas and liquid chromatography coupled to mass spectrometry, advanced spectroscopies, and highly selective immunoassays – enable the detection of these contaminants at trace levels. Reliable measurements are the cornerstone of regulatory enforcement, risk assessment, and the design of effective cleanup strategies. By integrating these strategies – tailored to the specific pollutant profile and site conditions -industries, governments, and communities can move toward safer, more sustainable e-waste management. Accurate monitoring, informed policy, and continued technological innovation will be key to protecting both the environment and public health from the silent threat of e-waste.

References

[1] Quinto S, Law N, Fletcher C, Le J, Antony Jose S, Menezes PL, Exploring the e-waste crisis: Strategies for sustainable recycling and circular economy integration. Recycling, 2025, 10, 72.

[2] Alabi OA, Adeoluwa YM, Huo X, Xu X, Bakare AA, Environmental contamination and public health effects of electronic waste: An overview. J Environ Health Sci Eng, 2021, 19, 1209–1227.

[3] Owusu-Sekyere K, Aladago DA, Leverenz D, Oteng-Ababio M, Kranert M, Environmental impacts on soil and groundwater of informal e-waste recycling processes in Ghana. Sustainability, 2024, 16, 4347.

[4] Martínez A, Toxicity of persistent organic pollutants: A theoretical study. J Mol Model, 2024, 30, 97.

[5] Das K, Masud MAA, Sarker A, Arafa RA, Patel M, Unveiling the sustainable and biological remediation of heavy metals contaminations in soils and water ecosystems through potential microbes – A review. Sustainability, 2025, 17, 7357.

[6] Ostovar M, Muñana S, Galdames A, Berganza J, Orueta M, Esteban JJ, Brettes P, Vilela JLV, Rubio LR, Advancements in biopile-based sustainable soil remediation: A decade of improvements, integrating bioremediation technologies and AI-based innovative tools. Environ Sci Pollut Res, 2025, 32, 22766–22793.

[7] Odeyingbo OA, Deubzer OK, Ogunmokun OA, Assessment of the impact of the revised national e-waste framework on the informal e-waste sector of Nigeria. Recycling, 2025, 10, 117.

[8] Njoku A, Agbalenyo M, Laude J, Ajibola TF, Attah MA, Sarko SB, Environmental injustice and electronic waste in Ghana: Challenges and recommendations. Int J Environ Res Public Health, 2023, 21, 25.

[9] Seif R, Salem FZ, Allam NK, E-waste recycled materials as efficient catalysts for renewable energy technologies and better environmental sustainability. Environ Dev Sustain, 2023, 1–36.

[10] Mathew JT, Inobeme A, Azeh Y, Musah M, Shaba EY, Tanko E, Etsuyankpa MB, Muhammad AI, Musa ST, Ndejiko MJ, Hussaini J, Adetunji CO, Mamman A, Total reflection X-ray fluorescence: Technological developments and expanding applications, In Singh VK, Ed, X-Ray Fluorescence Spectroscopy and Chemometrics. Springer, Cham, 2025a, 303–323.

[11] Mohamed D, Fayad A, Mohamed A-MO, Al Nahyan MT, The role of e-waste in sustainable mineral resource management. Waste, 2025, 3, 27.

[12] Dennison MS, Paramasivam SK, Wanazusi T, Sundarrajan KJ, Erheyovwe BP, Marshal Williams AM, Addressing plastic waste challenges in Africa: The potential of pyrolysis for waste-to-energy conversion. Clean Technol, 2025, 7, 20.

[13] Faibil D, Asante R, Agyemang M, Addaney M, Baah C, Extended producer responsibility in developing economies: Assessment of promoting factors through retail electronic firms for sustainable e-waste management. Waste Manag Res, 2023, 41, 117–142.

[14] Anderson L, Yu E, Chen W-T, Chemical recycling of mixed plastics in electronic waste using solvent-based processing. Processes, 2022, 10, 66.

[15] Grdić D, Despotović I, Ristić N, Grdić Z, Ćurčić GT, Potential for use of recycled cathode ray tube glass in making concrete blocks and paving flags. Materials, 2022, 15, 1499.

[16] Islam MT, Iyer-Raniga U, Lithium-ion battery recycling in the circular economy: A review. Recycling, 2022, 7, 33.

[17] Fatema K, Hassan MN, Hasan S, Roy H, E-waste recycling in an optimized way for copper recovery by leaching and a case study on e-waste generation and management in Dhaka city. Heliyon, 2024, 11, e41453.

[18] Srivastava AN, Sikarwar VS, Bisen D, Fathi J, Maslani A, Lopez Nino BN, Barmavatu P, Kaviti AK, Pohořelý M, Buryi M, E-waste unplugged: Reviewing impacts, valorization strategies and regulatory frontiers for efficient e-waste management. Processes, 2025, 13,. 2014.

[19] Ghiraldelli Miranda R, Ferreira Sampaio C, Gomes Leite F, Duarte Maia F, Junqueira Dorta D, Flame retardants: New and old environmental contaminants. IntechOpen, 2022.
[20] Wang Y, Qian H, Phthalates and their impacts on human health. Healthcare, 2021, 9, 603.
[21] Xue J, Xiao Q, Zhang M, Li D, Wang X, Toxic effects and mechanisms of polybrominated diphenyl ethers. Int J Mol Sci, 2023, 24, 13487.
[22] Montano L, Pironti C, Pinto G, Ricciardi M, Buono A, Brogna C, Venier M, Piscopo M, Amoresano A, Motta O, Polychlorinated biphenyls (PCBs) in the environment: Occupational and exposure events, effects on human health and fertility. Toxics, 2022, 10, 365.
[23] Montano L, Baldini GM, Piscopo M, Liguori G, Lombardi R, Ricciardi M, Esposito G, Pinto G, Fontanarosa C, Spinelli M, Palmieri I, Sofia D, Brogna C, Carati C, Esposito M, Gallo P, Amoresano A, Motta O, Polycyclic aromatic hydrocarbons (PAHs) in the environment: Occupational exposure, health risks and fertility implications. Toxics, 2025, 13, 151.
[24] Zubair M, Adrees A, Dioxins and furans: Emerging contaminants of air. IntechOpen, 2019.
[25] Samova S, Doctor H, The silent threat of BPA: Its pervasive presence and impact on reproductive health. Environ Health Insights, 2025, 19, 11786302251330774.
[26] Hiromori Y, Yui H, Nishikawa J, Nagase H, Nakanishi T, Organotin compounds cause structure-dependent induction of progesterone in human choriocarcinoma Jar cells. J Steroid Biochem Mol Biol, 2016, 155, 190–198.
[27] Gałązka A, Jankiewicz U, Endocrine disrupting compounds (nonylphenol and bisphenol A): Sources, harmfulness and laccase-assisted degradation in the aquatic environment. Microorganisms, 2022, 10, 2236.
[28] Kudzin MH, Piwowarska D, Festinger N, Chruściel JJ, Risks associated with the presence of polyvinyl chloride in the environment and methods for its disposal and utilization. Materials, 2023, 17, 173.
[29] Nevondo V, Okonkwo OJ, Status of short-chain chlorinated paraffins in matrices and research gap priorities in Africa: A review. Environ Sci Pollut Res, 2021, 28, 52844–52861.
[30] Ohoro CR, Adeniji AO, Okoh AI, Okoh S, Polybrominated diphenyl ethers in the environmental systems: A review. J Environ Health Sci Eng, 2021, 19, 1229–1247.
[31] Genuis SK, Birkholz D, Genuis SJ, Human excretion of polybrominated diphenyl ether flame retardants: Blood, urine, and sweat study. Biomed Res Int, 2017, 2017, 3676089.
[32] Mathew JT, Inobeme A, Azeh Y, Musah M, Shaba EY, Tanko T, Etsuyankpa MB, Musa ST, Hussaini J, Adetunji CO, Muhammad AI, Mamman A, X-ray fluorescence spectroscopy for analyzing food products coupled with chemometrics, In Singh VK, Ed, X-Ray Fluorescence Spectroscopy and Chemometrics. Springer, Cham, 2025b, 255–273.
[33] Mathew JT, Inobeme A, Azeh Y, Musah M, Shaba EY, Tanko E, Etsuyankpa MB, Musa ST, Ndejiko JM, Hussaini J, Adetunji CO, Muhammad AI, Mamman A, X-ray fluorescence in analyzing agricultural contamination via chemometric tools, In Singh VK, Ed, X-Ray Fluorescence Spectroscopy and Chemometrics. Springer, Cham, 2025c, 233–254.
[34] Sandu MA, Preda M, Tanase V, Mihailescu D, Virsta A, Ivanescu V, Trends in polychlorinated biphenyl contamination in Bucharest's urban soils: A two-decade perspective (2002–2022). Processes, 2025, 13, 1357.
[35] Patel AB, Shaikh S, Jain KR, Desai C, Madamwar D, Polycyclic aromatic hydrocarbons: Sources, toxicity, and remediation approaches. Front Microbiol, 2020, 11, 562813.
[36] Sahoo BM, Ravi Kumar BVV, Banik BK, Borah P, Polyaromatic hydrocarbons (PAHs): Structures, synthesis and their biological profile. Curr Org Synth, 2020, 17, 625–640.
[37] Hishamuddin NH, Khan MF, Suradi H, Siraj BMZ, Islam MT, Sairi NA, Tajuddin HA, Jamil AKM, Akanda MJH, Yusoff S, The sources of polycyclic aromatic hydrocarbons in road dust and their potential hazard. Sustainability, 2023, 15, 12532.
[38] Abdulai PM, Bede-Ojimadu O, Onyena AP, Frazzoli C, Mogborukor NA, Ekhator OC, Udom GJ, Nwanaforo E, Orisakwe OE, Public health effects of polycyclic aromatic hydrocarbons exposure

through air, water, soil, and food in Ghana: Possible economic burden. Environ Health Insights, 2025, 19, 11786302251343767.
[39] Viteri F, Sánchez NE, Alexandrino K, Determination of polycyclic aromatic hydrocarbons (PAHs) in leaf and bark samples of *Sambucus nigra* using high-performance liquid chromatography (HPLC). Methods Protoc, 2023, 6, 17.
[40] Nassikas NJ, McCormack MC, Ewart G, Balmes JR, Bond TC, Brigham E, Cromar K, Goldstein AH, Hicks A, Hopke PK, Meyer B, Nazaroff WW, Paulin LM, Rice MB, Thurston GD, Turpin BJ, Vance ME, Weschler CJ, Zhang J, Kipen HM, Indoor air sources of outdoor air pollution: Health consequences, policy, and recommendations – An official American Thoracic Society workshop report. Ann Am Thorac Soc, 2024, 21, 365–376.
[41] Mathew JT, Hussaini J, Inobeme A, Azeh Y, Musah M, Shaba YE, Etsuyankpa MB, Tanko E, Ndejiko JM, Adetunji CO, Muhammad AI, Musa ST, Mamman A, X-ray fluorescence analysis of vegetation tissues via chemometric tools, In Singh VK, Ed, X-Ray Fluorescence Spectroscopy and Chemometrics. Springer, Cham, 2025d, 209–231.
[42] Minale YF, Gajdoš I, Štefčák P, Dulebová Ľ, Jachowicz T, Szabó T, Ádámné Major A, Marossy K, Thermodynamic and technological compatibility of polyvinyl chloride, thermoplastic polyurethane, and bio-plasticizer blends. Polymers, 2025, 17, 1149.
[43] An J, Roh -H-H, Jeong H, Lee K-Y, Rhim T, Rapid assessment of di(2-ethylhexyl) phthalate migration from consumer PVC products. Toxics, 2024, 12, 7.
[44] De Paula L, Alves C, Food packaging and endocrine disruptors. J Pediatr, 2024, 100, S40–S47.
[45] Giuliani A, Zuccarini M, Cichelli A, Khan H, Reale M, Critical review on the presence of phthalates in food and evidence of their biological impact. Int J Environ Res Public Health, 2020, 17, 5655.
[46] Gorini F, Tonacci A, Sanmartin C, Venturi F, Phthalates and non-phthalate plasticizers and thyroid dysfunction: Current evidence and novel strategies to reduce their spread in food industry and environment. Toxics, 2025, 13, 222.
[47] Moghazy M, Papathanasiou M, Tzoupis H, Papavasileiou KD, Xing C, Lauschke VM, Afantitis A, Melagraki G, A systematic literature review of reproductive toxicological studies on phthalates. Int J Mol Sci, 2025, 26, 8761.
[48] Shumbula P, Maswanganyi C, Shumbula N, Type, sources, methods and treatment of organic pollutants in wastewater. IntechOpen, 2022.
[49] Nhung NTH, Nguyen XT, Long VD, Wei Y, Fujita T, A review of soil contaminated with dioxins and biodegradation technologies: Current status and future prospects. Toxics, 2022, 10, 278.
[50] Kim K, The role of endocrine disruption chemical-regulated aryl hydrocarbon receptor activity in the pathogenesis of pancreatic diseases and cancer. Int J Mol Sci, 2024, 25, 3818.
[51] Lu S, Zang E, Clean Air Act 1990 and its long-term impact on mortality risk among older adults in the United States. Innov Aging, 2024, 8, 968.
[52] Mathew JT, Shaba EY, Musah M, Azeh Y, Muhammad EB, Tanko E, Tanko MS, Muhammad AI, Hussaini J, Mamman A, Aerogel nanocomposites: Properties and applications, In Inobeme A, Mathew J, Adetunji C, Eds, Advances, Applications, and Future Perspectives of Aerogels. IGI Global Scientific Publishing, 2025e, 25–62.
[53] Ergül HA, Erdem T, Pelin EG, Tonay AM, Aksan S, Öztürk B, Assessment of pesticide, dioxin-like polychlorinated biphenyl, and polycyclic aromatic hydrocarbon existence around Galindez Island, and comparison with the organic pollution bibliography of the Antarctic Peninsula. Sustainability, 2022, 14, 3994.
[54] Souza MCO, Domingo JL, Dioxins and the One Health paradigm: An interdisciplinary challenge in environmental toxicology. Toxics, 2025, 13, 964.
[55] Kiki C, Daniel Ahlonsou E, Chabi K, Bernice Eloise Adéoyé A, Persistent organic pollutants from plastics: Sources, environmental monitoring, toxicity, and remediation strategies. IntechOpen, 2025.

[56] Xi P, Sun F, Tang X, Fan X, Cong G, Lu Z, Zhuo Q, Effect mechanism of phosphorus-containing flame retardants with different phosphorus valence states on the safety and electrochemical performance of lithium-ion batteries. Processes, 2025, 13, 2248.
[57] Kasiński S, Dębowski M, Municipal solid waste as a renewable energy source: Advances in thermochemical conversion technologies and environmental impacts. Energies, 2024, 17, 4704.
[58] Liu R, Tian Z, Cheng H, Zhou H, Wang Y. Organic acid leaching was an efficient approach for detoxification of metal-containing plant incineration ash. Environ Sci Pollut Res Int. 2021. 31, 1–22.
[59] Mathew JT, Azeh Y, Adetunji CO, Monday M, Hussaini J, Tanko MS, Catalytic application of aerogels, In Inobeme A, Mathew J, Adetunji C, Eds, Advances, Applications, and Future Perspectives of Aerogels. IGI Global Scientific Publishing, 2025f, 147–178.
[60] Deary ME, Griffiths SD, The impact of air pollution from industrial fires in urban settings: Monitoring, modelling, health, and environmental justice perspectives. Environments, 2024, 11, 157.
[61] Bertone M, Stabile L, Buonanno G, An overview of waste-to-energy incineration integrated with carbon capture utilization or storage retrofit application. Sustainability, 2024, 16, 4117.
[62] Mathew JT, Hussaini J, Azeh Y, Musah M, Shaba EY, Tanko E, Muhammad AI, Etsuyankpa MB, Musa TS, Mamman A, Environmental and health impact assessment on aerogel application, In Inobeme A, Mathew J, Adetunji C, Eds, Advances, Applications, and Future Perspectives of Aerogels. IGI Global Scientific Publishing, 2025g, 179–202.
[63] Thi Hong Hoa N, Hoan NT, Nguyen Trong N, Linh NTN, Quy BM, Pham TTH, Vq N, Nguyen VP, Nguyen VD, Enhancement of leaching copper from printed circuit boards of discarded mobile phones using ultrasound–ozone integrated approach. Metals, 2023, 13, 1145.
[64] Landrigan PJ, Raps H, Cropper M, Bald C, Brunner M, Canonizado EM, Charles D, Chiles TC, Donohue MJ, Enck J, Fenichel P, Fleming LE, Ferrier-Pages C, Fordham R, Gozt A, Griffin C, Hahn ME, Haryanto B, Hixson R, Ianelli H, Dunlop S, The Minderoo-Monaco Commission on plastics and human health. Ann Glob Health, 2023, 89, 23.
[65] Porowski R, Kowalik R, Ramiączek P, Bąk-Patyna P, Stępień P, Zielecka M, Popielarczyk T, Ludynia A, Chyb A, Gawdzik J, Application assessment of electrical cables during smoldering and flaming combustion. Appl Sci, 2023, 13, 3766.
[66] Mogashane TM, Mokoena L, Tshilongo J, A review on recent developments in the extraction and identification of polycyclic aromatic hydrocarbons from environmental samples. Water, 2024, 16, 2520.
[67] Li Z, Jia P, Zhao F, Kang Y, Mercury pollution, treatment and solutions in spent fluorescent lamps in mainland China. Int J Environ Res Public Health, 2018, 15, 2766.
[68] Cayumil R, Khanna R, Rajarao R, Ikram-ul-haq M, Mukherjee PS, Sahajwalla V, Environmental impact of processing electronic waste: Key issues and challenges. InTechOpen, 2016.
[69] Eze C, Vinken M, E-waste: Mechanisms of toxicity and safety testing. FEBS Open Bio, 2024, 14, 1420–1440.
[70] Atia DM, Hassan AA, El-Madany HT, Eliwa AY, Zahran MB, Degradation and energy performance evaluation of mono-crystalline photovoltaic modules in Egypt. Sci Rep, 2023, 13, 13066.
[71] Wang W, Kannan K, Leaching of phthalates from medical supplies and their implications for exposure. Environ Sci Technol, 2023, 57, 7675–7683.
[72] Gaur N, Dutta D, Jaiswal A, Dubey R, Vrat Kamboj D, Role and effect of persistent organic pollutants to our environment and wildlife. IntechOpen, 2022.
[73] Khoury D, Millet M, Jabali Y, Delhomme O, Occurrence of polycyclic aromatic hydrocarbons and polychlorinated biphenyls in fogwater at urban, suburban, and rural sites in northeast France between 2015 and 2021. Atmosphere, 2024, 15, 291.
[74] David E, Niculescu VC, Volatile organic compounds (VOCs) as environmental pollutants: Occurrence and mitigation using nanomaterials. Int J Environ Res Public Health, 2021, 18, 13147.

[75] Peivasteh-Roudsari L, Barzegar-Bafrouei R, Sharifi KA, Azimisalim S, Karami M, Abedinzadeh S, Asadinezhad S, Tajdar-Oranj B, Mahdavi V, Alizadeh AM, Sadighara P, Ferrante M, Conti GO, Aliyeva A, Mousavi Khaneghah A, Origin, dietary exposure, and toxicity of endocrine-disrupting food chemical contaminants: A comprehensive review. Heliyon, 2023, 9, e18140.
[76] Predieri B, Iughetti L, Bernasconi S, Street ME, Endocrine disrupting chemicals' effects in children: What we know and what we need to learn?. Int J Mol Sci, 2022, 23, 11899.
[77] Mathew JT, Musah M, Azeh Y, Shaba EY, Muhammad EB, Hussaini J, Tanko MS, Mamman A, Exploring future perspectives in aerogel-based research, In Inobeme A, Mathew J, Adetunji C, Eds, Advances, Applications, and Future Perspectives of Aerogels. IGI Global Scientific Publishing, 2025h, 203–228.
[78] Okoye ONN, Ijoma IK, Okechukwu Okeke D, Health risk assessment of polycyclic aromatic hydrocarbons in common food crops within and outside selected mining areas of Ebonyi State, Nigeria. Sci Rep, 2025, 15, 24982.
[79] Volf M, Vučemilović A, Dobrović Ž, Enhancing environmental and human health management through the integration of advanced revitalization technologies utilizing artificial intelligence. Toxics, 2024, 12, 847.
[80] Singh V, Cortes-Ramirez J, Toms L-M, Sooriyagoda T, Karatela S, Effects of polybrominated diphenyl ethers on hormonal and reproductive health in e-waste-exposed population: A systematic review. Int J Environ Res Public Health, 2022, 19, 7820.
[81] Inobeme A, Mathew JT, Devolli A, Adetunji CO, Sharma N, Maliki N, Hussaini J, Metal components in industrial wastes and methods for metal ions recovery, In Metal Value Recovery from Industrial Waste Using Advanced Physicochemical Treatment Technologies. Elsevier, 2025, 1–15.
[82] Manzoor MF, Tariq T, Fatima B, Sahar A, Tariq F, Munir S, Khan S, Nawaz Ranjha MMA, Sameen A, Zeng XA, Ibrahim SA, An insight into bisphenol A, food exposure and its adverse effects on health: A review. Front Nutr, 2022, 9, 1047827.
[83] da Silva RC, Teixeira MP, De Paiva LS, Miranda-Alves L, Environmental health and toxicology: Immunomodulation promoted by endocrine-disrupting chemical tributyltin. Toxics, 2023, 11, 696.
[84] Cseresznye A, Hardy EM, Ait Bamai Y, Cleys P, Poma G, Malarvannan G, Scheepers PTJ, Viegas S, Martins C, Porras SP, Santonen T, Godderis L, Verdonck J, Poels K, Silva MJ, Louro H, Martinsone I, Akūlova L, Van Dael M, Van Nieuwenhuyse A, Covaci A, HBM4EU e-waste study: Assessing persistent organic pollutants in blood, silicone wristbands, and settled dust among e-waste recycling workers in Europe. Environ Res, 2024, 250, 118537.
[85] Mathew JT, Adetunji CO, Inobeme A, Musah M, Azeh Y, Etsuyankpa MB, Otori AA, Chado IM, Shaba EY, Akoji JN, Mamman A, Oyewole OA, Application of nanochitosan for the biodegradation of agricultural wastes, In Nanochitosan Applications for Enhanced Crop Production and Food Security. Wiley, 2025, 361–382.
[86] Pineda-Vásquez T, Rendón-Castrillón L, Ramírez-Carmona M, Ocampo-López C, From e-waste to high-value materials: Sustainable synthesis of metal, metal oxide, and MOF nanoparticles from waste printed circuit boards. Nanomaterials, 2023, 14(1), 69.
[87] Mehri SM, Farajzadeh MA, Mogaddam MRA, Matrix cleanup assisted extraction of phenolic pollutants from diverse wastewaters using a magnetic core-shell adsorbent. Sci Rep, 2025, 15, 39154.
[88] Chen Y, Li H, Huang H, Zhang B, Ye Z, Yu X, Shentu X, Recent advances in non-targeted screening of compounds in plastic-based and paper-based food contact materials. Foods, 2023, 12, 4135.
[89] Deschamps E, Calabrese V, Schmitz I, Hubert-Roux M, Castagnos D, Afonso C, Advances in ultra-high-resolution mass spectrometry for pharmaceutical analysis. Molecules, 2023, 28. 2061.
[90] De la Hera O, Alonso RM, Contribution of gas chromatography–mass spectrometry to the volatile organic compound profile of Vespa velutina nigrithorax larvae. Chemosensors, 2025, 13, 175.

[91] Dalmia A, Cudjoe E, Jalali J, Qin F, LC-MS/MS method with electrospray ionization and atmospheric pressure chemical ionization for analysis of pesticides in hemp. J Cannabis Res, 2021, 3, 50.

[92] Mathew JT, Adetunji CO, Inobeme A, Otori AA, Musah M, Azeh Y, Shaba EY, Muhammad IA, Mamman A, Oyewole OA, Relevance of nanochitosan in food and food-packaging sectors: Current trends, In Nanochitosan Applications for Enhanced Crop Production and Food Security. Wiley, 2025, 345–360.

[93] Al-Amin K, Kawsar M, Mamun MTRB, Sahadat Hossain M, Fourier transform infrared spectroscopic technique for analysis of inorganic materials: A review. Nanoscale Adv, 2025, 7, 6677–6702.

[94] Mathew JT, Inobeme A, Adetunji CO, Oyewole OA, Recent applications of chitosan-based nanoformulated metal and metal oxide nanoparticles, In Chitosan-Based Nanoparticles for Biomedical Applications. Elsevier, 2025, 573–589.

[95] Temwani K, Pan D, Wu Z, Zhang Y, Lan H, Advances in miniaturized liquid chromatography for detection of organic pollutants in food, environmental, and biological samples. Separations, 2025, 12, 312.

[96] Xu S, Wang J, Deng D, Sun Y, Wang X, Zhang Z, Pretreatment combining matrix solid-phase dispersion with dispersive liquid–liquid microextraction for PBDEs in vegetables by GC-MS/MS. RSC Adv, 2023, 13, 15772–15782.

[97] Guirette-Barbosa OA, Durán-Muñoz HA, Cruz-Domínguez O, Carrera-Escobedo JL, Celaya-Padilla JM, Castañeda-Burciaga S, Management system according to ISO/IEC 17025: Method validation. Appl Sci, 2024, 14, 4114.

[98] Inobeme A, Adetunji CO, Ajai AI, Inobeme J, Mathew JT, Obar A, Maliki M, Nwakife N, Eziukwu C, Chemical nanosensors for monitoring environmental pollution, In Advanced Application of Nanotechnology to Industrial Wastewater. Springer, 2023, 93–103.

[99] Bala S, Garg D, Thirumalesh BV, Sharma M, Sridhar K, Inbaraj BS, Tripathi M, Recent strategies for bioremediation of emerging pollutants: A review. Toxics, 2022, 10, 484.

[100] Inobeme A, Mathew JT, Adetunji CO, Agbugui MO, Osarenren EJ, Oti C, Ngonso BF, Inobeme J, Oyedolapo MB, Bernard E, Jibrin NA, Biocomposites with graphene derivatives, In Advances in Biocomposites and Their Applications. Elsevier, 2024, 149–166.

[101] Sánchez-Castro I, Molina L, Prieto-Fernández MÁ, Segura A, Past, present and future trends in remediation of heavy-metal contaminated soil: Techniques applied in real contamination events. Heliyon, 2023, 9, e16692.

[102] Wang M, Kong D, Liu L, Wen G, Zhang F, In situ conductive heating for thermal desorption of volatile organic-contaminated soil based on solar energy. Sustainability, 2024, 16, 8565.

[103] Dagnew M, Xue Q, Zhang J, Wang Z, Zhou A, Li M, Zhao C, Review of advanced oxidation techniques for pesticide degradation in aqueous environments. Sustainability, 2025, 17, 4710.

[104] Yadav P, Petrella A, Todaro F, De Gisi S, Vitone C, Petti R, Notarnicola M, Ex situ stabilization and solidification of marine sediments using green cement admixtures. Materials, 2024, 17, 3597.

[105] Triozzi M, Binetti MS, Campanale C, Uricchio VF, Massarelli C, Integrated assessment of smart passive bioventing for remediation of persistent organic pollutants. Sustainability, 2023, 15, 3764.

[106] Sharma M, Agarwal S, Agarwal Malik R, Kumar G, Pal DB, Mandal M, Sarkar A, Bantun F, Haque S, Singh P, Srivastava N, Gupta VK, Recent advances in microbial engineering approaches for wastewater treatment: A review. Bioengineered, 2023, 14, 2184518.

[107] Cramer R, Parker BL, Stapleton JM, Remediation geology and process-based conceptual site models for optimizing groundwater remediation. Sustainability, 2025, 17. 2027.

Ashwani Kumar, Ankush Kumar Tangra*, Amanpreet Singh

Electronic waste pollution: a global challenge with focus on end-of-life solar cell waste

Abstract: Electronic waste (e-waste) is one of the solid waste streams that is escalating the rapidest in the world. The e-waste presents significant difficulties for the environment, the economy, and community health. Solar cell development has opened up new avenues for the invention of sustainable and green energy, but it has also made end-of-life (EoL) waste management a progressively pressing issue. The deployment of solar cells has expanded significantly due to the worldwide push for sustainable and eco-friendly energy, and forecasts show that solar cell waste will increase exponentially over the next few decades. In addition to dropping the chance to recover precious elements like rare earth, indium, and silver, improper handling of these minerals might result in the release of dangerous compounds like cadmium, lead, and other heavy metals. The growing problem of solar cell waste, which is a subset of the larger issue of e-waste, is examined in this study with an emphasis on developing nations where the use of solar cells is increasing but waste management infrastructure is still lacking. Recycling and material recovery are two examples of circular economy initiatives that greatly lessen environmental consequences. The results highlight the critical need for long-term national planning. Coordinated regulation and investment in recycling businesses to maintain the clean energy transition, social justice, and environmental sustainability.

Keywords: Solar cell, e-waste, recycling, planning, end of life

1 Introduction of e-waste: a global issue

One of the most crucial environmental problems of the twenty-first century is e-waste. E-waste is the solid waste source with the fastest rate of growth in the world. E-waste is defined as unwanted electrical and electronic equipment (EEE) that has outlived its useful life, including the lot from computers, laptops, televisions, generators, DVDs, mobile phones, and home appliances to renewable energy technology like photovoltaic modules or has become outdated as a result of the quick development of

***Corresponding author: Ankush Kumar Tangra,** Department of Physics, Research and Incubation Center, Rayat Bahra University, Kharar, Punjab, India, e-mail: aktangra@gmail.com
Ashwani Kumar, University Center for Research and Development, Chandigarh University, Gharuan, Mohali 140413, Punjab, India
Amanpreet Singh, Department of Chemistry, Research and Incubation Center, Rayat Bahra University, Punjab, India

 | https://doi.org/10.1515/9783112248072-006

technology and rising customer demand. Global e-waste extents have improved at an exceptional rise in electronic consumption, shorter device lifespans, and rapid innovation cycles [1]. The world's largest and fastest-growing industry sector is the electronics sector, which includes telecommunications. This is due to people's dependence on modern technology for a luxury lifestyle [2]. There is a great deal of electrical waste as a result of the enormous growth in new technologies brought about by the growing demand and consumption rate of electric equipment worldwide. E-waste is especially challenging to address because of its dual nature. On the one hand, e-waste contains toxic elements such as mercury, lead, cadmium, and brominated flame retardants, which can seriously endanger human health and the environment if not properly handled [3]. However, it is also important for reservoirs of valuable materials, viz., rare earth metals, copper, gold, and silver, which can be extracted and reincorporated into manufacturing processes to promote a circular economy. This paradox puts e-waste in the middle of three big issues: preserving the environment, generating income and jobs, and ensuring equitable treatment of communities [4]. Over the past 20 years, the quantity of e-waste produced has increased at an alarming rate. Global e-waste was anticipated to be 41.8 million tons in 2014, but by 2019, it had increased to 53.6 million tons, and by 2030, it is expected to reach approximately 74 million tons [1] (Table 1). This shows that the quantity of e-waste produced worldwide is growing at a startling pace of almost 2 million tons annually. The amount of e-waste generated annually was 57.4 million tons in 2021 alone. Global e-waste patterns are depicted in Figure 1, which shows a consistent rise between 2017 and 2030. Europe has the highest recycling rates, while Asia produces the most e-waste overall. The composition of e-waste is dominated by both big and small equipment. Global recycling rates are still low compared to generation, which highlights the need for better recycling and waste management programs. There are several reasons for this growing issue. Economic expansion, shifting lifestyles, and rapid technological developments mean that new gadgets with the latest features are always hitting the market, forcing people to throw away older but often still usable products. Additionally, in order to announce new technology, both the public and private sectors frequently replace outdated electronic devices [2, 5]. This complexity is increased by the growth of renewable energy technologies, like solar cells, which, while environmentally valuable in terms of clean energy, will also add significantly to the burden of e-waste as panels reach the end of their 20–30-year lifespan. As a result, the problem of e-waste is not limited to user electronics but also encompasses the infrastructure supporting the shift to sustainable energy sources. E-waste is now the waste stream with the greatest rate of growth in the world, surpassing municipal solid trash and posing noteworthy obstacles to waste administration [6].

The distribution of e-waste is not identical worldwide. Because of their high consumption and quick cycles of device replacement. The inappropriate recycling and disposal of e-waste contributes significantly to pollution and greenhouse gas emissions, which have a negative impact on the environment and human health. Traditionally, a

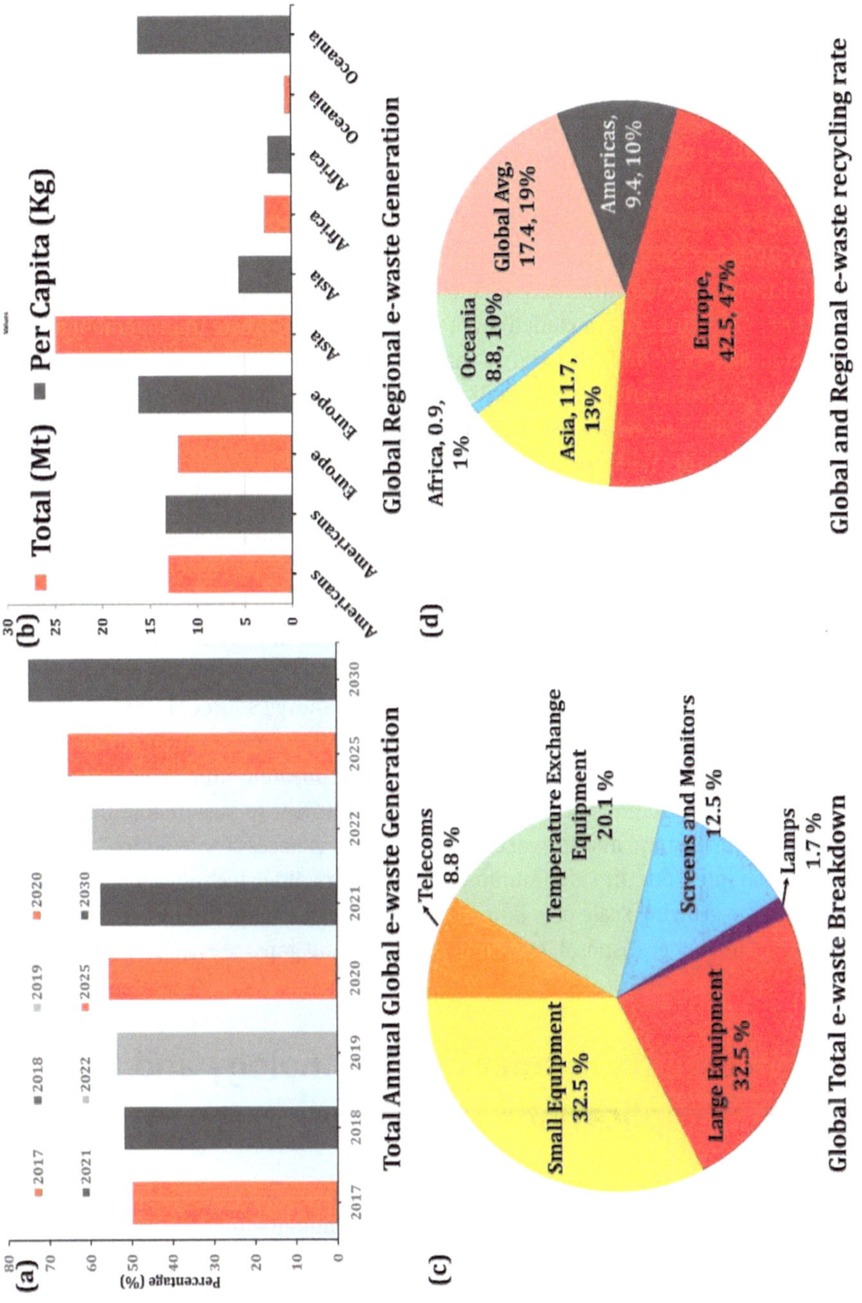

Figure 1: Global e-waste generation, composition, and recycling rates. (a) Annual global e-waste growth, (b) regional e-waste generation, (c) breakdown by equipment type, and (d) regional recycling rate comparison.

popular way to dispose of e-waste, landfilling, produces large volumes of methane, a powerful greenhouse gas, along with pollution, risks from PV metals, and land degradation. Solar photovoltaic (PV) panels are one of the many electric components that are included in the problem of e-waste, which is becoming more and more acknowledged as a new waste stream. If proper waste management solutions are not implemented, solar panels are predicted to contribute considerably to waste formation, given their typical lifespan of around 25 years and the growing installation capacity worldwide [7]. In reality, PV components were added as a discarded electronic device in July 2012, when the European Union formally updated its waste electrical and electronic equipment (WEEE) guideline. This included them in the electronic waste management system for collection and recycling. This shows that in order to guarantee a circular economy and promote environmental sustainability, efficient waste management and recycling techniques are very necessary [8]. The emergence of e-waste as a worldwide problem highlights the pressing need for combined approaches that include financial, technological, and regulatory remedies. E-waste must be reframed as a resource management challenge that calls for circular economy thinking rather than only being seen as an environmental concern. Key modules of the answer include increased enforcement of international agreements, capacity building in poor nations, eco-design principles in product development, and technological advancements in recycling [9–11]. Furthermore, a new frontier in waste management is presented by the confluence of conventional e-waste streams and solar e-trash, as renewable energy sources like solar PV take centre stage in the global energy landscape. Although they are essential for decarbonization, solar panels will increase the amount of complicated, multilateral trash produced worldwide by millions of tons. A strong commitment to sustainable innovation, international cooperation, and forward-thinking policy is necessary to address this dual issue. In this chapter, our aim is to summarize the photovoltaic technology and material composition, the growing scale of solar cell waste, the assessment of environmental and health risks of improper disposal, recycling practices, and gaps.

2 Advancements in solar cell technology and their material implications

An essential and quickly growing clean energy source for the world's shift to a decarbonized future is solar PV technology. It is simple to implement, eco-friendly, and has seen a significant drop in installation costs over time, around 50% since 2010. First applied in space, PV systems are widely used in residential, commercial, and utility-scale application. This technology is praised for being abundant, clean, efficient, non-polluting, and becoming more and more economical [12–16]. Carbon emissions from solar power sources are modest; silicon-based solar panels, for instance, have been shown to emit very little CO_2 when compared to fossil fuels [17]. The PV effect, which occurs when photons from sunlight excite electrons in a semiconductor material to produce an electric current. Solar cells have undergone tremendous development over the last 60 years, emerging as a key component of the clean energy revolution and providing a substitute for power generation derived from fossil fuels. But this industry's volatile growth also brings with it problems related to material usage, environmental impact, and panel EoL management. Solar cell development is frequently explained in terms of three generations. With over 90% of the market, crystalline silicon (c-Si), the foundation of the first generation, continues to rule the industry today. The typical lifespan of these modules is 25–30 years, with a power conversion efficiency (PCE) of 20–25% [11]. High-purity silicon and intricate manufacturing procedures are necessary for the resource and energy-intensive fabrication of crystalline silicon cells, despite their maturity and durability. Thin film technologies, including amorphous silicon (a-Si), copper indium gallium selenide (CIGS), and cadmium telluride (CdTe), are included in the second generation. These devices typically have a lower efficiency of about 10–18% but they are lighter and less expensive, and they are frequently appropriate for flexible applications. Environmental issues are also brought up by their use of rare and occasionally hazardous elements like indium and cadmium [18]. Emerging technologies such as tandem configurations, dye-sensitized solar cells (DSSCs), and perovskite solar cells (PSCs) make up the third generation. Among these, perovskite has garnered a lot of interest due to its ability to achieve efficiencies above 25% in just ten years of development. Issues including stability, long-term durability, and lead-related toxicity continue to be obstacles to commercialization despite this quick development [19]. From first-generation silicon cells to sophisticated third-generation nanocrystal and PSCs, Figure 2 illustrates the development of solar cell technology. In addition to addressing waste and recycling issues, it highlights the advantages of sustainable energy generation while identifying hazardous elements like lead, cadmium, and arsenic, as well as important recoverable resources like silicon and aluminum.

However, handling the future waste load related to millions of tons of EoL PV modules comes with the promise of sustainability [11]. With installed capacity reaching almost 400 GW by the end of 2017 and an 90% increase in installations over the previous ten years, from 104 GW to 1,053 GW, the adoption of PV systems has significantly accelerated globally. The total PV capacity worldwide exceeded 971 GW by 2021 and 600 GW by 2023. By 2024, solar and wind power will account for around 21% of all electricity generation worldwide, making solar energy a crucial source to fulfill the world's growing electricity demands [1, 12–16]. Solar power is expected to develop at an average annual rate of 20–25% until 2030, according to the International Renewable Energy Agency (IRENA), making it a key component of the world's renewable energy mix. According to life-cycle studies, however, the manufacturing phase is the environmental "hotspot" accounting for the majority of energy use, greenhouse gas emissions, and material depletion, despite efficiency advances. The manufacture of glass, the purifying of silicon, and the addition of hazardous or rare materials all in-

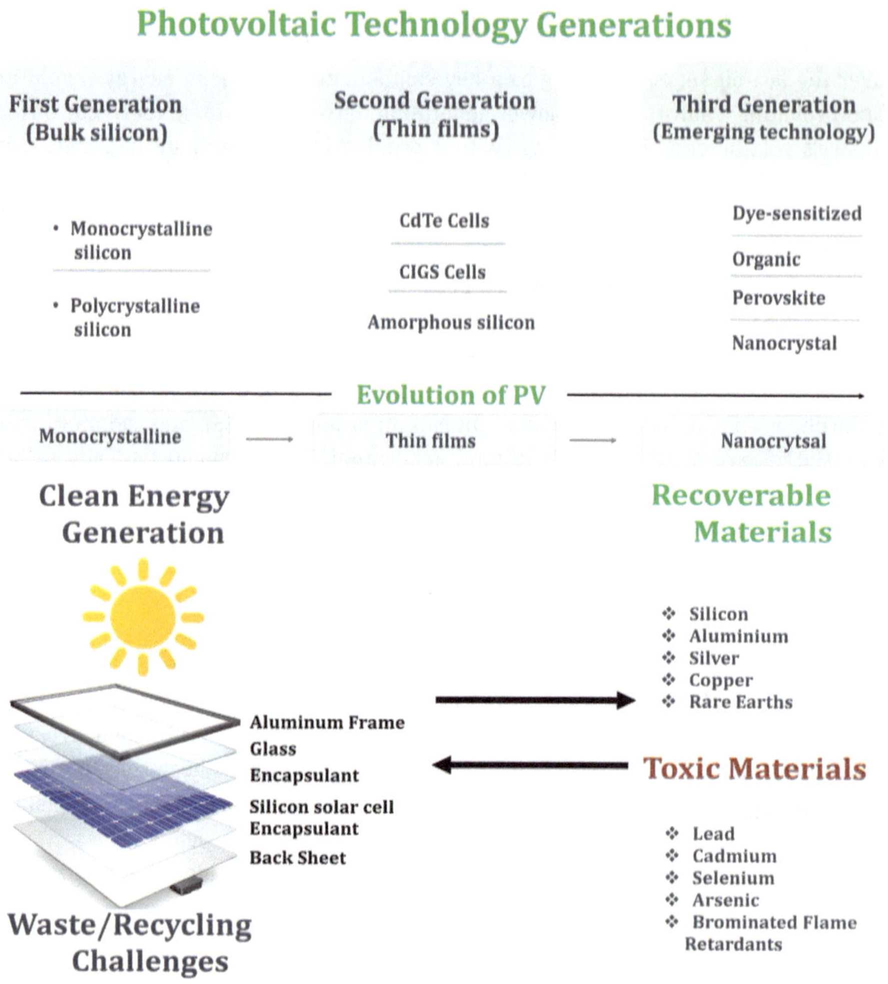

Figure 2: The evolution of photovoltaic technology generations and recycling challenges.

crease the environmental impact of solar systems. The value and intricacy of solar modules are highlighted by their composition. An aluminum frame, a glass front cover, semiconductor layers made of silicon or substances like gallium selenide and cadmium telluride, and encapsulants like copper, silver, and ethylene vinyl acetate (EVA) make up a standard panel [20]. If panels are disposed of incorrectly, these components provide a risk of environmental pollution in addition to chances for resource recovery through recycling. Global differences are highlighted by the regional dispersion of solar adoption as installations continue to rise dramatically. Typical module stacks and technology families. Crystalline silicon (C-Si) modules make up the majority of installed modules. Environmental stressors, viz., heat, UV, humidity, dust, and hail, can cause corrosion, discoloration, and delamination over time. This is described by a durability study and field experience. Different semiconductor stacks (Such as Cd/Te or Cu-In-Ga-Se compounds) are used by thin film modules (most notably CdTe, CIGS, and a-Si), which have an impact on recycling pathways and hazard profiles [21]. According to comparative overviews, composition-aware EoL techniques are necessary for thin film devices because they contain materials of economic relevance and/or regulatory concern (e.g., Cd, Te, Ga, and Se) [20]. High recovery potential for tellurium and cadmium through a series of crushing, separation, and hydrometallurgical stages is demonstrated by historical and industrial experience with CdTe recycling [22]. Newer technologies like perovskite promise shorter energy payback periods, but it's important to carefully consider their eco-toxicity and recycling issues. There are unique EoL concerns with emerging perovskite PV. Reviews contend that lead is frequently present in perovskite cells and that they may have precious metal connections (Au/Ag), which present both a risk (Pb) and an opportunity at EoL. They urge eco-design that accounts for disassembly and value-adding processes [23]. Consequences for EoL procedures, Delamination is a crucial phase in any technology family because of the EVA bond. The following methods have been documented: chemical (solvent or acid/base leaching), mechanical (shredding/screening), and thermal (e.g., 500–650 °C to remove EVA); they are frequently combined to balance energy use, emissions, throughput, and output purity [17, 20]. EoL care is still a major concern in spite of these advancements; usually lasting 25–30 years, solar panels are removed when their efficiency starts to drop. By 2050, it is anticipated that the total volume of solar waste produced worldwide will be around 78 million tons [11]. These panels have the potential to greatly increase the already severe e-waste problem in the absence of efficient recycling infrastructure. Some technologies use hazardous materials, including lead, cadmium, and arsenic, which raise disposal concerns [24]. To guarantee that solar technology continues to play a role as a genuinely sustainable energy source.

3 The growing scale of solar panel waste

Solar cell technology's explosive growth has created previously unheard of possibilities for the production of clean energy, but it has also brought with it the difficulty of handling enormous amounts of waste from solar panels. According to predications, the number of retired and EoL panels will rise significantly over the next few decades as the global few decades as the global solar industry continues to grow. According to estimates from the IRENA, depending on deployment and replacement scenarios, PV waste might amount to ~60–78 million tons by 2050. This figure reflects both "regular loss" panels, which fail too soon because of damage, manufacturing flaws, or replacement with newer, more efficient technologies, and "regular loss" panels, which have reached the end of their anticipated 25–30-year lifespan [12]. The geographic dispersion of solar installations is reflected in the global distribution of solar waste. According to projections, China, the largest manufacturer and installer of PV modules worldwide, will produce the most waste by the middle of the century, up to 20 million tons under early loss scenarios. Together, the United States, India, Japan, and Germany will contribute tens of millions of tons of discarded modules, making them other major contributors. According to IRENA, for instance, the United States alone may produce 7.5–10 million tons of PV waste by 2050, while Japan and India may each contribute over 6 million tons. Germany is likely to have severe disposal challenges in Europe by 2050, as more than 4 million tons of solar waste are anticipated. The fast increase in PV installations worldwide over the last decade is a major contributor to this expansion. With yearly additions topping 240 GW, global capacity has increased from just 400 GW in 2017 to over 1.2 TW in 2022. With total capacity predicted to surpass 8,500 GWs by 2050, this trend is anticipated to continue [30]. The amount of decommissioning that will occur as solar panels get older will increase significantly, producing a parallel waste stream that may compete with or even outnumber more conventional types of electronic waste. The logistical and environmental issues of end-of-life management are exacerbated by the fact that solar panels, in contrast to small consumer electronics, are huge, heavy, and multilayered goods that typically weigh between 20 and 30 kg [23, 31].

Problems with aging and technical replacement contribute to the size of waste from solar panels. Despite solar modules of 2–3 decade lifespans, many system owners replace their panels quicker than planned due to rapid efficiency increases and declining costs. This phenomenon, which is also referred to as the early loss scenario, increases the unpredictability of waste streams and speeds up trash formation. In many instances, panels may be decommissioned because newer technologies can generate more power with fewer units, not because they have failed. These characteristics make predicting more difficult and necessitate infrastructures for recycling and garbage management that are adaptable and agile. Solar waste that is not properly managed poses serious environmental risks. Silver, copper, and other metals embedded in crystalline silicon modules are among the valuable but dangerous materials found in panels, along with lead, cadmium, selenium, and arsenic in some technologies [32]. These materials can leak into soil and groundwater if they are disposed of inappropriately in landfills, posing long-term risks to the environment and human health. However, there is a chance for urban mining with these same materials. Reusing discarded panels to recover high-purity silicon, silver, and aluminum can improve supply chains, lessen reliance on primary raw materials, and promote the circular economy [12, 24]. The global solar waste challenge will be further shaped by regional differences in waste management capacity. Under its WEEE Directive, Europe has taken the first step by requiring up to 80% of EoL panels to be recycled [11]. Conversely, nations with quickly expanding solar markets, like India, are just starting to create regulatory frameworks and lack extensive infrastructure for recycling solar waste. Existing e-waste systems in China and the United States, which are already under stress due to the exponential growth of electronic and battery waste streams, will be put to the test by the volume of waste that is expected. This disparity in readiness highlights how urgent it is to establish international collaboration and standardized methods for managing PV waste [25]. There are also policy and economic concerns about the amount of solar waste. Approximately 2 billion new PV modules could be produced by 2050, because of the recovery of materials from recycled panels, which would have significant economic value. However, attaining this goal necessitates significant expenditure on collection systems, recycling technologies, and recovery incentive policy mechanisms. Without such steps, there is a chance that solar panels, which are a representation of clean energy, will contribute to hazardous waste and environmental damage [9, 33]. In the end, the increasing amount of waste from solar panels worldwide presents both a problem and an opportunity. The estimated tens of millions of tons of discarded modules raise concerns about the possibility of overburdening waste systems and compromising the environmental advantages of solar technology. On the other hand, the same forecast highlights how solar waste has the potential to develop into a useful resource stream that can help new industries, lessen dependency on virgin materials, and close loops in the renewable energy industry. It will take a coordinated effort that incorporates policy reform and technological innovation to address this problem. The problem of solar panel waste can be turned into a basis for sustainable development if it is handled appropriately, guaranteeing that solar energy will continue to have a promising and responsible future [34–38].

Table 1: Reveals the projected solar panel waste by country/region.

Region/country	2030 projection	2050 projection	References.
United Kingdom (UK)	~30–200 kt	~1–1.2 Mt	[26, 27]
India	~50–320 kt	~4.4–7.5 (scenario-dependent) ~1.8 Mt (national review)	[20, 26, 27]
China	–	~19.9 Mt (early-loss IRENA) ~22.34 Mt (Marcuzzo et al.)	[28]
Japan	~2.1 Mt	~6.5–7.6 Mt (IRENA) ~3.22 Mt	[26, 28]
United States	–	~7.5 – 10 Mt (IRENA) ~5.29 Mt (Marcuzzo et al.)	[26, 28]
European Union (EU)	–	~9.5 Mt	[28]

4 Environmental and health risks of improper disposal

One of the primary factors of the global switch to renewable energy is the quick and extensive adoption of solar PV technology. Although PV systems provide clean power with low carbon emissions during their active lifetime, managing them beyond their EoL poses an increasing socio-environmental problem. While they contain toxic elements, improper disposal of retired PV panels and other electronic trash presents serious threats to human health and the integrity of the environment. If not managed correctly, what is a sign of clean energy can become a source of toxicity and long-term pollution (Figure 3). Multiple studies have confirmed that improper disposal of e-waste and solar PV waste poses significant health and environmental risks due to the release of harmful emissions and the leaching of toxic substances. Toxic gases like dioxins, particulate matter, brominated compounds, and hydrocarbons are released into the atmosphere. When PV modules and electronic components are disassembled, shredded, or burned without the right controls. Research has reported that dioxin concentrations from open burning can reach levels 15 to 56 times above the WHO-recommended safe limits, with accumulation documented in human milk, placentas, and hair [39–42]. This exposure highlights the long-term health hazards that communities residing close to unofficial recycling facilities or landfills face, such as increased risks of cancer, tumors, and respiratory ailments. Workers in the unorganized sector are especially at risk as they handle PV and electronic debris without safety gear, which exposes them to toxins through direct skin contact and inhalation. As time goes on, this exposure contributes to immunological suppression, respiratory conditions, and, in extreme situations, cancer. Electronic devices like solar panels are intricate goods made of a variety of components, including dangerous compounds, such as lead, cadmium, and arsenic, as well as precious metals like copper and silver. More than 90% of a conventional crystalline silicon (c-Si) PV panel, which makes up the majority of the worldwide market, is made of nonhazardous elements including aluminum, polymers, and glass [43, 44]. The compounds in the remaining portion, however, have the potential to be hazardous if discharged into the environment. Likewise, potentially dangerous metals, including cadmium, tellurium, indium, gallium, and selenium, are present in thin film panels.

The following are the important hazardous elements found in PV panels and e-waste [45–48]:

(i) Lead (Pb): The solder used to make electrical connections in c-Si panels and other devices contains lead. Strong neurotoxins like lead can harm children's neurological and reproductive systems, impede renal function, and create lasting cognitive deficiencies. Rainwater may leach anywhere from 13% to 90% of the lead in a damaged photovoltaic unit, according to studies. Under some circumstances, leached lead quantities from c-Si modules have been reported to surpass regulatory limits.

(ii) Cadmium (Cd): An essential element in NiCd batteries and CdTe thin film solar cells. In addition to being a proven carcinogen, cadmium can seriously harm bones, kidneys, and lungs. Although the compound CdTe is less soluble and more stable than the forms of cadmium present in batteries, public cadmium concerns have been a major problem for CdTe technology. Similar to lead, quantities of cadmium from discarded thin-film modules may exceed toxicity thresholds.

(iii) Arsenic (As): Gallium arsenide (GaAs) solar cells use arsenic (As), which is extremely toxic and a proven carcinogen; arsenic may affect the skin and neurological system. Research indicates that amounts of arsenic that is far greater than regulatory limits can leak out of GaAs cells.

(iv) Selenium (Se): One element included in CIGS thin film panels is selenium. Selenium is hazardous at greater doses and can bioaccumulate in food chains, endangering animals, even though it is an important mineral in low amounts.

(v) Flame retardants and polymers: Polymers containing fluorinated plastics or brominated flame retardants are frequently used to make backsheets, encapsulants, and electronic casings for photovoltaic panels. It is not possible to recycle these plastics. These materials emit corrosive gases like hydrofluoric acid and extremely hazardous and persistent organic pollutants like dioxins and furans when they are inappropriately disposed of by open burning, a frequent practice in informal recycling sectors. Dioxins have been associated with a higher risk of cancer [5].

These dangers are increased by the estimated volume of PV waste. This might result in thousands of tons of dangerous substances, like lead and cadmium, possibly entering the environment if improperly managed [20, 26, 47, 48].

Landfilling: The main danger of inappropriate disposal, such as landfilling, is that hazardous chemicals will seep into the ecosystem. Rainwater and other natural elements can break down the components of PV panels and other e-waste and release hazardous compounds when they are illegally buried in landfills. Leaching is a process that may pollute groundwater and soil, opening doors for pollutants to infiltrate ecosystems and human water sources. Once in the soil, these chemicals can be absorbed by crops, establishing exposure pathways up the food chain that can cause kidney damage and immunological delays in people. It has been demonstrated that circuit boards and chips containing phosphorus compounds and brominated flame retardants in aquatic settings may poison aquatic life and acidify water bodies, upsetting ecosystems.

Air pollution from informal recycling: In order to recover important metals like copper, informal recycling activities, which are prevalent in underdeveloped countries, frequently use fundamental techniques like open burning of panels. When plastic parts, like the backsheet and encapsulant, burn, a mixture of harmful chemicals is released into the atmosphere. Dioxins, furans, polycyclic aromatic hydrocarbons, volatile organic compounds, and small dust particles carrying heavy metals are some examples of these emissions. These contaminants harm the quality of the air, can result in acid rain, and seriously endanger the respiratory health of the local population and employees [5, 45]. Another factor contributing to climate change is improper solar PV recycling methods. Global warming can be made worse by the open burning and landfilling of PV waste, which can emit greenhouse gases such as carbon dioxide, methane, hydrochlorofluorocarbons (HCFCs), and hydrofluorocarbons (HFCs). According to studies, the concentrations of precious metals (gold, silver, and palladium) in one metric ton of discarded circuit boards can be up to 800 times higher than those found in natural ores. This means that burning crude not only wastes resources but also increases emissions [10]. Lastly, recycling procedures themselves might be dangerous if they are not adequately planned. PV panel crushing and thermal treatment frequently produce dust that is tainted with silicon, lead, and other substances. If breathed, this dust can result in silicosis and other long-term respiratory conditions. In order to prevent secondary contamination, chemical dissolution procedures for ethylene-vinyl acetate (EVA) encapsulants in solar panels must be carefully managed since they can produce hydrofluoric acid, nitric oxides, and toxic liquid wastes.

Loss of biodiversity and ecosystem disruption: Natural processes are disturbed, and wildlife is harmed when heavy metals and persistent organic pollutants contaminate soil and water habitats. As they move up the food chain and reach larger quantities in predators like humans, these chemicals have the ability to bioaccumulate in living things. In many animal species, this can result in mortality, developmental abnormalities, and reproductive problems, which eventually lower biodiversity. Cadmium, for instance, is known to be harmful to fish and other species, and it is dangerous when it builds up in the food chain [49].

Residues from even cooperative EoL therapy must be handled carefully. An integrated c-Si flow including disassembly, glass separation, cutting + incineration (energy recovery), acid leaching, electrolysis, and neutralization is modeled by a frequently referenced life cycle assessment example [20]. About 686 kg of cleaned glass (out of 700 kg), Al scrap, about 0.50 kg of Ag scrap, about 1.08 kg of Cu scrap, as well as about 44 kg of bottom ash, about 2 kg of fly ash (hazardous; landfilled), and about 306 kg of liquid waste plus about 50 kg of sludge that needs to be disposed of carefully are among the outputs; process emissions include about 2 kg of NO_x. Implication: formal recycling lowers environmental risk compared to landfills, but it still necessitates hazardous residue management and air emission controls, which is another reason to plan capacity and standards as quantities increase [20, 50].

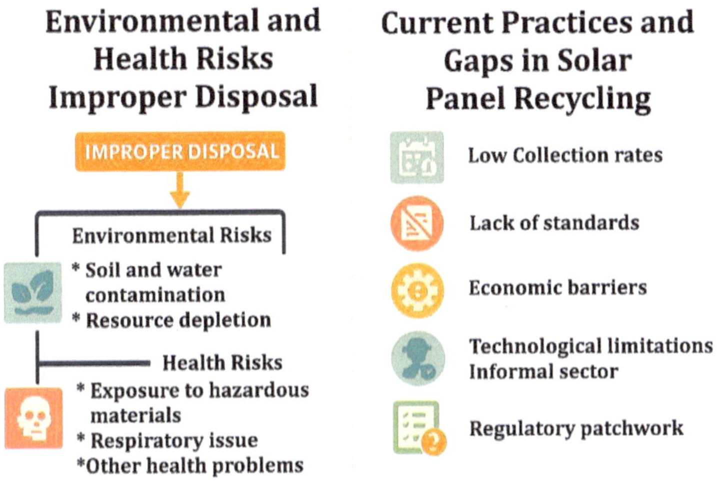

Figure 3: Various environmental and health risks of improper disposal and gaps in solar panel recycling.

5 Current practices and gaps in solar panel recycling

Over the past 20 years, solar energy has quickly expanded globally, with crystalline silicon PV modules leading the industry. Around 400 GW of PV capacity was added worldwide in 2017, and by 2050, it is expected to have grown to 4,500 GW. The enormous number of EoL solar panels is a new environmental concern, even if this expansion is necessary to meet climate targets. Solar panels installed during the recent boom will soon become a large waste stream because they typically have a lifespan of 25–30 years. According to projections, the amount of PV waste generated worldwide may increase to 78 million tons by 2050 and as much as 8 million tons by 2030 [12, 41]. However, despite improvements in manufacturing efficiency, EoL panel management systems are still disjointed and undeveloped. A combination of mechanical, thermal, and chemical processes is used in solar panel recycling today, but these methods have serious problems with cost, scalability, and environmental sustainability. Even though the issue has started to be addressed by pilot projects and specialist recyclers in Europe, the United States, and Japan, there are still gaps in global infrastructure, legislation, and commercial feasibility.

Solar panel recycling is a complicated procedure because of its multilayered, composite structure, which is intended to last for many years. The three main categories of current recycling techniques are chemical, thermal, and physical/mechanical treatments; they are frequently combined to optimize material recovery.

The most popular method for recycling crystalline silicon modules is bulk recycling. The recovery of glass and aluminum frames, which together make up over 70% of the panel weight, is its main goal. To disrupt the ties between layers, the laminate is first mechanically treated by shredding and breaking it into tiny pieces. After that, the crushed materials are sorted using a number of methods: (i) sieving and density-based separation are used to separate glass fragments from the lighter polymer fractions and denser metallic/silicon parts; (ii) conductive materials (silicon and metals) and nonconductive materials (glass and plastic) can be separated using electrostatic separation; and (iii) ferrous and nonferrous metals are separated using magnetic separators and eddy current [18]. However, this approach produces a low-value output with little incentives for recyclers since it ignores rare metals like silver, copper, and indium as well as high-value semiconductor components. Although semi-high value recycling is an advance, it frequently ignores other embedded components and focuses on the selective recovery of silicon wafers or metallic layers. The goal of closed-loop recycling, the most sophisticated technique, is to immediately reintroduce recycled silicon and other components into the manufacturing of new modules. In line with the ideas of the circular economy, Deutsche Solar AG in Europe has experimented with reusing recovered silicon cells in standard module manufacture [11]. Thermal recycling, primarily used to remove the EVA encapsulant, represents a significant challenge in PV recycling (Table 2):

- One important thermal technique is pyrolysis, which involves heating the panel laminate to high temperature (usually 500–600 °C) without the presence of oxygen. Glass and complete silicon cells may be easily separated thanks to this technique, which breaks down the EVA into gases and oils. After recovery, the silicon wafers can be cleaned and used again. Although efficient, pyrolysis requires a lot of energy and, if not properly managed, can result in hazardous pollutants.
- Although it is also used to burn off polymers, incineration, which entails burning in the presence of oxygen, carries a greater risk of producing harmful pollutants and oxidizing important metals, making their recovery more challenging.
- Recycling of chemicals, chemical techniques are used to recover dangerous and valuable compounds with high purity that are hard to separate mechanically or thermally.

Chemical recycling techniques concentrate on using solvent treatments or acid leaching to recover high-purity silicon, silver, and other important materials. These techniques are more effective at recovering materials than mechanical or thermal procedures, but they pose serious environmental risks since they employ powerful acids like nitric and hydrofluoric acid. Nitric acid dissolution, for instance, may effectively remove metal connections and EVA, but it may also harm wafers, decreasing the recovery rate of undamaged silicon. Some innovations, such as vacuum blasting, have been proposed to reduce chemical reliance, but these remain in experimental stages. In general, chemical processes emphasize the balance between ecological effect and recovery efficiency. Even with advances in technology, recycling solar panels is still not economically viable. Especially in areas where waste quantities are still relatively low, the expenses of setting up and running specialized recycling facilities are sometimes more than the value of the recovered materials. Recycling has more environmental advantages than landfilling, according to life cycle assessments (LCAs), including lower emissions and energy use, although profitability is still low. At small sizes (less than 4,000 tons annually), it has been demonstrated that a streamlined recycling procedure consisting of deframing, shredding, and electrostatic separation is more lucrative than thorough recovery techniques. Such simplified methods might not be enough to maximize resource recovery, though, when the amount of PV waste produced worldwide rises to millions of tons [11].

Regional differences in current regulatory frameworks are significant; the European Union's WEEE Directive offers the most sophisticated example. Extended producer responsibility (EPR) mandated by the EU requires manufacturers to pay for the collection and recycling of PV modules that have reached the end of their useful lives [12, 51, 52]. The United States, on the other hand, does not have a federal framework, although states like California and Washington have passed laws designating PV modules as "universal waste" and mandating recycling programs from manufacturers. Japan has also evolved, investing in improved separation processes and recycling infrastructure, while India has just recently included solar photovoltaics under its e-waste management Rules (2022) [25, 52]. Technical, financial, and policy-related issues can be used to categorize the gaps in the present recycling methods. Due to their multilayered structure, crystalline silicon modules are technically difficult to recycle, and current techniques either waste valuable resources or create additional environmental hazards. Without subsidies or legal requirements, recycling is not economically appealing due to the lack of significant market incentives and the low value of recovered glass. Effective design of recycling infrastructure is hampered by policy gaps such as the absence of standardized international standards, lax enforcement of current regulations, and limited data on PV waste creation and locations [11, 12, 52].

6 Future prospective

Technological innovation, circular economy frameworks, and changing regulatory processes that work together to turn waste difficulties into opportunities will dictate how solar panel e-waste is managed in the future. EoL modules will be in mass at previously unheard of levels as the installed capacity of PV is predicted to surpass several terawatts by the middle of the century. E-waste might exceed 78 million tons globally by 2050, according to projections, highlighting the need for long-term plans that account for complete material recovery and reintegration into production cycles in addition to disposal. The development of technology has a significant impact on expectations for

Table 2: Comparison of current solar panel recycling methods.

Recycling method	Key process	Advantages	Limitations/gaps	References
Mechanical (shredding, crushing, milling, and eddy current separation)	Panels are dismantled, frames removed, glass and metals separated	Low-cost, relatively simple, scalable; recovers bulk materials (glass, aluminum, copper)	High-value silicon and silver are lost; recovered glass is sometimes of poor quality (used for filler or insulation rather than new panels); closed-loop recycling is not achieved.	[11, 12]
Thermal (pyrolysis and heating to decompose polymers)	Encapsulants (EVA) are eliminated by controlled heating, which also splits silicon wafers and ribbons	Can recover upto ~ 90% of module weight; effective removal of EVA and polymers	Extremely energy intensive; emits harmful gases; expensive emission control is necessary; carbon footprint is large.	[11, 29]
Chemical (acid leaching and solvent treatments)	Wafers and metals are released when strong acids (such as HF and HNO_3) break down encapsulants and metal connections	High-purity silicon, silver, and indium may be recovered with greater efficiency than mechanical	High operating expenses, the possibility of wafer damage, the production of hazardous waste (toxic fumes, liquid effluents), and safety issues.	[11, 29]
Combined/hybrid (mech. + thermal + chemical)	Sequential processes (e.g., mechanical deframing, thermal EVA removal, chemical wafer recovery)	Increased material purity maximizes recovery rates and gets closer to closed-loop recycling	Complicated, costly, and only available for test projects; not yet industrially scaled	[11, 12]

Emerging eco-design approaches (design for recycling panels, detachable encapsulates)	Panels designed for easy disassembly and recycling	Decreases the environmental effect, boosts material recovery, and lowers recycling costs	Still at the experimental stage; not yet used in large-scale PV production	[11]

Gaps and Priority Actions:

(i) Delamination bottleneck: EVA removal still drives time and cost; solvent methods increase the burden of chemical management. Research and development of eco-design for disassembly and low-energy delamination is still a top goal [17, 26, 27].

(ii) Markets and output purity: Without higher quality routes, glass markets might become saturated; mechanical lines provide mass but lower purity, which lowers pricing.

(iii) The finance and clarity of policy: Recycling acceptance is low when rules lack PV-specific categories or objectives; EPR with incentives (landfill process, eco-modulation) can improve economics [28].

(iv) Logistics and capacity: Despite expanding waste streams, nations lacking specialized infrastructure, such as India in earlier evaluations, face gaps in their capability for collection and processing [20].

(v) Data transparency: Recycling is made more difficult by thousands of panel configurations; separation and recovery would be enhanced by long-lasting labeling and connecting BoM data to panels [26].

(vi) Safety and residues: Thermal stages necessitate emissions management; even compliance lines produce ash, sludge, and liquid pollutants that need to be handled carefully [17, 20].

the future. Although mechanical recovery of glass and aluminum is a major emphasis of current recycling procedures, research is moving toward high-value and closed-loop recycling. In order to lessen dependency on the extraction of virgin resources, future recycling systems seek to recover semiconductor materials like silicon, silver, and indium with high purity. Solvent-based delamination and controlled pyrolysis are two examples of chemical and thermal advancements that are being refined to increase material yields while reducing environmental emissions [29]. Distributed "micro-recycling" approaches, which decentralize waste processing, reduce transportation costs, and allow customized recovery procedures near generating sites, are equally promising. PV modules are positioned as a component of a larger sustainable materials ecosystem through these strategies, which show a shift from waste minimization to resource cycling. Future results will also be defined by regulatory and policy frameworks. By requiring EPR and encouraging the construction of specialized PV recycling facilities, the European Union has created a comparatively developed system under the WEEE Directive. Other areas, including the US and India, are currently developing their capability and adjusting their legislative frameworks to accommodate the expected increase in PV waste. To promote compliance and accountability, the combination of EPR standards, periodic reporting requirements, and fines for landfill disposal is probably going to spread around the world. These frameworks will also incentivize manufacturers to create panels that are recyclable, which will facilitate layer separation and reduce the need for hazardous ingredients. Long-term regulations may place an emphasis on modularity and eco-design, tying product innovation to EoL tactics to guarantee a circular supply chain.

There is a lot of economic potential in the future of managing solar panel e-waste from a market and resource standpoint. A significant portion of the metals and glass needed to produce new modules might be obtained from secondary raw materials recovered from waste streams, which would save prices and supply chain concerns. According to studies, recycling EoL PV might yield billions of dollars in recovered materials, generating green employment and lowering carbon emissions linked to primary production and mining. Building a parallel system of sustainable waste management guarantees that the spread of clean energy does not result in an environmental burden of hazardous wastes, as solar energy becomes increasingly important in the pursuit of decarbonization goals.

7 Conclusion

The world's will to move toward a sustainable, low-carbon future is demonstrated by the quick uptake of solar energy. However, as this chapter has demonstrated, the advantages of solar technology are inextricably linked to the escalating problems associated with solar cell e-waste. During their useful lives, solar panels provide clean energy, but their complicated materials composition and growing EoL volumes pose serious health, financial, and environmental issues. PV technology advancements have increased efficiency and worldwide deployment, but they also bring with them elements like lead, cadmium, and rare metals that make recycling and disposal more difficult. By the middle of the century, solar panel waste is expected to reach tens of millions of tons, making it one of the waste streams with the greatest rate of growth in the world. Due to groundwater pollution, toxic emissions, and heavy metal leaching, improper disposal endangers both human health and ecosystems. Recycling methods are still scarce today, with the majority of efforts concentrated on recovering bulk materials like aluminum and glass while ignoring valuable metals and semiconductors. Technical capabilities, financial incentives, and legislative frameworks continue to differ significantly, especially in poorer nations where recycling infrastructure is still in its infancy.

Coordination of international action is crucial for the future. More effective material recovery may be possible with investments in cutting-edge mechanical, thermal, and chemical recycling techniques in conjunction with eco-design techniques. To promote accountability and scale solutions, policy instruments like national recycling laws, extended producer responsibility, and standardized international standards will be essential. Promoting the ideas of the circular economy may also turn waste into a resource, lowering reliance on imported goods and opening up new markets. In conclusion, EoL plans need to be just as strong as the clean energy transition. Society can guarantee that solar energy fulfills its potential in powering sustainability without leaving behind new environmental legacies by incorporating innovation, regulation, and circularity into solar panel waste management.

References

[1] Heacock M, Kelly CB, Asante KA, Birnbaum LS, Bergman AL, Brune MN, Suk WA, E-waste and harm to vulnerable populations: A growing global problem. Environ Health Perspect, 2016, 124(5), 550–555.

[2] Seif R, Salem FZ, Allam NK. E-waste recycled materials as efficient catalysts for renewable energy technologies and better environmental sustainability: R. Seif et al. Environment, development and sustainability. 2024 Mar;26(3):5473–508.

[3] Weckend S, Wade A, Health GA, End of life management: Solar photovoltaic panels (No. NREL/TP-6A20-73852; T12-06: 2016). National Renewable Energy Laboratory (NREL), Golden, CO (United States).

[4] Tangra AK, Sharma M, Zainudeen UL, Singh LG, Investigation of Inorganic electron–hole transport material for high efficiency, stable and low-cost perovskite solar cell. J Mater Sci Mater Electron, 2020, 31(16), 13657–13666.

[5] Jain M, Kumar D, Chaudhary J, Kumar S, Sharma S, Verma AS, Review on e-waste management and its impact on the environment and society. Waste Manag Bull, 2023, 1(3), 34–44.

[6] Fawole AA, Orikpete OF, Ehiobu NN, Ewim DRE, Climate change implications of electronic waste: Strategies for sustainable management. Bull Natl Res Cent, 2023, 47(1), 147.

[7] Tasnim SS, Rahman MM, Hasan MM, Shammi M, Tareq SM. Current challenges and future perspectives of solar-PV cell waste in Bangladesh. Heliyon. 2022; 8(2), e08970.
[8] Xu Y, Li J, Tan Q, Peters AL, Yang C, Global status of recycling waste solar panels: A review. Waste Manag, 2018, 75, 450–458.
[9] Heacock M, Kelly CB, Asante KA, Birnbaum LS, Bergman AL, Bruné MN, Suk WA, E-waste and harm to vulnerable populations: A growing global problem. Environ Health Perspect, 2016, 124(5), 550–555.
[10] Fawole AA, Orikpete OF, Ehiobu NN, Ewim DRE, Climate change implications of electronic waste: Strategies for sustainable management. Bull Natl Res Cent, 2023, 47(1), 147.
[11] Chen PH, Chen WS, Lee CH, Wu JY, Comprehensive review of crystalline silicon solar panel recycling: From historical context to advanced techniques. Sustainability, 2024, 16(1), 60.
[12] Ndalloka ZN, Nair HV, Alpert S, Schmid C, Solar photovoltaic recycling strategies. Sol Energy, 2024, 270, 112379.
[13] Chowdhury MS, Rahman KS, Chowdhury T, Nuthammachot N, Techato K, Akhtaruzzaman M, Amin N, An overview of solar photovoltaic panels' end-of-life material recycling. Energy Strateg Rev, 2020, 27, 100431.
[14] Babaei A, Nasr Esfahani A, A review of photovoltaic waste management from a sustainable perspective. Electricity, 2024, 5(4), 734–750.
[15] Lotey GS, Tangra AK, Kanoun MB, Goumri-Said S, Kumar S, Garg MP, Sun NX, Potential of low-cost inorganic $CaFeO_3$ as transporting material for efficient perovskite solar cells. Mater Today Commun, 2023, 35, 105956.
[16] Xu Y, Li J, Tan Q, Peters AL, Yang C, Global status of recycling waste solar panels: A review. Waste Manag, 2018, 75, 450–458.
[17] Lotey GS, Sun NX, Tangra AK, Kanoun MB, Goumri-Said S, Kanoun AA, Garg MP, Boosting efficiency and stability of perovskite solar cells via integrating engineered Li/Na-ferrite-based inorganic charge transport layers: A combined experimental and theoretical study. J Nanopart Res, 2023, 25(9), 179.
[18] Cellura M, Luu LQ, Guarino F, Longo S, A review on life cycle environmental impacts of emerging solar cells. Sci Total Environ, 2024, 908.
[19] Seo B, Kim JY, Chung J, Overview of global status and challenges for end-of-life crystalline silicon photovoltaic panels: A focus on environmental impacts. Waste Manag, 2021, 128, 45–54.
[20] Rathore N, Panwar NL, Strategic overview of management of future solar photovoltaic panel waste generation in the Indian context. Waste Manag Res, 2022, 40(5), 504–518.
[21] Nain P, Kumar A, Understanding the possibility of material release from end-of-life solar modules: A study based on literature review and survey analysis. Renew Energy, 2020, 160, 903–918.
[22] Fthenakis VM, End-of-life management and recycling of PV modules. Energy Policy, 2000, 28(14), 1051–1058.
[23] Akulenko ES, Hadadian M, Santasalo-Aarnio A, Miettunen K. Eco-design for perovskite solar cells to address future waste challenges and recover valuable materials. Heliyon. 2023; 9(2), e13584
[24] Sahajwalla V, Hossain R, Rethinking circular economy for electronics, energy storage, and solar photovoltaics with long product life cycles. MRS Bull, 2023, 48(4), 375–385.
[25] Chatterjee B, Das S, Sustainable management of discarded solar Photovoltaic material: Enhancing electronic waste recycling in India, In National Conference on Technological Advancements in Waste Management: Challenges and Opportunities. 2023, 542, 57–71.
[26] Ali A, Malik SA, Shafiullah M, Malik MZ, Zahir MH, Policies and regulations for solar photovoltaic end-of-life waste management: Insights from China and the USA. Chemosphere, 2023, 340, 139840.

[27] https://www.irena.org/publications/2016/Jun/End-of-life-management-Solar-Photovoltaic-Panels
[28] Kwok KH, Savaget P, Fukushige S, Halog A, The necessity for end-of-life photovoltaic technology waste management policy: A systematic review. J Clean Prod, 2024, 461, 142497.
[29] Bhattacharjee J, De A, Kamila B, Mandal A, Recent scenario of e-waste recycling: Chemical engineering. Int J Chem React Eng, 2025, 23(6), 675–700.
[30] Kumari M, Kaur M, Tangra AK, Numerical optimization of HTL-free Perovskite solar cells: A pathway to enhanced efficiency. ICCK Trans Adv Funct Mater Process, 2025, 1(1), 25–31.
[31] Sah D, Kumar S, Recovery and analysis of valuable materials from a discarded crystalline silicon solar module. Sol Energy Mater Sol Cells, 2022, 246, 111908.
[32] Gahlot R, Mir S, Dhawan N, Recycling of discarded photovoltaic solar modules for metal recovery: A review and outlook for the future. Energy Fuels, 2022, 36(24), 14554–14572.
[33] Tangra AK, Kumar A, Kumari M, Arya S, Aljaloud K, Shkir M, AlFaify S, Emerging trends in inorganic hole transport materials for high-efficiency PSCs. Inorg Chem Commun, 2025, 182, 115643.
[34] Lesniewska F, Renewable energy, waste management and the circular economy in the EU: Solar PV and wind power, In: Research Handbook on EU Energy Law and Policy. 2017, 2017, 460–468.
[35] Matheri AN, Nabadda E, Mohamed B, Sustainable and circularity in the decentralized hybrid solarbioenergy system. Environ Dev Sustain, 2023, 26, 16987–17011.
[36] Akhter M, Al Mansur A, Islam MI, Lipu MH, Karim TF, Abdolrasol MG, Alghamdi TA, Sustainable strategies for crystalline solar cell recycling: A review on recycling techniques, companies, and environmental impact analysis. Sustainability, 2024, 16(13), 5785.
[37] Tao J, Yu S, Review on feasible recycling pathways and technologies of solar photovoltaic modules. Sol Energy Mater Sol Cells, 2015, 141, 108–124.
[38] Ceballos DM, Dong Z, The formal electronic recycling industry: Challenges and opportunities in occupational and environmental health research. Environ Int, 2016, 95, 157–166.
[39] Beaucham CC, Brueck SE, Kawamoto MM, Evaluation of exposure to metals at an electronic scraprecycling facility. 2014, 1, 1–44.
[40] Chakraborty P, Syed JH, Chandra S, Hande S, Pokhrel B, Islam E, Miah MAH, Electronic wasterecycling in South Asia: Overview of associated risks from a cocktail of micro-pollutants and recommendations for sustainable e-waste management. J Hazard Mater Adv, 2025, 18(3), 100715.
[41] Adeola FO, Global impact of chemicals and toxic substances on human health and the environment, In: Handbook of Global Health. 2020, 978-3-030-05325-3, 1–30.
[42] Preet S, Smith ST, A comprehensive review on the recycling technology of silicon based photovoltaic solar panels: Challenges and future outlook. J Clean Prod, 2024, 448, 141661.
[43] Allouhi A, Rehman S, Buker MS, Said Z, Up-to-date literature review on solar PV systems: Technology progress, market status and R&D. J Clean Prod, 2022, 362, 132339.
[44] Emetere ME, Omotoso AW, Afolalu SA, Amusan L, Jen TC, Hossain KM, Naturally occurring radioactive materials and heavy metals in photovoltaics: Insights into short-and long-term waste hazards. J Hazard Toxic Radioact Waste, 2024, 28(3), 05024001.
[45] Rawat S, Verma L, Singh J, Environmental hazards and management of e-waste, In: Environmental Concerns and Sustainable Development: Volume 2: Biodiversity, Soil and Waste Management. 2019, 1, 381–398.
[46] Tasnim SS, Rahman MM, Hasan MM, Shammi M, Tareq SM, Current challenges and future perspectives of solar-PV cell waste in Bangladesh. Heliyon, 2022, 8(2), e08970.
[47] Nain P, Kumar A, Identifying issues in assessing environmental implications of solar PVs-related waste, In Recent Developments in Waste Management: Select Proceedings of Recycle. 2020, 57, 71–90.
[48] Chowdhury PR, Goswami S, Choudhary R, Choudhury M, Electronic wastes, its impact on wildlife and biodiversity, In: Electronic Waste. CRC Press, 2025, 65–71.

[49] Hansen UE, Nygaard I, Dal Maso M, The dark side of the sun: Solar e-waste and environmental upgrading in the off-grid solar PV value chain, In: The Dark Side of Innovation, 2022, 1, 35–55.
[50] Andersen T, A comparative study of national variations of the European WEEE directive: Manufacturer's view. Environ Sci Pollut Res, 2022, 29(14), 19920–19939.
[51] Serpe A, Purchase D, Bisschop L, Chatterjee D, De Gioannis G, Garelick H, Verbeek S, 2002–2022: 20 years of e-waste regulation in the European Union and the worldwide trends in legislation and innovation technologies for a circular economy. RSC Sustain, 2025, 3(3), 1039–1083.
[52] Sivagami K, Bose S, Vinayak AK, Sreenivas M, Ghosh A, Narasimhan M, Gurumoorthy AV, Solar panel recycling from circular economy viewpoint: A review. Appl Sol Energy, 2024, 60(2), 328–345.

Sayed Md Abrar Labib†, Ashraful Hossain Shikder†, Monjur Mourshed*

Innovative e-waste recycling technologies, systematic management strategies, and challenges

Abstract: This chapter examines the latest developments in e-waste recycling technology, such as mechanical separation, hydrometallurgical, pyrometallurgical, and bioleaching methods. It highlights the need for digital monitoring and automation in streamlining processes while discussing methodical management techniques that incorporate collection, segregation, material recovery, and safe disposal. Critical analysis is done on issues including technological constraints, the dominance of the informal sector, economic viability, and environmental compliance. In order to scale up creative ideas, the chapter also discusses the necessity of industry, academia, and government cooperation. Sustainable e-waste recycling may be accomplished by fusing technical innovation with strategic management, promoting a circular economy while reducing negative effects on the environment and human health.

Keywords: E-waste, circular economy, management strategy, environmental sustainability, waste battery

1 Introduction

Global electronic waste (e-waste) is among the fastest-growing waste streams in the world, fueled by the speed of innovation, short product lifetimes, and consumer demand for new technology. The Global E-waste Monitor of the United Nations estimated that 53.6 million tons of e-waste were produced worldwide in 2019, and is estimated to reach 74.7 million tons by 2030 if this growth rate continues [1]. Despite the large resource potential that has been estimated to be over US$57 billion worth of recoverable materials in 2019, only approximately 17.4% of this waste was collected formally and recycled [1]. This highlights the growing recycling technologies gap in which developed countries use advanced hydrometallurgical, pyrometallurgical, and biotechnological methods, while developing countries are still relying on rudimentary methods with a poor recovery yield and devastating environmental costs [2, 3].

*Corresponding author: **Monjur Mourshed,** Department of Mechanical Engineering, Rajshahi University of Engineering and Technology, Rajshahi 6204, Bangladesh,
e-mail: monjurmourshed@me.ruet.ac.bd
Sayed Md Abrar Labib, Ashraful Hossain Shikder, Department of Mechanical Engineering, Rajshahi University of Engineering and Technology, Rajshahi 6204, Bangladesh

 | https://doi.org/10.1515/9783112248072-007

Improper management of e-waste has created widespread environmental and social damage. Informal recycling practices that may release toxicants, including lead, cadmium, mercury, and brominated flame retardants in air, soil, and groundwater if not properly disposed of, are open burning of wires, acid baths for metal extraction, and landfills without protective lining [4]. These pollutants are associated with respiratory diseases, neurological disorders, and developmental problems among exposed populations [5]. Informal recycling facilities like Agbogbloshie (Ghana) or Guiyu (China) have become case studies of environmental injustice, where low-income communities have to carry added health risks and work in unsafe conditions, without access to protective equipment or supervision [4].

At the same time, the economic and ecological inefficiencies of current systems are provoking. For example, printed circuit boards (PCBs) contain up to 30% metals, including copper, gold, silver, and palladium in some places at levels much higher than natural ores [6]. Yet much of this value is lost because of the lack of effective recovery practices or loss through leakage to informal markets. Globally, the informal sector takes precedence in e-waste collection in many developing economies and is unregulated, adding to the pollution and loss of resources [3]. This underlines not only a lack of technological capacity but also a governance challenge: a lack of harmonized legislation, extended producer responsibility (EPR) enforcement, and good integration between formal and informal systems [6].

Against this background, this current study comprehensively presents the e-waste recycling and recovery technologies, as well as sheds light on the potential management strategies and potential challenges. First of all, it overviews some recent advances of e-waste recycling technologies, including hydrometallurgical and pyrometallurgical approaches, bioleaching, and mechanical separation, evaluating their scalability, efficiency, and environmental implications. Second, it looks at systematic management strategies such as models of collection, segregation strategies, safe disposal, and digital tracking systems, and EPR and take-back schemes to enhance accountability. Third, it identifies ongoing challenges: high costs, dominance of the informal sector, lack of consumer awareness, and regulatory weaknesses, and explores pathways to sustainable solutions. By focusing on collaboration between governments, industries, academia, and communities, the chapter is designed to bring these integrated approaches for e-waste management into the spotlight; approaches that will mitigate the harm to nature's resources and maximize the use of these materials in a circular economy [1, 6].

2 E-waste recycling technologies

2.1 An overview

E-waste recycling includes a range of technologies with differences in principle, efficiency, and environmental effects. The principal ones, in general, include mechanical separation, pyrometallurgy, hydrometallurgy, and bioleaching. A comparative analysis of e-waste recycling technologies has been presented in Table 1. Physical techniques used in mechanical techniques include: dismantling, shredding, magnetic separation, density-based sorting, and electrostatic separation. Pyrometallurgical techniques use mechanical treatment of high temperature (smelting, burning, and refining) in the form of e-waste to alloy metals or slags lacking e-waste. Hydrometallurgy includes operations such as leaching, solvent extraction, purification, and electrowinning, which are used to destroy and reclaim metals using water. Bioleaching uses the solubilizing ability of microorganisms (bacteria, fungi) to solubilize metals in a more environmentally friendly way, perhaps in conjunction with hydrometallurgical operations. Both methods involve a trade-off as it relates to cost, power consumption, recovery rate, environmental impact, and elimination of various types of e-waste.

Table 1: Comparative chart of e-waste recycling technologies [1, 3, 6, 7].

Technology	Process steps	Recovery efficiency	Cost	Environmental impact	Maturity
Mechanical separation	Dismantling → shredding → magnetic/density/ electrostatic separation	Moderate (good for bulk metals, poor for Au/ Ag/REEs)	Low/ moderate	Generates dust; limited chemical use; safe if controlled	Commercial, widely used (mainly as pretreatment)
Pyrometallurgy	High-temperature smelting → refining → slag disposal	High for base and precious metals; up to 95% in advanced smelters	High	High CO_2, toxic gas emissions; slag disposal required	Commercial, mature (large smelters in EU/ China)
Hydrometallurgy	Leaching (acids/ cyanide/etc.) → solvent extraction → electrowinning/ precipitation	Very high (Cu, Au, Ag often >90%; REEs moderate)	Moderate/ high	Risk of toxic effluents; requires wastewater management	Semi-commercial, lab → industrial pilot

Table 1 (continued)

Technology	Process steps	Recovery efficiency	Cost	Environmental impact	Maturity
Bioleaching (biotechnology)	Microbial leaching (e.g., *Acidithiobacillus*, fungi) → metal recovery (precipitation/ electrowinning)	Moderate to high (Cu, Co, Ni >80% under lab conditions; Au/Ag lower)	Low but very time-consuming	Environmentally friendly, minimal chemicals; slow kinetics	Lab and pilot scale; limited industrial deployment

However, studies conducted in hydrometallurgy tend to prove less harmful to the environment than pyrometallurgy in some types of metal recovery work, especially where acid leaching and proper treatment of chemicals are viable. Another example of such a life-cycle assessment is one that studied copper and gold recovery processes on electronic waste, which found that hydrometallurgical processes were less prone to produce greenhouse gas emissions and less energy, whereas pyrometallurgical processes have difficulties in managing acidic effluents and assuring a safe environment [7]. Pyrometallurgy, on the other hand, is more advanced and able to handle mixed, unsorted waste. The recycling method is energy-intensive, less sensitive to heterogeneity, and results in both slag and gaseous emissions that must be treated.

Mechanical separation is particularly a pretreatment for all the other methods. It aids size reduction, plastic, glass separation, metals separation, de-bulking of nonmetallic components, and shallower concentration of the valuable fractions like circuit boards, metals, etc. Its benefits include reduction of energy input (when compared to thermal or chemical processes), less complicated infrastructure, and accelerated throughput. Nevertheless, the mechanical processes alone tend to be least effective in recovering precious metals and cannot extract metals deposited within complex materials, in mixed plastics, or stratified materials [8].

The next method is bioleaching, also called bio-hydrometallurgy, which has a reduced chemical and thermal imprint. Recent reviews indicate that base metals in printed circuit boards (electric, magnetic, and auxiliary) and used batteries can be recovered reasonably well at acidophilic bacteria *Acidithiobacillus* and *Sulfobacillus* with optimization of the bioleaching process (pH, temperature, and aeration) [8–10]. As an illustration, when applying a two-stage method, which involves using acidophilic iron/sulfur-oxidizing microorganisms, followed by soluble precious or rare metal solubilizers, more total recycling can be achieved. The disadvantages can usually be a lengthy time period, expandability complications, and responsiveness to waste structure and microbial health.

In brief, initial findings show that hydrometallurgy is good in terms of recovery and environmental considerations and poor in chemical handling; pyrometallurgy is high in throughput, but low in environmental considerations and mechanical separation is high in front-end process but the low in terms of an independent process to high value metals; bioleaching is high in terms of environmental considerations and is low in terms of throughput and industrial implementation.

2.2 Hydrometallurgical recycling

E-waste has valuable metals, where the extraction can be carried out, provided that the correct sound recycling technology is applied [10]. As the extraction of primary resources persists to fulfill demand needs, the sheer amount of valuable resources, especially precious and critical metals that are buried in discarded devices, has placed e-waste as an important urban mine. Hydrometallurgical recycling is one of the major processes, which are used to retrieve any of these resources and can be considered a high-purity, scalable chemical procedure, which is usually applied to the metal concentrates that are produced through the mechanical pretreatment of e-waste material, including shredded PCBs. This is carried out using aqueous solutions which selectively dissolve, separate, and recover individual metals. Compared to pyrometallurgy, in which metals are obtained by means of high-temperature smelting, hydrometallurgy is more selective and consumes less energy, and can be used in particular to recover low-concentration and valuable metals. Metal recovery normally follows a multistep approach comprising three basic steps shown in Figure 1(a). The very beginning one is acid leaching. It is a simultaneous and most vital step in which the target metals are separated out of the solid e-waste blend into a liquid form (a lixiviant). The most frequent agents of leaching include strong acids such as sulfuric (H_2SO_4), hydrochloric (HCl), and nitric (HNO_3), or complexing agents such as cyanide and thiosulfate. In the case of base metals such as copper, the sulfuric acid solution is usually very simple and sufficient usually though precious metals (Au and Ag) need more aggressive and selective agents. It is aimed at the maximum dissolution of the target metal and minimum dross dissolution of the impurities [11].

Once the leachate is prepared, then the metal-rich solution has an amalgamation of dissolved metals. These metals are separated using solvent extraction in order to be selected. It is a procedure that requires blending the aqueous lean with an immiscible organic solvent, which also has a specified extractant. The extractant constructs a complex with the ions of the target metal, removing it from the aqueous phase and into the organic phase. This process is repeated with varying solvents and pH in order to isolate high-purity streams of single metal streams, a requirement of the last recovery process. The ultimate pure concentrated metal solution is then sent to the recovery procedure known as electrowinning, which is a type of electroplating in which an electric current is applied through the solution, and the pure metal serves to plate onto a cathode plate. It is the high purity of copper (Cu) and lead (Pb) recovery standard in the industry. Alternatively, the recovery of metals may be achieved through precipitation, which entails the addition of chemical substances that will be filtered and refined into an insoluble solid compound [12].

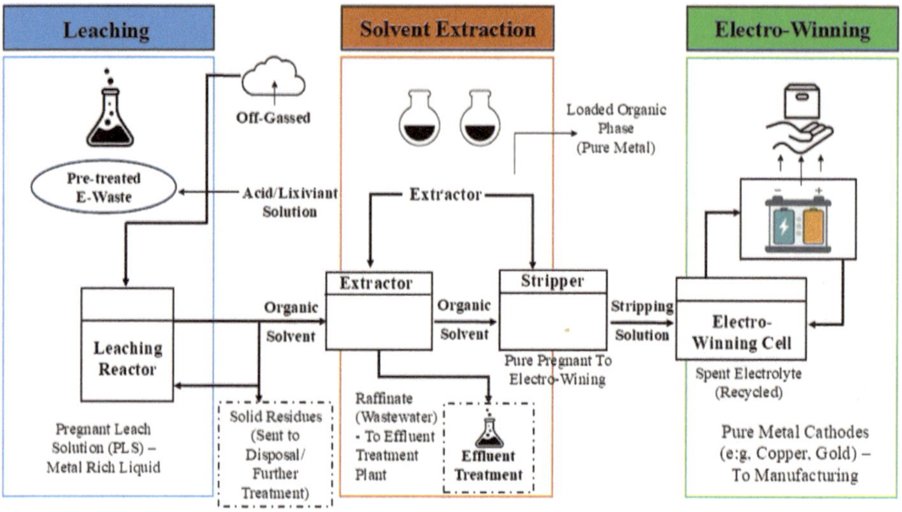

(a)

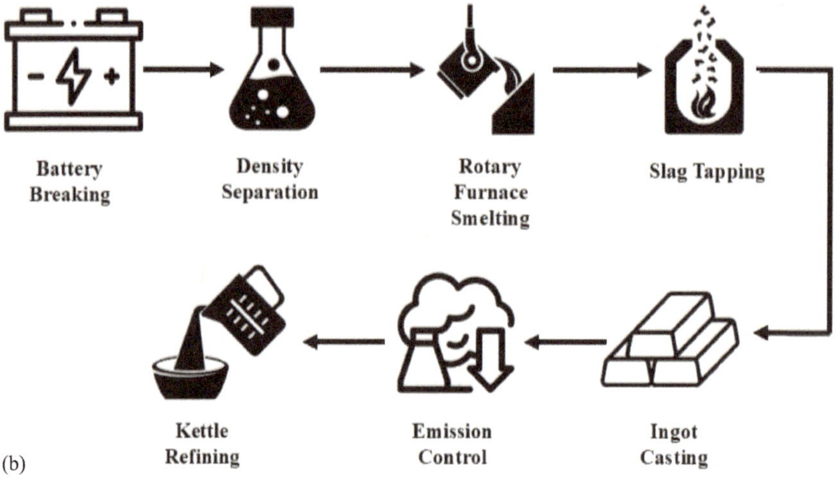

(b)

Figure 1: E-waste recycling processes: (a) hydrometallurgical, (b) pyrometallurgical, (c) bioleaching, and (d) mechanical separation processes.

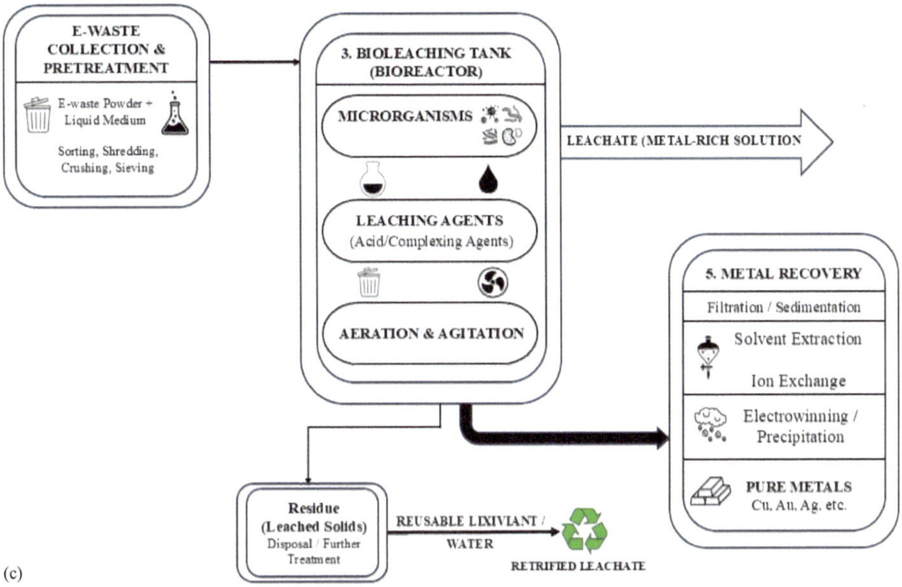

(c)

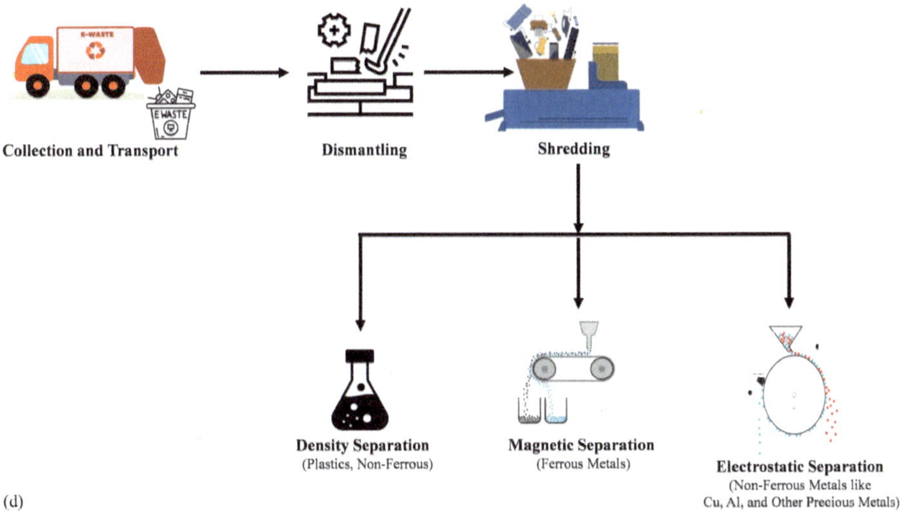

(d)

Figure 1 (continued)

Hydrometallurgy has also proven to be a very useful process due to its capability of obtaining high recovery efficiencies of primary metals contained in e-waste. Extensive studies have always proved that there is high performance in different metal groups. Metals recovery of copper is commonly reported as high, with an average above 99% recovery rate when using optimized sulfuric acid leaching and electrowinning circuits (as recorded in a number of studies examining the use of effective Cu recovery of PCBs) [10]. The efficiency of recovery of precious metals such as gold (Au) and silver (Ag) is usually above 95%. Although the traditional cyanidation process is very effective, newer and more environmentally friendly processes have been developed through the use of different lixiviants such as thiourea or thiosulfate and solvent extraction or adsorption to ensure high recovery and reduce the risk of toxicity. Extraction of rare earth elements (REEs), which include yttrium, europium, and neodymium in e-waste such as fluorescent light tubes or hard drive magnets, poses a more complicated problem. Multistep hydrometallurgical processes, in working groups, however, are capable of a high degree of REE recovery, as high as 90, but this is an expensive process since the process requires several highly selective separation steps. Numerous studies, including the systematic investigation of sustainable e-waste practices, offer detailed information about the processes, and some of them indicate that under optimal conditions, close to a hundred percent of metal could be recovered. As an example, high Cu, Au, and Ag yields are characterized in processes that are oriented at cleaner production techniques [13]. Although the hydrometallurgical process was highly efficient and more preferable compared to pyrometallurgy in certain applications, it does not lack serious deficiencies in regard to the environment. The major problem is that there are large quantities of toxic effluents and process wastewater released. The chemicals employed in dissolving the metals are highly corrosive and toxic, requiring complex and expensive treatment of wastewater (H_2SO_4, HCl, and other dangerous reagents). The liquid wastes of solvent extraction (and the spent leachates) include traces of the strong lixiviants, fecal effluents of heavy metals (such as Pb, Cd, and Cr) that were unintentionally washed away or not recaptured. These toxic effluents pose a serious environmental hazard and should be managed safely because, in most cases, the environment is contaminated, especially water sources and soil, which has been an embarrassing issue in informal e-waste recycling facilities in countries around the world [14, 15]. The expensive nature and sophistication of a formal, environmentally sound hydrometallurgical effluent treatment infrastructure can also pose a serious impediment to the popularization of formal hydrometallurgical plants. Table 2 represents different hydrometallurgical leaching efficiencies for different E-waste metals.

Table 2: Hydrometallurgical leaching efficiencies for key metals from e-waste.

Metal	Typical recovery (%)	Primary reagent used	Environmental drawback	References
Copper (Cu)	90–99	Sulfuric acid (H_2SO_4), nitric acid (HNO_3), ammonia (NH_3) solutions, or Fe^{3+}/peroxide	Requires neutralization of large volumes of acidic/ammoniacal solutions; generation of metal-contaminated sludge; NO_x fumes from HNO_3.	[16]
Gold (Au)	95–99	Cyanide (NaCN or KCN), thiourea, thiosulfate, halogen (e.g., Cl_2 or HCl/H_2O_2)	Cyanide has high toxicity and strict safety/disposal protocols. Thiourea as a reagent degrades rapidly, complex regeneration. Halogens are highly corrosive, potential for toxic gas evolution.	[17]
Silver (Ag)	90–98	Nitric acid (HNO_3), sulfuric acid (H_2SO_4) + oxidant, thiosulfate H_2SO_4	Nitric acid generates toxic NO_x gases; thiosulfate process requires precise control and is prone to side reactions.	
Nickel (Ni)	90–99	Sulfuric acid (H_2SO_4), hydrochloric acid (HCl)	Acid leaching generates large volumes of acidic effluent requiring extensive neutralization and sludge management (often containing heavy metals).	[18]
Cobalt (Co)	90–98	Sulfuric acid (H_2SO_4) + reductant (H_2O_2 or Na_2SO_3)	Similar to Ni, the process involves large volumes of acid and the potential for heavy metal contamination in the residual solution and sludge.	
Rare earth elements (REEs)	85–95	Strong acids (HCl, HNO_3, and H_2SO_4)	Leaching with strong mineral acids is highly corrosive and creates highly acidic wastewater rich in other leached metals, necessitating complex separation steps.	[19]

2.3 Pyrometallurgical recycling

The most industrialized and mature thermal process to extract valuable metals from e-waste is pyrometallurgical recycling, also known as smelting, especially various high-volume concentrates such as shredded PCBs [20]. The result of the process is characterized by the high-temperature smelting in specially designed furnaces, in which e-waste is used as a secondary fuel after its organic (plastic/resin) content [6].

The pyrometallurgical method follows multiple steps, which are shown in Figure 1(b). During this process of high heat treatment, the metals are melted and segregated according to their chemical characteristics. Copper is a great scavenger, which produces precious metals (gold, silver, palladium, and platinum) in a molten copper alloy (called copper matte or black copper), and other materials that are not metallic (plastics, ceramics, and fiberglass) and have low value are oxidized and released into the lightweight slag phase [20]. Further refining of the copper alloy is then done by electro-refining the copper alloy to remove the pure copper and the precious metals.

The conventional method has discrete strengths, mostly in its high ability to recover base and precious metals, with a regular recovery of over 95% of Cu, Au, Ag, and it is highly scalable and able to handle very large, and often heterogeneous, e-waste streams [14]. It is the paradigm of the massive operations, and it goes smoothly into the already existing nonferrous metallurgical industry. The industry in which this technology finds considerable use includes notable industrial plants such as Umicore, doing business in Belgium, Aurubis, doing business in Germany, and DOWA in Japan, which handle some of the largest tonnages of e-waste and secondary feedstocks on an annual basis [2].

However, there are major drawbacks of the pyrometallurgical process to the environment, as mentioned in Figure 1(b). Such a process is energy-intensive and consumes a large amount of fuel to keep the temperatures at extreme values, thus leaving a large carbon footprint [14]. In addition to this, the incineration of plastic and organic materials in the e-waste feedstock results in emission of toxic gases like sulfur dioxide, nitrogen oxides, and, in certain cases, persistent organic pollutants (POPs) such as dioxins and furans in case of uncontrolled emission, and this necessitates expensive flue gas purification systems [20]. It is also very energy-consuming and a form of energy is used to sustain the required high temperatures to a huge amount of fuel meaning that it has a big carbon footprint. Moreover, plastic and organic materials in the e-waste undergo burning, resulting in the release of harmful gases, such as sulfur dioxide, nitrogen oxides, and in some cases, POPs, such as dioxins and furans, if emissions are not strictly regulated, and a complex and expensive flue gas purification system is required. Lastly, precious metals such as Lithium and some REEs are commonly oxidized and lost to the last, voluminous slag, a solid waste by-product that must be disposed of safely and over a long period of time, which is a consistent loss of resources [6].

2.4 Bioleaching and biotechnological approaches

Bioleaching offers a clean and cost-efficient substitute to the conventional recycling techniques of e-waste, such as pyrometallurgy and hydrometallurgy, which are generally costly and result in harmful emissions. As a biotechnological solution, microorganisms like bacteria and fungi are used to extract useful metals like copper, gold, and silver in e-waste by converting them into soluble forms. It works under lower temperatures and pressure, which greatly minimizes energy and toxic substances discharge. Although bioleaching is a good resource recovery method, it remains unlikely to be implemented on a large scale by industries due to issues in kinetic speed, e-waste toxicity to microorganisms, and the difficulty of scaling up laboratory-based processes. Consequently, the development of the future work is oriented to the optimization of the conditions of the process and the examination of the hybrid technologies to enhance the efficiency and make bioleaching a system of sustainable e-waste management [21–23].

Microbial metal leaching or bioleaching is a process shown in Figure 1(c) in which microorganisms are used to dissolve metals in a wide range of materials, e.g., ores and electronic waste. Two significant groups of pertinent microbes include *acidophilic chemolithotrophic* microbes, e.g., *Acidithiobacillus ferrooxidans*, and *heterotrophs*, e.g., *fungi*. The *Acidithiobacillus* species are most suitable since they can multiply in low pH and have the ability to oxidize ferrous iron into ferric iron or elemental sulfur into sulfuric acid. These biological acids and ferric ions are effective lixiviants and dissolve the solid material of metals. However, Fungi have another mode of attack, and they synthesize organic acids such as citric, oxalic, and gluconic acids to dissolve metals by acidolysis, complexolysis, and redoxolysis. The method is specifically helpful with nonsulphide substances and can be performed at a significant variety of pH levels. This may be accomplished by direct means, such as the microbes sticking to the material, or indirectly, such as the leaching by the metals of the microbial metabolic by-products, as in a two-step or spent medium process to prevent the toxic impact of the waste on the microorganisms themselves [24].

Bioleaching organisms and their effectiveness are represented in Table 3. Bioleaching is emphasized as being ecologically friendly and a low-cost alternative to conventional recycling of e-waste. It is also environmentally friendly because it can be conducted at the ambient temperature and pressure, thereby consuming relatively low energy and carbon footprint as a result of such ambient temperature and pressure in comparison with high-temperature pyrometallurgical methods. Bioleaching does not produce toxic gases such as dioxins and furans, unlike pyrometallurgy, which emits these gases. It also consumes less hazardous chemicals compared to the hydro metallurgical processes, which reduces the chances of toxic effluents that may contaminate the environment (air, water, and soil). It rather uses microorganisms that are naturally found and their metabolic by-product to dissolve metals. The low cost of the process is explained by the fact that it has a simple infrastructure and does not consume much energy, which makes it economically feasible and can be considered by developing countries. The technique also has the capability of using cheap resources, such as agricultural waste as a source of nutrients to the microbes, thus lowering the cost of operation [23]. Bioleaching has always been termed as a low-cost substitute for the traditional ones. It is much less costly in terms of capital investment than pyrometallurgy since it does not need expensive and high-temperature furnaces and sophisticated gas treatment plants. This renders it more appropriate for the small-scale operations and more accessible to the developing world. The operating cost, however, may change as the cost is a determinant of the requirement of nutrients, particularly among the heterotrophic organisms. As an example, one study has established that the cost of glucose can hit up to 44% of the overall cost of a bioleaching process. This indicates that there are more affordable sources of nutrients, like agricultural refuse or other organic derivatives, that would make bioleaching far more profitable. A techno-economic study of copper recovery indicated that a profit margin of 21% would be obtained should 10,000 tons of black mass be processed annually with the help of a bio-lixiviant generated using corn stover [25].

Although bioleaching has many benefits, there are serious drawbacks that make it difficult to apply in a large-scale industry. A significant weakness is that the field is very slow concerning kinetics, and therefore, it can be protracted, in some cases, weeks or even months, to get good yields of metal. This is very contrary to the traditional processes, such as pyrometallurgy, which provides a significantly higher throughput. The low reaction rate is usually described by the fact that microorganisms have to adapt to the e-waste toxic environment, which may slow down their growth and metabolism and produce lesser amounts of metal-solubilizing agents. In this slowness, it is hard to apply on a commercial basis, where efficiency and speed are the most important factors [26].

The other major constraint is that it is quite hard to scale up experiments at the laboratory scale and translate them to the industry scale. The majority of effective bioleaching experiments are done in the controlled laboratory, and it is complicated to apply such findings to large-scale bioreactors or heap leaching processes. The difficulties of scale-up are to provide the optimal and constant conditions, e.g., pH, temperature, and the nutrient supply across extensive amounts of material. The e-waste is heterogeneous and contains deactivating elements that complicate the process of maintaining uniform microbial activity of a large-scale system. Thus, although the technology has shown promising outcomes in small-scale research, it remains in an infancy phase when it comes to large-scale and real-world implementation [26]. Bioleaching has other severe drawbacks besides slow kinetics and scaling issues. The complexity and heterogeneity of e-waste are one of the biggest problems because it is not a homogeneous substance. It is filled with a large variety of metals, nonmetals, and plastics, and usually complex multi-layered structures. Such a mixture implies that one microorganism or leaching agent is not often going to be sufficient to recover all useful materials. An example is that a microbe may be an effective copper remover, but not of gold or REEs, and this would need a multi-stage or hybrid operation, making it both more complex and expensive. Physical barriers of access to the metals by the microbes can also be provided by the presence of nonmetallic materials, including flame retardants and heavy plastics [27]. The other weakness is that secondary waste can be produced. Although bioleaching is usually cleaner than other processes, it is capable of producing liquid waste that requires proper handling. The leachate that is produced may be harmful unless there is proper neutralization and filtering, which is a threat to contaminating the environment. This causes the shift of the environmental burden of air pollution to liquid waste treatment, which will need certain and frequently expensive infrastructure. Moreover, the process is also vulnerable to changes in the quality of the e-waste feedstock, which can be fatal to the microorganisms due to high concentrations of some of the toxic metals, which will lead to failure of the whole process.

Table 3: Bioleaching organisms and their effectiveness.

Microorganism (bacteria/fungi)	E-waste component source	Target metal(s) recovered	Recovery percentage (%)	Mechanism/note
Acidithiobacillus ferrooxidans	PCBs	Copper (Cu) and nickel (Ni)	Cu: 100, Ni: 100	Redoxolysis (iron oxidation)
Acidithiobacillus ferrooxidans	WEEE shredding dust	Cerium (Ce), europium (Eu), and neodymium (Ne)	Ce, Eu, Ne: 99	Redoxolysis (recovery of rare earth elements)
Acidithiobacillus ferrooxidans	WEEE shredding dust	Lanthanum (La) and yttrium (Y)	La, Y: 80	Redoxolysis (recovery of rare earth elements)
Acidithiobacillus ferrooxidans + *A. thiooxidans* (mixed culture)	Mixed WEEE	Aluminum (Al), copper (Cu), nickel (Ni), and zinc (Zn)	90	Combined acidolysis and redoxolysis
Acidithiobacillus thiooxidans	PCBs	Copper (Cu)	Near 100 (complete mobilization)	Acidolysis (sulfuric acid production)
Leptospirillum ferrooxidans + *A. ferrooxidans* + *A. thiooxidans* (mixed culture)	TV PCBs	Cu	89	Combined acidolysis and redoxolysis
Aspergillus niger (fungus)	PCBs	Cu	High recovery (implied)	Organic acid production (e.g., citric and malic acid)
Pseudomonas putida	WEEE shredding dust	Au	48	Cyanogenic bioleaching
Sulphobacillus Thermosulfidooxidans + *acidophilic isolate*	PCBs	Cu, Al, Ni, and Zn	Cu: 89, Al: 79, Ni: 81, Zn: 83	Thermophilic chemolithotrophy

Table 3 (continued)

Microorganism (bacteria/fungi)	**E-waste component source**	**Target metal(s) recovered**	**Recovery percentage (%)**	**Mechanism/note**
Chromobacterium violaceum	Precious metals (from e-waste)	Au and Ag (implied)	Not specified	Cyanogenic bioleaching (HCN production)
Penicillium simplicissimum (fungus)	Various E-wastes	Cu, Ni, and Zn (implied)	Not specified	Organic acid production
Acidiphilium acidiphilium	Shredded PCBs	Cu and Ni	Leached Cu and Ni	Acidophilic heterotrophic leaching

2.5 Mechanical separation processes

One of the most basic and most popular methods of recycling e-waste is mechanical separation. It is based on the variations in the physical characteristics of the materials, like size, density, magnetism, and conductivity, to be used to extract valuable fractions of the discarded electronic equipment. Figure 1(d) shows multiple steps to perform mechanical separation. It usually starts with dismantling, and it can be done either by hand or by semi-automated and robotic solutions. Dismantling is necessary as it can ensure that the hazardous parts, such as batteries, capacitors, and mercury lamps, are not mixed with the main part of the waste and therefore would not contaminate the downstream processes. Robotic dismantlers, including the Apple Daisy, have been implemented in more advanced recycling plants, which are more efficient and safer by removing reusable components [28].

After the dismantling process is done, the e-wastes are then shredded. Shredding refers to the mechanical reduction in size through hammer milling, granulators, or rotary shredding to reduce the waste into small sizes. This process releases metals, plastics, and ceramics, which are frequently closely contained in electronic assemblies. It is also found that shredding raises the surface area of the materials, making it easier to separate further. It, however, involves proper dust management since the fine particles produced in this phase might include dangerous heavy metals. Some of the materials have a difference in specific gravity, and the density separation can be applied after Shredding. Lighter fractions like plastics and foams can be separated by air classification systems, whereas heavier metallic fractions can be separated by water-based sink-float separation, allowing the plastics to float and the metals to be sunk. Further improved equipment, such as hydrocyclons, can be used to provide high efficiencies for complex mixture separation. Fractions are further purified by using magnetic and electrostatic separation techniques. The magnetic separation is directed to ferromagnetic metals (iron, nickel, and steel), which can be easily remanent with the aid of a drum or over-belt magnet. Electrostatic separation, on the other hand, is the separation of the conductive and nonconductive material using electric fields. This enables the retrieval of precious nonferrous metals like copper and aluminum, as well as precious metals, out of plastics and ceramics.

Recent developments in the field of automation have changed mechanical separation during e-waste recycling. Robotics is finding more applications for automated dismantling and picking high-value or dangerous components. Machine vision, along with artificial intelligence and near-infrared (NIR) sensors, allows automated sorting of the materials that have been shredded in terms of shape, color, and spectral properties [29]. Moreover, data analytics through Internet of Things (IoT) can be used to track the efficiency of the plants in real-time and schedule maintenance of the equipment. They are currently developing innovations that will lead to a fully automated smart recycling system, which will combine the shredding, robotic dismantling, AI sorting, and advanced separation technologies into a single system. Using the power of deep learning and computer vision allows real-time and high-speed identification and classification of materials based on a variety of sensor inputs (e.g., NIR and XRF) despite the problem of material heterogeneity and contamination [30]. AI-controlled robotic arms perform tasks that are high-dexterity (such as selective disassembly, e.g., battery removal), or high-throughput (e.g., 80+ picks per minute), with high precision and consistency, which are far safer to operate and far surpass human labor, making them more valuable than micro-refinements or purification can restore [31].

3 Systematic e-waste management strategies

An effective, good end-of-life electronic strategy must begin with secure and convenient collection sites which deliberately connect formal and informal networks, which bring together producer accountability and infrastructures of cities, as well as facilitate consumer engagement. Third world experience indicates that the initiation of the e-waste collection program into systems (deposit/drop-off bins, material-recovery facilities, and location-retailer take-back points) should be supported by an appreciation that in most cases the informal collectors and backyard dismantlers are the existing collection network and that formal programs are best used in a situation where they create an economic opportunity to enter that informal sector [32, 33]. Researchers proposed the model of the integrated e-waste management (IEWM) scheme (as shown in Figure 2), where the collection can be viewed as the segment of the municipal solid-waste system, and the selection of collection schemes (drop-off, deposit containers, sorted/MRFs, or mixed/MRFs) and is proposed to be raised in case of the need to save money and prevent the dangerous mixing of fractions [34].

Formal take-back and subsidy schemes have proven to increase rapidly and be adequately funded: in China national old-for-new rebate schemes and producer funded collection systems have shown that consumer returns can sky-rocket when retailers or producers can offer explicit economic incentives and convenient drop-off routes; however, the formal sector must be competitive at informal price and must be able to achieve logistical distribution of subsidy and haulage benefits to keep the flows going to licensed end-users [32, 35]. On a broader basis, empirical studies of large-scale programs (e.g., appliance rebate/old-for-new pilots) indicate that appropriately financed formal programs can attract tens of millions of units of products in a relatively short period, but concerns emerge as to the amount of optimal subsidy amount and sharing of responsibilities, as well as sustainability [14]. After gathering, it must be sorted using systematic segregation to ensure a more effective and safe recovery of materials to safeguard health, and to recover the material. Its best approach is to separate sources (where possible at the generation point) and sort at MRFs: presort initial (large hazardous) material (e.g., batteries and CRTs) should be removed manually or with automated methods, downstream metals recovery should receive mixed material freed by shredding/granulating, and a clean feedstock should be obtained by physical sorting of ferrous (e.g., by use of magnetic separation) and nonferrous (e.g., eddy-current and density separation) and the sorting of waste by generation [34].

According to consumer behavior research, it has been found that the factors that have significant effects on whether a household will approach a formal collector or hoard or sell to informal buyers are derived in the form of convenience, perceived economic gains, and trust to use the take-back channels as a channel of disposing of e-waste materials. Consequently, system designers should employ easy pickup/drop-off, clear-cut pricing/vouchers, and social awareness measures in tandem so that the stored stocks may be transformed into material recovery feedstock. The lack of such incentives and ease of disposal also leads to the study of consumer disposal behavior, pointing out that these incentives would merely leave many consumers holding on to obsolete electronics in their homes over the years, at the limit of collections [36].

Digital tracking technologies such as the IoT and blockchain would be essential to improving the integrity of a systematic management approach by providing an end-to-end audit trail of trust to all electronic devices. IoT sensors (e.g., RFID tags, GPS, and smart waste bins with fill-level sensors) are used to track and give real-time data concerning the waste production, position, and transit. This data line can be used to support operational efficiency by optimizing logistics, including dynamic routing of collection cars to minimize expenses and carbon footprints. The blockchain technology, in its turn, serves as the noncentralized, nonmanipulable registry and documents all the transactions and transfers of the e-waste object since it was produced to the eventual treatment (recycling or disposal). Blockchain can guarantee the immutability of data and automate compliance checks with registration of material attributes and use of smart contracts, which can help avoid problems with recycling fraud, fraudulent

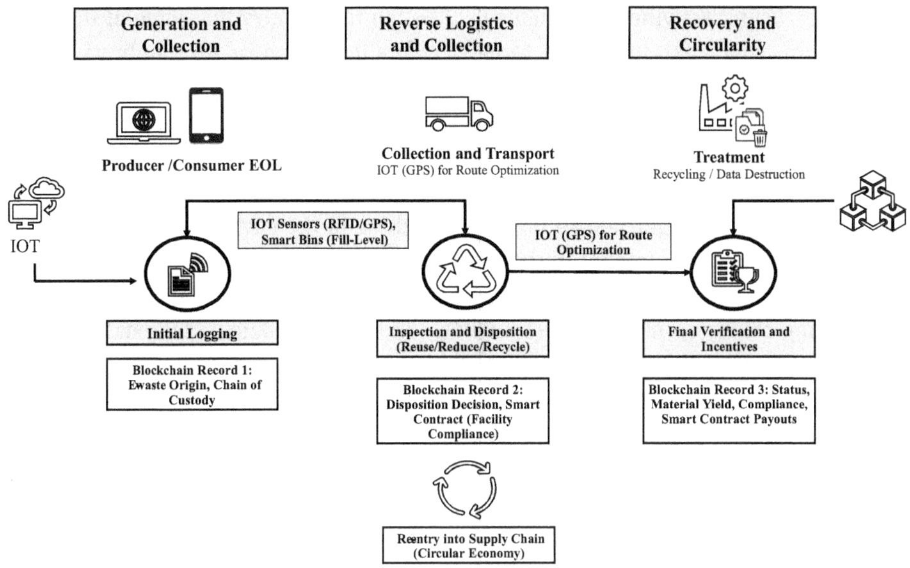

Figure 2: Systematic e-waste management strategies (integrated digital tracking).

compliance checks, compliance with ethical standards of disposal, and issue insurmountable certificates of disposal or data destruction. It is an integrated digital system that enhances transparency and accountability of all parties, including producers and recyclers, and allows them to achieve incentive-based schemes to promote responsible disposal of consumers [37]. In Figure 2, an efficient e-waste management strategy is shown, which requires convenient (formal/informal sectors) collection, progressive segregation (initial and mechanical), and digital tracking (IoT). All this in unity seals material-processing cycles, reduces environmental and health risks of informal processing, as well as increases economic feasibility of urban mining, subsidy targeting, and stakeholder education.

4 Key Challenges in recycling and management

The waste electrical and electronic equipment (e-waste) issue around the globe is characterized by its enormous size, speedy accumulation, and multi-dimensional, systemic obstacles that restrict the effective usage of the resource to a highly debilitating extent. It is the fastest-growing type of domestic waste across the globe, and e-waste production in the world has been the highest ever in 2022, at 62 million tons (Mt) and is predicted to increase to 82 million tons (Mt) by 2030, increasing five times faster than all recycling activities documented. Although the country has some of the most valuable raw materials estimated to be US$62 billion, even in 2022, only a quarter (22.3%) of this was captured in the form of formally collected and properly recycled amounts [38]. The mismatch creates a wide untapped economic potential, at the same time, reminding us of the inherent weaknesses of the existing management systems [3].

The fact that the informal sector is far-reaching in the developing countries, and especially in Asia predominant in the production of e-waste, is a fundamentally complicated issue. Approximately 18 Mt of e-waste are handled through loosely regulated but existing collection networks in most low- and lower-middle-income nations with poor infrastructure, which collect high quantities of materials but lack the proper regulation to control their proper management [34]. This efficiency is considerably paid for with a terrible price: the environmental risks and human health dangers as the outcome are enormous. Informal sources, e.g., primitive backyard recycling and open burning (as formerly witnessed in such regions as Guiyu, China) [14, 39], emit toxic gases, such as lead, mercury, cadmium, and POPs into the air, soil, and water. These exposures are very dangerous occupational and communal health hazards, especially to vulnerable groups such as children and women.

A question on technology maturity and cost used in the formal sector is an additional hindrance to sustainable management. Although recycling on an industrial scale is required to maintain environmental sustainability in terms of environmentally friendly material retrieval, it has cost challenges dealing with sorting, material heterogeneity, and the costly nature of acquiring a sophisticated hydrometallurgical or pyrometallurgical treatment method [3, 41]. Formal recyclers frequently will find it difficult to compete with the cheap operation of the uncontrolled informal sector that uses cheap labor to do the manual dismantling. This, combined with the poor infrastructure and slow legislative development, saw the reignition of an e-waste policy in only 81 countries in 2023. This slows down the transition of global practices to sustainable circles [41]. Ultimately, the lack of consumer awareness and interest in any formal system is a pernicious factor that destroys its success. It has been found that most consumers are not ready to dispose of e-waste in the proper manner because they would prefer to store old items at their homes or dispose of them as ordinary trash, which constitutes a clear point of congestion in the reverse supply chain [36, 41]. Further, there is the extensive systemic obstacle at the first stage of the product life cycle, the design of the electrical and electronic equipment (EEE). The latest technologies are growing nonmodular, complicated, and with a short life-cycle, and it is one of the main challenges that do not provide efficient recovery [41]. This lack of longevity, as well as the few possibilities to repair it, increases the speed of the e-waste stream development [38]. This complication in disassembly is also extended by the scattering of valuable substances throughout complex units, which complicates the formal recovery of essential raw materials (such as the rare earth) and discourages the use of further treatment technology. This has been a dynamic, together with the policy fragmentation and the lack of enforcement, especially on the illegal transboundary movement of e-waste, a major governance concern. Despite the international regulations such as the Basel Convention [39], every year, some millions of tons of unwanted goods in the developed countries are transported to developing countries under the banner of used goods or donation, and put aside the responsibility of cleaning them up, and compound the health and environmental disasters in the recipient countries.

The difficulty of sustainable e-waste management is a complex goal that is a result of a complex systemic failure, behavioral (resistance), and economic limitations, as generalized by the worldwide research. The supremacy of the informal sector is the most pressing and threatening problem, as it is the highest priority problem with a score of 9.5. This is in reference to the dire and direct health and environmental dangers of crude and uncontrolled recycling frameworks (such as open burning) that are being undertaken across the world, and dismantle formal ones [14]. Tightly connected is the decisive impediment of ineffective regulation and enforcement (score: 8.5); a poorly executed legislation, even in the countries that have laws (only 78 countries have e-waste laws), serves as a systemic failure point, as the compliant recycler is not able to compete with the illegal operation [40]. The third significant challenge is that of low awareness and participation (score: 8.0) by the general population, in that consumer indifference to dispose of e-waste appropriately, chokes the formal recycling pipeline with feedstock [36]. Lastly, the technological cost (score: 7.0) of the compatible, state-of-the-art recycling operations offers a considerable financial challenge, though it is second last in priority, as compared to addressing the structural and behavioral crises that would lead to the e-waste in the facilities in the first place.

5 Role of stakeholders in sustainable management

The concept of sustainable e-waste management is in itself a group effort, in which the cooperation and collaboration of all the stakeholder groups are necessary to create a perfect cycle of economy and achieve net-zero results [42] as presented in Figure 3(a). The government regulatory bodies are of paramount importance to shape the basic framework for developing and actively implementing environmental regulations, including the restriction of hazardous substances directive, and issuing subsidies or tax incentives to formal recyclers to minimize high operational costs and stimulate the transition to informal and pollution practices [14]. Most importantly, the government should also provide stability in the policies, and the legal system that supports cross-sector cooperation should be strong [43].

The industry, comprising the manufacturers and producers, is the key financial and physical contributor by means of the EPR schemes. Within the framework of EPR, it is the manufacturers who will have an incentive to go beyond end-of-life collection to incorporate green design principles, which means that products will be easier to break down, repair, and recycle, increasing the efficiency of material recovery and decreasing the degree of embedded toxicity [41]. The producers should also invest in and make use of formalized recycling avenues to make the treatment of the collected waste environmentally sound. In the meantime, academia is a driving force of innovation and capacity building that drives research and development to other cost-effective, nontoxic pretreatment and recycling technology, including advanced hydrometallurgy [40]. Also, universities and technical institutions should be tasked with the job of training a specific workforce to formalize the e-waste industry and offer the talented labor that is required to carry out the intricate, contemporary recycling processes. Last, the communities (the group of consumers) will play a vital role in completing the loop by actively engaging in the formal collection systems and increasing their awareness of the disposal procedures. Since it has been established that a great majority of consumers are not ready to properly dispose of e-waste, it is better to hoard products or discard them as simple garbage [36]. Involvement and a population education campaign are imperative to raise the collection rates and ensure the overall system runs as expected [44].

- **Community to industry (material and demand flow):** Community starts the loop by creating e-waste and disposing of it properly through take-back schemes, which are later processed by an Industry. The community also contributes to the market by ensuring that the products supplied by industry preferably possess green design or environmentally friendly properties.
- **Industry loop (material and information flow):** The Industry accepts the collected e-waste and recycles it (recycling/refurbishment), and utilizes the collected material again in its manufacturing process to produce new or eco-designed EEE and hence completes the material loop.
- **Regulatory role of government (top-down influence):** Government sets the terms, which are obligatory to be followed by the industry in terms of the measures of EPR and the financial resources, such as subsidies or fines. It contributes to academia too and trains the community.
- **Enabling role (knowledge flow) of academia:** Academia gives the scientific and technical base through the R&D on superior recycling technologies and promotes the specialized workforce required by the industry. It also provides evidence-based recommendations to the Government in order to refine the policies.
- **Feedback and reporting (bottom-up information):** The industry provides reporting of its collection and recycling performance data to the Government that complies with the EPR targets and reports its technical difficulties to academia to guide research work. Essential behavioral data/feedback is given by the community to the rest of the stakeholders.

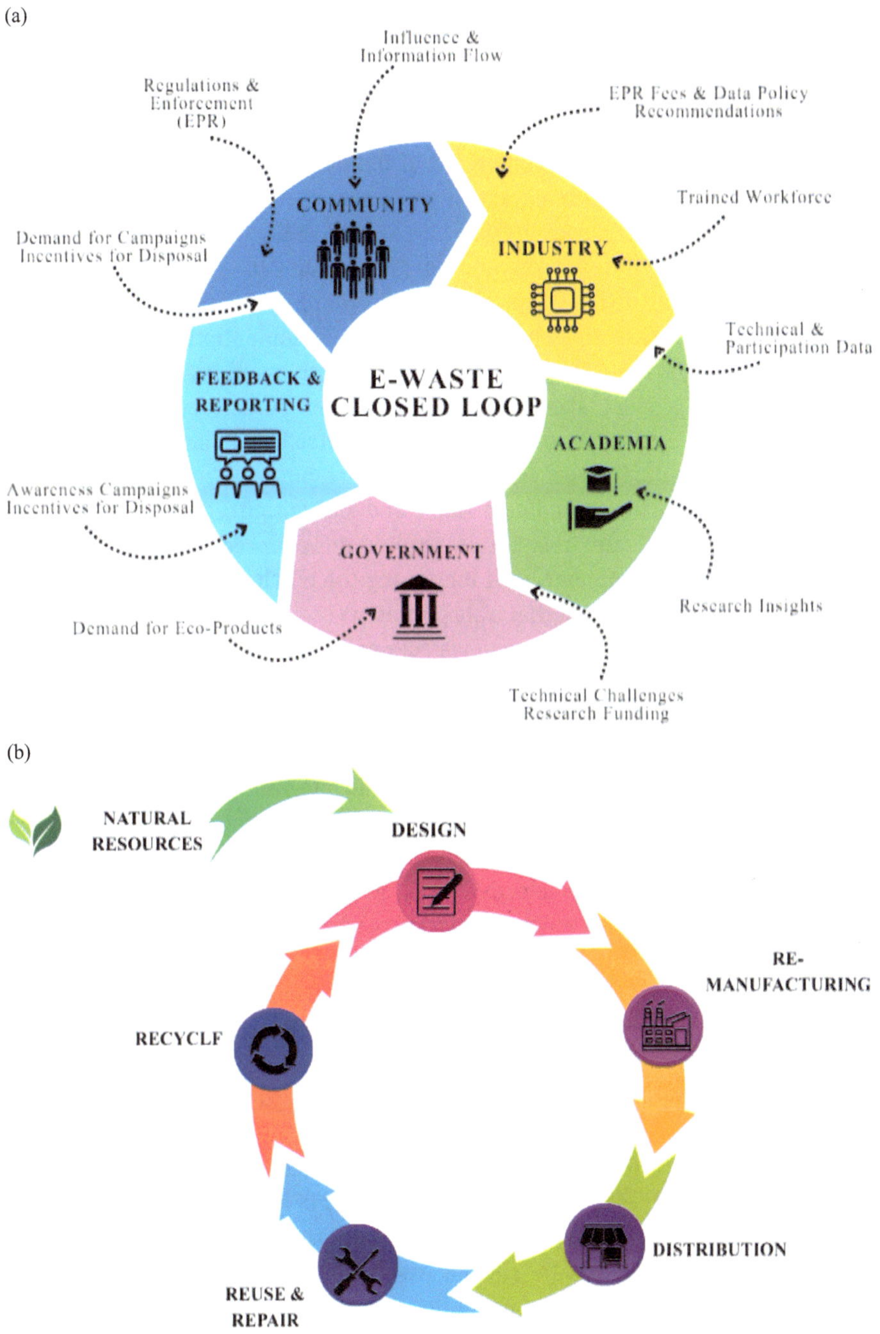

Figure 3: (a) Stakeholder interaction: closed-loop e-waste management, and (b) e-waste circular economy model.

6 Pathways for sustainable recycling and circular economy

To truly achieve an e-waste-based circular economy (as shown in Figure 3(b)), there must be a radical and purposeful alteration of the old linear antiquity of the linear take-make-dispose model, where e-waste is no longer seen as garbage, but as a high-value raw material of secondary materials [45]. Sustainability is dependent on technological change, systemic integration, and advanced policy to ensure the utmost resource usefulness and to divorce economic growth from primary resource usage.

The key component in this change is the direct combination of recycling and manufacturing, which creates a strong system of industrial symbiosis [46]. This would no longer be a mere product take-back model wherein an infrastructure and logistical connection between the recycling Company and manufacturing plant has been established. Through high-purity, recovered resources are put back into the production cycle, allowing manufacturers to obtain stable and locally sourced material availability at the same time, reducing reliance on virgin mining procedures that have negative impacts on the environment, thus increasing the efficiency of overall resource utilization and cutting down material transportation costs [41].

In an attempt to seal the very complicated amalgamation of precious metals and rare raw materials embraced by contemporary electronics, the industry has to balance gearing with the application of hybrid methods in the treatment of materials. Although the standard pyrometallurgy and bulk mechanical separation are and will be vital in the processing of volumes, the end state is integrating mechanical pretreatment (obtaining material liberation and sorting) with the advent of innovative refinement methods [47]. One such possibility is the application of bioleaching, the use of specialized microorganisms to selectively dissolve metals, which is exciting, low-energy, and environmentally benign compared to other routes of hydrometallurgical processing, which are highly acidic or energetically consuming. It is a particularly promising method of source recovery of metals (such as gold and palladium) of shredded circuit boards, which is a cleaner and cost-effective source, and can even be viable in underdeveloped areas with infrastructural limitations [3]. After all, it is these sophisticated recycling plans that are continually optimized thanks to the intensive utilization of life cycle assessment to guarantee genuine environmental positive impact, which are the basic driving forces that will turn the circular economy theoretical concept into the concrete, economically efficient reality of resource savings.

In addition, to achieve a fully operational circular economy, one will have to deal with the issue at the front-end of the product design. Going beyond recycling, the circular model requires manufacturers to adopt the principle of the so-called Green Design, and pay more attention to the notion of modularity, durability, and repairability to increase the useful life of EEE devices [41]. This involves using materials that are nonhazardous and ensuring that the products are designed in such a manner that they can be disassembled easily, reducing the amount of energy and complexity involved in recovering the end-of-life products significantly. Simultaneously, the government policies will be required to alter further than raising the collection rates in an attempt to promote a higher degree of repair and reuse, which has the potential to increase the EPR system to reimburse longevity and recycling. Nevertheless, the public cannot be afforded without any technical or systemic pathway, and a large amount of e-waste is still stored or disposed of in a poor manner due to the poor participation and awareness of consumers [36]. This is why effective take-back logistics, financial incentives (deposit refund schemes), and education of the population should be the elements of successful circular models to guarantee the input of recoverable products into the formal and sophisticated cycles of recycling required by modern technologies.

7 Conclusion

The overall analysis of e-waste management reveals the emergent necessity of the combined strategy of applying innovative technologies in recycling with the approach of system management strategies. Recycling processes such as hydrometallurgical, pyrometallurgical, bioleaching, and mechanical separation processes were discussed in the chapter, as these can improve the recovery of materials and help reduce the environmental impact.

Well-developed management systems, through optimized collection, segregation, recovery of materials, and disposal which is safe disposal, are obligatory, and digital tracking and automation are major optimizable of such systems. Nevertheless, there are a number of challenging issues that need to be overcome, and these are technological restrictions, overwhelming prevalence of the informal sector, economic viability, and environmental overregulation.

Eventually, the integration of efforts is necessary so that sustainable e-waste recycling can be achieved and a circular economy can be built. To deal with these challenges, on the way to scalable and innovative solutions, effective collaboration among industry, academia, and government is required to acquire valuable resources and successfully resolve the environmental and health risks facing the poorly managed electronic waste.

References

[1] Dutta D, Goel S, Understanding the gap between formal and informal e-waste recycling facilities in India. Waste Manag, 2021 Apr, 125, 163–171.

[2] MDK I, MSTS Khatun, Mourshed M, An in-depth analysis and review of management strategies for E-waste in the south Asian region: A way forward towards waste to energy conversion and sustainability. Heliyon, 2024 Apr, 10(7), e28707.

[3] Shahabuddin M, Uddin MN, Chowdhury JI, Ahmed SF, Uddin MN, Mofijur M, et al, A review of the recent development, challenges, and opportunities of electronic waste (e-waste). Int J Environ Sci Technol, 2023 Apr 31, 20(4), 4513–4520.
[4] Adanu SK, Gbedemah SF, Attah MK, Challenges of adopting sustainable technologies in e-waste management at Agbogbloshie, Ghana. Heliyon, 2020 Aug, 6(8), e04548.
[5] Owusu-Sekyere K, Aladago DA, Leverenz D, Oteng-Ababio M, Kranert M, Environmental impacts on soil and groundwater of informal e-waste recycling processes in Ghana. Sustainability, 2024 May 21, 16(11), 4347.
[6] Yong YS, Lim YA, Ilankoon IMSK, An analysis of electronic waste management strategies and recycling operations in Malaysia: Challenges and future prospects. J Clean Prod, 2019 Jul, 224, 151–166.
[7] Priya A, Hait S, Comparative assessment of metallurgical recovery of metals from electronic waste with special emphasis on bioleaching. Environ Sci Pollut Res, 2017 Mar 14, 24(8), 6989–7008.
[8] Blumbergs E, Shishkin A, Markus K, Serga V, Goljandin D, Klauson A, et al, Economic aspects of mechanical pre-treatment's role in precious metals recovery from electronic waste. Metals, 2024 Jan 13, 14(1), 95.
[9] Tezyapar Kara I, Kremser K, Wagland ST, Coulon F, Bioleaching metal-bearing wastes and by-products for resource recovery: A review. Environ Chem Lett, 2023 Dec 22, 21(6), 3329–3350.
[10] Moursheda M, Khatun S, Islam K, Masuk NI, Masud MH, Generation, composition, collection, and treatment of e-waste, In: Management of Electronic Waste. Wiley, 2024, 39–72.
[11] Gulliani S, Volpe M, Messineo A, Volpe R Recovery of metals and valuable chemicals from waste electric and electronic materials: A critical review of existing technologies. RSC Sustainability [Internet]. 2023 Jan 1 [cited 2025 Nov 4];1(5):1085–108. Available from: https://www.sciencedirect.com/org/science/article/pii/S2753812523000083
[12] Ashiq A, Kulkarni J, Vithanage M, Hydrometallurgical recovery of metals from e-waste, In: Electronic Waste Management and Treatment Technology. Elsevier, 2019, 225–246.
[13] Kikuchi-Uehara E, Nakatani J, Hirao M, Analysis of factors influencing consumers' proenvironmental behavior based on life cycle thinking. Part II: Trust model of environmental information. J Clean Prod, 2016 Jul, 125, 216–226.
[14] Duan H, Hu J, Tan Q, Liu L, Wang Y, Li J, Systematic characterization of generation and management of e-waste in China. Environ Sci Pollut Res, 2016 Jan 26, 23(2), 1929–1943.
[15] Widmer R, Oswald-Krapf H, Sinha-Khetriwal D, Schnellmann M, Böni H, Global perspectives on e-waste. Environ Impact Assess Rev, 2005 Jul, 25(5), 436–458.
[16] Rode J, Heinz N, Cornelissen G, Le Menestrel M, How to encourage business professionals to adopt sustainable practices? Experimental evidence that the 'business case' discourse can backfire. J Clean Prod, 2021 Feb, 283, 124618.
[17] Eryildiz B, Ozbey-Unal B, Gezmis-Yavuz E, Koseoglu-Imer DY, Keskinler B, Koyuncu I, Flux-enhanced reduced graphene oxide (rGO)/PVDF nanofibrous membrane distillation membranes for the removal of boron from geothermal water. Sep Purif Technol, 2021 Nov, 274, 119058.
[18] Yang F, Hao L, Zhu Y, Liu H, Chen S, Shao Y, Preparation of graphene modified melamine sponge and solar-assisted cleanup of heavy oil spills. J Environ Chem Eng, 2022 Jun, 10(3), 107779.
[19] Batuecas E, Serrano D, Horvat A, Abelha P, Sustainable conditions for waste tires recycling through gasification in a bubbling fluidized bed. J Clean Prod, 2023 Aug, 415, 137839.
[20] Harvey J, Khalil M, Chaouki J, Pyrometallurgical processes for recycling waste electrical and electronic equipment, In Electronic Waste. Wiley, 2022, 135–164.
[21] Adetunji AI, Oberholster PJ, Erasmus M, Bioleaching of metals from e-waste using microorganisms: A review. Minerals, 2023 Jun 18, 13(6), 828.
[22] Jaiswal M, Srivastava S, A review on sustainable approach of bioleaching of precious metals from electronic wastes. J Hazard Mater Adv, 2024 May, 14, 100435.

[23] Arya S, Kumar S, Bioleaching: Urban mining option to curb the menace of E-waste challenge. Bioengineered, 2020 Jan 14, 11(1), 640–660.
[24] Mishra S, Panda S, Akcil A, Dembele S, Biotechnological avenues in mineral processing: Fundamentals, applications and advances in bioleaching and bio-beneficiation. Miner Process Extr Metall Rev, 2023 Jan 2, 44(1), 22–51.
[25] Advancements and use of OMIC technologies in the field of Bioleaching: A review. Biointerface Res Appl Chem, 2020 Oct 13, 11(3), 10185–10204.
[26] Adetunji AI, Oberholster PJ, Erasmus M, Bioleaching of metals from e-waste using microorganisms: A review. Minerals, 2023 Jun 18, 13(6), 828.
[27] Jaiswal M, Srivastava S, A review on sustainable approach of bioleaching of precious metals from electronic wastes. J Hazard Mater Adv, 2024 May, 14, 100435.
[28] Apple expands global recycling programs – Apple [Internet]. [cited 2025 Oct 28]. Available from: https://www.apple.com/newsroom/2019/04/apple-expands-global-recycling-programs/
[29] Wang L, Jiang S, Shi Y, Du X, Xiao Y, Ma Y, et al, Blockchain-based dynamic energy management mode for distributed energy system with high penetration of renewable energy. Int J Electr Power Energy Syst, 2023 Jun 1, 148, 108933.
[30] Vermeșan H, Tiuc AE, Purcar M, Advanced recovery techniques for waste materials from IT and telecommunication equipment printed circuit boards. Sustainability, 2019 Dec 20, 12(1), 74.
[31] Baker NA, Stehr J, Handmann U, E-waste recycling gets smarter with digitalization, In: 2023 IEEE Conference on Technologies for Sustainability (SusTech). IEEE, 2023, 205–209.
[32] Kibria MDG, Masuk NI, Safayet R, Nguyen HQ, Mourshed M, Plastic waste: Challenges and opportunities to mitigate pollution and effective management. Int J Environ Res, 2023 Feb 20, 17(1), 20.
[33] Islam MDR, Ruponti SA, Rakib MDA, Nguyen HQ, Mourshed M, Current scenario and challenges of plastic pollution in Bangladesh: A focus on farmlands and terrestrial ecosystems. Front Environ Sci Eng, 2023 Jun 15, 17(6), 66.
[34] Ikhlayel M, An integrated approach to establish e-waste management systems for developing countries. J Clean Prod, 2018 Jan, 170, 119–130.
[35] Mourshed M, Masud MH, Rashid F, Joardder MUH, Towards the effective plastic waste management in Bangladesh: A review. Environ Sci Pollut Res, 2017 Dec 27, 24(35), 27021–27046.
[36] Gilal FG, Shah SMM, Adeel S, Gilal RG, Gilal NG, Consumer e-waste disposal behaviour: A systematic review and research agenda. Int J Consum Stud, 2022 Sep 2, 46(5), 1785–1803.
[37] Sosunova I, Porras J, IoT-enabled smart waste management systems for smart cities: A systematic review. IEEE Access, 2022, 10, 73326–73363.
[38] The Global E-waste Monitor 2024 – E-Waste Monitor [Internet]. [cited 2025 Oct 29]. Available from: https://ewastemonitor.info/the-global-e-waste-monitor-2024/
[39] Widmer R, Oswald-Krapf H, Sinha-Khetriwal D, Schnellmann M, Böni H, Global perspectives on e-waste. Environ Impact Assess Rev, 2005 Jul, 25(5), 436–458.
[40] Widanapathirana S, Perera IJJU, Bellanthudawa BKA, Electrical and electronic waste (e-waste) recycling and management strategies in South Asian region: A systematic review from Sri Lankan context. Waste Dispos Sustain Energy, 2023 Dec 24, 5(4), 559–575.
[41] Fan G, Wang M, Dang J, Zhang R, Lv Z, He W, et al, A novel recycling approach for efficient extraction of titanium from high-titanium-bearing blast furnace slag. Waste Manag, 2021 Feb, 120, 626–634.
[42] Cao D, Puntaier E, Gillani F, Chapman D, Dewitt S, Towards integrative multi-stakeholder responsibility for net zero in e-waste: A systematic literature review. Bus Strategy Environ, 2024 Dec 14, 33(8), 8994–9014.
[43] Kaushal RK, Nema AK, Multi-stakeholder decision analysis and comparative risk assessment for reuse–recycle oriented e-waste management strategies: A game theoretic approach. Waste Manag Res, 2013 Sep 15, 31(9), 881–895.

[44] Pandey J, Kaushal RK, Shukla SP, Behavioural and structural analysis of stakeholders in E-waste management. Mater Today Proc, 2022, 51, 416–421.
[45] Le MH, Lu WM, Chang JC, Recycling e-waste and the sustainable economy: A bibliometric exploration. Sustainability, 2023 Nov 20, 15(22), 16108.
[46] Quinto S, Law N, Fletcher C, Le J, Antony Jose S, Menezes PL, Exploring the e-waste crisis: Strategies for sustainable recycling and circular economy integration. Recycling, 2025 Apr 14, 10(2), 72.
[47] Van Yken J, Boxall NJ, Cheng KY, Nikoloski AN, Moheimani NR, Kaksonen AH, E-waste recycling and resource recovery: A review on technologies, barriers and enablers with a focus on Oceania. Metals, 2021 Aug 19, 11(8), 1313.

Arefin Sultana, Monjur Mourshed*

Policy framework and ethics for sustainable e-waste management: a global scenario

Abstract: Global policy frameworks and ethical issues in sustainable e-waste management are thoroughly examined in this chapter. It looks at national laws, international agreements, and extended producer responsibility programs, emphasizing differences in their efficacy and enforcement. Environmental justice, worker safety, and fair resource allocation are among the ethical aspects that are severely examined. Comparative case studies from rich and developing nations highlight potential and shortcomings in policy alignment with sustainable development goals. In addition to providing policy proposals to support equitable, transparent, and efficient e-waste management globally, the chapter highlights the necessity of integrated governance models that strike a balance between economic growth, environmental protection, and human rights.

Keywords: E-waste, sustainable management, policy framework, environmental protection

1 Introduction

The twenty-first century has seen a sharp increase in the global proliferation of electronic and electrical equipment (EEE) due to advancements in digital technologies, declining consumer electronics prices, and growing global connectedness. Significant socioeconomic advantages have resulted from this expansion, including advancements in industrial productivity, healthcare, education, and communication [1]. However, it has also led to an additional problem: the exponential rise of electronic garbage, or "e-waste." With an estimated 53.6 million metric tons (Mt) produced in 2019, e-waste is currently recognized as the solid waste stream with the greatest rate of growth in the world. If present trends continue, this amount is expected to surpass 74 Mt by 2030 [2, 3].

E-waste includes a complex mixture of dangerous compounds, including lead, cadmium, mercury, brominated flame retardants, and persistent organic pollutants, as well as precious resources like gold, palladium, copper, and rare earth elements [4, 5].

***Corresponding author: Monjur Mourshed,** Department of Mechanical Engineering, Rajshahi University of Engineering and Technology, Rajshahi 6204, Bangladesh,
e-mail: monjurmourshed@me.ruet.ac.bd
Arefin Sultana, Varendra University, Rajshahi 6204, Bangladesh; Department of Public Health Sciences, Stockholm University, Stockholm, Sweden

 | https://doi.org/10.1515/9783112248072-008

These materials cause pollution and long-term ecological harm when they are improperly treated and seep into the soil, water, and air [6, 7]. Severe health effects, such as neurodevelopmental abnormalities, renal damage, respiratory illnesses, and reproductive toxicity, can result from human exposure to hazardous components through open burning, acid leaching, and informal recycling [8, 9]. The severe environmental and occupational risks connected to uncontrolled e-waste processing are shown by informal recycling centers in areas like Agbogbloshie (Ghana), Guiyu (China), Moradabad (India), and portions of Nigeria and Côte d'Ivoire [10, 11]. Strong governance systems are crucial in light of these dangers. Legal frameworks for controlling hazardous waste flows, encouraging environmentally responsible management, and limiting transboundary shipments from developed to developing nations are provided by international environmental agreements like the Basel Convention (1989), Stockholm Convention (2001), and Bamako Convention (1991) [11]. These governance commitments are operationalized at the domestic level by national and regional legislative instruments, such as the Waste Electrical and Electronic Equipment (WEEE) Directive of the European Union (EU), China's Circular Economy Promotion Law, Japan's Home Appliance Recycling Law, and India's E-waste Management Rules (2016, amended 2022) [11].

Significant differences still exist between the Global North and the Global South in spite of these developments. According to Ongondo et al. [12], developed nations often maintain sophisticated recycling infrastructure, standardized take-back programs, EPR procedures, and more robust enforcement capabilities. On the other hand, a number of nations in the Global South still confront substantial obstacles, such as restricted access to technology, insufficient infrastructure for regulations, lax enforcement, low consumer awareness, and economic reliance on unofficial recycling industries [13, 14]. In addition to raising grave ethical questions about worker safety, resource distribution equality, and the disproportionate burden placed on vulnerable populations, these systemic injustices perpetuate environmental injustice.

However, e-waste governance mechanisms are still dispersed, varied, and frequently ineffectual despite increased worldwide awareness. Although many nations have passed domestic laws and ratified international accords, the results of their implementation differ greatly. Regulatory compliance and accountability are nevertheless threatened by lax enforcement, a lack of institutional capacity, and widespread corruption [15]. According to Widmer et al. [16], policy incoherence among governmental entities also leads to administrative overlap, contradictory mandates, and ineffective resource allocation. These difficulties are exacerbated by ongoing data gaps on the production of e-waste, collection rates, material flows, and recycling results gaps that obstruct evidence-based policymaking and mask the actual scope of the issue [11, 14]. When taken as a whole, these obstacles worsen worldwide disparities in the e-waste management system and lead to persistent policy failures.

The policy frameworks and ethical issues supporting international e-waste governance are thoroughly examined in this chapter. In order to highlight the differences between official and informal recycling systems, it examines international treaties, national legal frameworks, and local regulatory procedures. Case studies from the EU, Japan, China, India, Ghana, and a few African countries are used. Socioeconomic dynamics, governance achievements, ongoing enforcement shortcomings, and the consequences of these gaps for public health, environmental integrity, and sustainable development are all examined in this paper. Finding prospects for integrated, multistakeholder governance models that uphold the values of the circular economy, improve institutional accountability, and bolster moral decision-making throughout the e-waste value chain is given special attention.

2 Global overview of e-waste policies

2.1 An overview

E-waste management has become a global challenge nowadays, and it has led to the necessity of creating international laws, policies, and frameworks on e-waste movement, treatment, and disposal. E-waste is now one of the rapidly growing waste streams that continues to expand with fast technological advancement [11, 14]. Moreover, it raises concerns on many important issues such as environmental protection, human health, resource recovery, and global fairness. As e-waste movement across borders is common now, countries have created some shared rules and regulations along with ethical standards. The purpose of these agreements is to protect both the people and the planet and to manage the e-waste stream in a responsible manner. Among these, the Basel Convention (1989), the Stockholm Convention (2001), and the Bamako Convention (1991) are the three main global agreements that lead to e-waste management [11]. The Basel Convention forms the core of international legal regulations for managing hazardous wastes and preventing developed countries' exports of hazardous waste (inclusive of e-waste) to developing countries. The Basel Ban Amendment significantly strengthened this protection further. The Stockholm Convention aims to reduce or eliminate toxic chemicals and pollutants found in e-waste. By doing so, it contributes to the protection of the environment and human health. Unlike the other treaties, the Bamako Convention completely bans the import of hazardous waste into Africa. It also reflects Africa's rejection of being used as a dumping ground for waste from richer nations and promotes regional cooperation for safe waste management.

2.2 The Basel Convention

As the industrial production was growing in the 1970s and 1980s, companies in wealthy nations were facing high disposal costs, advanced technology requirements, and strict environmental regulations to dispose of the hazardous and toxic wastes. Thus, they were in search of the cheapest way to dispose of e-waste, where laws would be weaker, and oversight would be limited. Many developing countries in the 1980s lacked strict environmental laws, enforcement capacity, and hazardous waste treatment facilities [11]. Also, exporting the waste to those developing countries was often far cheaper. This made them vulnerable to being targeted as cheap disposal sites. To address and halt this toxic trade, countries negotiated the Basel Convention among nations. The Basel Convention on the Control of Transboundary Movements of Hazardous Wastes and Their Disposal was adopted in 1989 and entered into force in 1992, under the United Nations Environment Programme (UNEP). Its aim was to control how hazardous and other wastes (including e-waste) move across borders. According to the convention, if one country wants to send hazardous waste to another, it must first notify the receiving country and get written permission, which is called Prior Informed Consent (PIC) [17]. Initially, the convention focused only on "hazardous" e-waste (like batteries, CRTs, or waste with toxic chemicals). But over time, new rules were added to cover more types of waste. Its overall goal is to ensure that waste is managed in an environmentally sound manner (ESM) everywhere [10, 17]. The treaty provides detailed technical manuals to help countries handle, recycle, and dispose of waste in safe ways. For e-waste, these guidelines describe safe recycling methods, limits on burning or landfilling, and measures to protect workers and communities. Several important amendments have strengthened the Basel Convention's relevance to e-waste management. The Ban Amendment was adopted in 1995 and entered into force in December 2019 to prohibit hazardous waste exports from OECD (Organisation for Economic Co-operation and Development) countries to other Parties that have accepted the amendment [18]. It thereby protected developing nations from becoming dumping grounds for toxic waste. The Plastic Waste Amendments have been effective since January 2021, and they introduced tighter controls on plastic waste exports. It is an important step as much e-waste is now mixed with plastics, and so most plastic scrap shipments now require PIC [19]. Most significantly, the E-waste Amendment was adopted in 2022 and is effective from January 2025, which expands coverage beyond hazardous materials to include all e-waste, whether hazardous or not [20]. This reform closes a long-standing loophole in which exporters declared e-waste as "non-hazardous secondhand goods," and it transforms the Basel Convention into a comprehensive global system for regulating cross-border movements of e-waste. The Basel Convention has 191 Parties that have fully ratified or acceded to the treaty. In addition, Haiti and the United States have signed the convention but have not yet ratified it, meaning they have not become full Parties [18].

2.3 The Stockholm Convention

During the late twentieth century, scientists discovered that persistent organic pollutants (POPs) were toxic, bioaccumulated, and highly persistent in the environment for decades. POPs were traveling across borders by water and air, and were being found in food chains, human breast milk, wildlife, and even in places where they had never been used. This showed that POPs were globally distributed and required a global solution. This led to the Stockholm Convention, a widespread agreement that the world needed to strictly control the invisible threat of these forever chemicals. The treaty was adopted in 2001 and became effective in 2004 under the UNEP. Over the years, the convention has expanded to cover many harmful chemicals found in electronic products, which include PCBs (2001), dioxins and furans (2001), flame retardants like penta- and octa-BDE (added in 2009), HBCD (2013), hexachlorobutadiene or HCBD (2015), and short-chain chlorinated paraffins or SCCPs (2017). Newer pollutants like PFOA (2019) and PFHxS (2022) have been added more recently [21]. These updates indicate that the convention is increasingly focusing on toxic chemicals linked to e-waste altogether. The Stockholm Convention is structured around three main annexes, and each one defines pathways to manage different types of persistent organic pollutants. Firstly, Annex A requires member countries to ban and systematically phase out substances such as PCBs, brominated diphenyl ethers (BDEs), and PFHxS. Secondly, Annex B deals with chemicals that still have limited acceptable uses but are otherwise strictly controlled. For example, DDT is still allowed for malaria control under certain conditions. Thirdly, Annex C targets POPs that are unintentionally produced. These include dioxins and furans, which are by-products of open burning, smelting, and waste incineration. Such activities are often linked to informal e-waste recycling. Together, these annexes ensure that the Convention bans harmful chemicals, controls restricted uses, and reduces unintended releases [21]. As of 2025, the Stockholm Convention has almost universal participation. In fact, a total of 186 Parties are committed to eliminating or restricting POPs. However, one important political gap remains, which is that the United States signed the Convention in 2001 but has not yet ratified it [21]. This means it is not fully bound by the treaty's formal obligations. This situation, therefore, shows that there is global agreement on the need to address POPs. At the same time, it highlights how domestic politics can affect international environmental governance.

2.4 The Bamako Convention

"Never again to toxic dumping": with this determination, African countries negotiated the Bamako Convention on the Ban of the Import into Africa and the Control of Transboundary Movement and Management of Hazardous Wastes within Africa. It was negotiated in Bamako, Mali, in 1991 under the Organization of African Unity (OAU, now

African Union) and entered into force in 1998 [22]. It was created in response to the weaknesses of the Basel Convention, as many African countries considered Basel too lenient because it allowed hazardous waste exports under the PIC system. High-profile toxic dumping scandals in the 1980s prompted urgent action. For instance, in 1988, Italian companies illegally shipped thousands of tons of toxic waste to Koko, Nigeria. Unlike Basel, the Bamako Convention completely prohibits the import of all hazardous and radioactive wastes into Africa, regardless of consent [22].

It also explicitly criminalizes such acts. The treaty requires Parties to strictly control the transboundary movement of hazardous waste within Africa. Moreover, it mandates that wastes should be managed in an environmentally sound way. It also applies a broader definition of hazardous waste than the Basel Convention, and this makes the Convention highly relevant to e-waste. By banning imports, Bamako seeks to prevent Africa from becoming a dumping ground for obsolete electronics and other hazardous wastes. It focuses mostly on self-reliance by ensuring that Africa manages its waste domestically. Moreover, it promotes capacity enhancement and technical cooperation among member states. It also addresses radioactive waste, which is not covered by the Basel Convention [23]. By 2025, 35 African countries, including major e-waste destinations such as Nigeria and Ghana, will be signatories to the Convention [22]. However, the enforcement of this convention is uneven owing to insufficient capacity, a weak monitoring system, and persistent informal recycling activities. Overall, the Bamako Convention constitutes a much stricter regional counterpart to the Basel Convention that reflects African countries' determination to protect themselves from toxic trade and its environmental and health consequences.

2.5 Strengths and weaknesses of existing frameworks

The main strength of the Basel Convention is that it provides clear obligations, such as the PIC system. This contributes to both reducing illegal dumping and promoting waste minimization at its origin, as well as supporting environmentally responsible management. The convention also provides institutional support, such as guidelines, technical assistance, and regional centers to build capacity. However, the Convention has notable weaknesses, including ineffective enforcement in many countries, resulting from limited monitoring capabilities. Furthermore, vague and unclear definitions cause loopholes in e-waste management. Consequently, North–South tensions persist, with developing countries often bearing the burden of toxic waste dumping. The liability framework of the Basel Convention falls short in holding parties accountable for e-waste-related environmental damage. Uneven application is also notable due to differences in laws, capacities, and priorities. On the other hand, the Stockholm Convention prioritizes protection from POPs and allows precautionary approaches for safer chemical management of e-waste. It provides time-limited exemptions and flexibility to ease transition for developing countries while reducing harmful chemical use. The convention is dynamic and expandable as it allows the listing of new POPs recognized by scientific evidence. But adding new chemicals can be slow and influenced by political considerations, resulting in delayed action. Temporary exemptions can reduce effectiveness by allowing continued use of hazardous substances. The focus is also POPs-centered and does not cover all hazardous chemicals broadly. Moreover, developing countries may feel burdened to meet obligations without sufficient technical and financial support. In contrast, the Bamako Convention takes a much stronger stance by completely prohibiting the import of hazardous waste into Africa without exception. It reinforces regional solidarity and also symbolizes moral authority and leadership from the Global South in setting stricter environmental standards. However, its effectiveness is limited as all African Union states have ratified or implemented it. Enforcement capacity is weak in many member states due to a lack of infrastructure and resources. Many countries still rely on global frameworks like Basel, and illegal dumping and e-waste imports remain persistent challenges. Therefore, the Bamako Convention is symbolically powerful but practically a little constrained. A summary of the international framework is presented in Table 1.

Table 1: International policy frameworks overview.

Convention/ treaty	Year	Scope	Effectiveness	Limitations	References
Basel Convention	1989	Controls transboundary movements of hazardous wastes and their disposal	Establishes prior informed consent (PIC), promotes environmentally sound management, and provides technical assistance and regional centers	Weak enforcement in many countries; ambiguities in definitions, especially for e-waste; limited liability coverage	[18]
Stockholm Convention	2001	Focuses on persistent organic pollutants (POPs) that threaten human health and environment	Precautionary approach allows chemicals to be listed without full certainty; time-limited exemptions for developing countries; dynamic list can expand	Listing new chemicals is slow and politicized; exemptions dilute effectiveness; limited to POPs, not all hazardous chemicals	[21]

Table 1 (continued)

Convention/ treaty	Year	Scope	Effectiveness	Limitations	References
Bamako Convention	1991	African regional treaty banning import of hazardous wastes	Stricter than Basel in prohibiting imports; reflects African countries' commitment to environmental protection	Uneven enforcement due to limited capacity; informal recycling continues; some countries have limited implementation	[22]
OECD Council Decision on Waste	2001 (revised)	Provides guidelines for the transboundary movement of wastes among OECD countries	Encourages waste minimization, environmentally sound management, and notification procedures	Applies only to OECD members; lacks binding enforcement mechanisms outside member states; limited impact on non-OECD countries	[24]

3 National regulations and extended producer responsibility

3.1 The EU WEEE Directive

The EU enacted the WEEE Directive in 2002, which was later updated and recast in 2012 [25]. The main objective of the WEEE Directive is to support environmentally sustainable production and consumption and to prevent the WEEE creation. It also promotes resource efficiency and the recovery of secondary raw materials. This is done through reuse, recycling, and other recovery methods. The Directive encourages better environmental practices and requires all actors involved in the EEE lifecycle to follow these improved practices [26]. EPR is the key principle of the directive that requires producers to fund the collection of WEEE. It also places responsibility on the producers to manage the processing of WEEE and to make sure that the waste is disposed of in an environmentally friendly way [26]. This responsibility includes providing consumers with the option to return WEEE at no cost. Thereby, it encourages proper waste management and helps reduce illegal disposal. The directive sets ambitious collection targets: Member states must collect 65% of the average weight of EEE sold in the previous three years. Alternatively, they must collect 85% of the WEEE generated within their own territory [27]. The main target is to recover resources and enhance the recycling rates. However, there are still many challenges as different member states experience different impacts, including unlawful exports of e-waste and regulatory problems for smaller producers [28]. Despite progress, challenges are also visible in achieving the goals. In 2022, about 14.4 Mt of EEE were placed on the EU market, but only 5 Mt of WEEE were collected. This amount corresponds to 11.2 kg per inhabitant and represents a collection rate of 40.1%. This rate is significantly below the 65% goal [27]. The data reveals shortcomings in enforcement and shows gaps in consumer participation. At the same time, it highlights the importance of regulatory structures aligned with collection schemes and strengthens the circular economy. Despite these obstacles, the WEEE Directive has had a major influence on global e-waste policy. It has encouraged many countries to adopt EPR systems and motivated them to move toward circular economy practices.

3.2 USA (state-level laws)

In the United States, electronic waste is mostly managed through state-level laws, resulting in a complex and fragmented policy framework. Unlike the EU's WEEE Directive, the United States has no federal-level EPR system for electronic waste. Instead, individual states have introduced their own legislation. These state-level efforts are aimed at tackling the environmental problems caused by outdated electronics. As of 2024, e-waste recycling is legally required in 25 states and the District of Columbia.

Many programs in these areas integrate EPR principles under which manufacturers bear responsibility for funding or managing recovery systems [14]. California was the first state to pass comprehensive e-waste legislation. It enacted the Electronic Waste Recycling Act of 2003 (SB 20/SB 50), which established a consumer-paid fee at the point of purchase. This fee supports authorized collection and recycling programs [29]. In 2022, California expanded its recycling framework through Senate Bill 1215. The law includes new product categories, such as solar panels and products with embedded batteries, in the recycling system. It also mandates that consumers pay a recycling fee when buying certain new or refurbished products starting January 1, 2026. In addition, the law requires manufacturers to label products with brand and battery chemistry information, while the California Department of Resources Recycling and Recovery (CalRecycle) administers the program [30]. Several US states have adopted producer-funded EPR programs to manage e-waste.

For example, Washington State operates the E-Cycle Washington program, where manufacturers of covered products like computers, monitors, laptops, and TVs must register every year and pay administrative fees. They also have to take part in approved recycling programs. Only manufacturers listed as "in compliance" are allowed to sell their products in the state. The rule increases accountability for recycling and collection in this way [31]. Oregon takes a similar approach through its E-Cycles program. Manufacturers of covered electronics must register with the Department of Environmental Quality, pay annual recycling fees, and set up collection sites in every city with more than 10,000 residents. This requirement helps ensure broad access to recycling services across the state [32]. Connecticut also requires manufacturers to register annually with the Department of Energy and Environmental Protection [33]. They are required to finance the collection and recycling of designated electronic devices and take part in the statewide recycling program. These requirements ensure that producers are responsible for the disposal of their products. These state initiatives show how local EPR policies can enforce producer responsibility, strengthen collection systems, and improve the sustainability of e-waste management, even without a national EPR law.

3.3 Japan's e-waste management and extended producer responsibility framework

Japan is often regarded as a leader in e-waste management, largely because it pairs strict environmental rules with a well-organized EPR system. The Home Appliance Recycling Law (HARL) of 2001 is a key example, as it was one of the first laws in the world to make producers directly responsible for e-waste. Under this law, companies must help fund and oversee the collection and recycling of major household appliances such as refrigerators, air conditioners, washing machines, and televisions. Consumers also share part of the responsibility by ensuring their old appliances are disposed of properly and by paying a portion of the recycling cost [34].

In 2009, an amendment to the law expanded its scope by including liquid-crystal and plasma televisions, as well as clothes dryers, among the appliances required to be recycled [35]. According to a joint press release from Japan's METI and MOE, about 15.26 million units across the four regulated appliance categories were collected in fiscal year 2021. They were all collected at designated sites, and all of them achieved recycle rates above the required legal thresholds. The Japanese government also reports that refrigerant fluorocarbons from appliances like air conditioners and refrigerators are being safely and properly recovered during the recycling process [36]. Under this law, retailers must accept used appliances from consumers. Meanwhile, producers must ensure that recycling is carried out at certified facilities. This creates a closed-loop system for recovering high-value materials. In 2013, Japan's "Act on Promotion of Recycling of Small Waste Electrical and Electronic Equipment" came into effect, which addresses smaller electronic devices [37]. This law extends EPR principles to computers, printers, and audio-visual equipment. Though it is not a full EPR scheme in the sense that manufacturers are not obliged to collect and recycle these small items financially or physically. Rather, they are to make efforts to reduce costs and design for recycling [38]. Both the Home Appliance Recycling Law (HARL) and the Act on Promotion of Recycling of Small Waste Electrical and Electronic Equipment place a strong emphasis on eco-design. They encourage manufacturers to create products that are easier to dismantle, recycle, and reuse. The system also relies on active consumer participation, with end-users paying a recycling fee when disposing of their electronics. Alongside this, highly coordinated logistics and certified recycling operators ensure that hazardous components are safely processed. Japan's EPR approach has achieved high recovery rates, and often these rates are above 80% for covered appliances [34]. It has created one of the world's most advanced material recovery systems. However, the framework faces some limitations, which include restricted product coverage, reliance on consumer-paid recycling fees, and challenges in addressing small-scale informal recycling outside regulated channels. Despite these challenges, Japan's model shows how a combination of producer responsibility, strong regulatory enforcement, eco-design, and consumer participation can create an efficient and sustainable e-waste management system. Together, these elements support the system for environmental protection and the circular economy.

3.4 China's circular economy approach

In this challenging era of waste management, China has created a strong policy framework for managing electronic waste that links e-waste management to its wider circular economy strategy. The main basis for this policy framework is the Circular Economy Promotion Law of 2008. This law aims to improve resource efficiency, reduce waste, and strengthen environmental protection across various industries [39]. Building on this foundation, the State Council released the Plan for the Implementation of the EPR System. This plan requires producers to take responsibility for the whole life cycle of their manufactured products, along with collecting used electronics. It also requires them to recycle these products properly and ensure environmentally sound disposal of end-of-life electronics [40]. Under this EPR framework, producers and importers must set up their recycling systems, either on their own or through certified recycling companies. The government supports these recycling systems through the WEEE Fund. The fund encourages formal recycling by providing subsidies to authorized recyclers that meet environmental and safety rules. This approach decreases the need for informal recycling, as certified recyclers receive financial support [41]. China's 14th Five-Year Plan of the year 2021 to 2025 strengthens the country's circular economy agenda. It aims to increase the use of secondary raw materials while promoting industrial symbiosis and eco-design [42]. China has intensified efforts to integrate informal recyclers into the formal system over the past few years. The country is also working to improve environmental standards in dismantling important regions such as Guiyu. Occupational health and safety standards for workers in these areas are also being strengthened.

3.5 India's 2016 and 2022 e-waste rules

India sets an example of how a developing country can embed EPR within its environmental policy framework by its E-Waste Management Rules from 2016 and 2022. The South Asian country has progressively developed a detailed e-waste management framework. The framework recognizes the environmental challenges posed by rapidly growing electronic waste and also addresses the social challenges associated with e-waste. India's e-waste (management) rules were introduced in 2016 to establish a legally binding EPR system. Under this system, producers are required to organize and finance the collection, recycling, and environmentally sound disposal of end-of-life electrical and electronic equipment (EEE) [43]. Producers are required to get EPR authorization from the Central Pollution Control Board (CPCB) and set up take-back systems or collection centers for their products. They also need to meet phased collection targets that correspond to the lifespan of each product. The Rules recognize the role of producer responsibility organizations (PROs) and refurbishers, and encourage them to work with the informal sector to make recycling more formal. The 2022 Rules broadened the scope to cover 107 types of electronic products and set recycling efficiency standards, supporting India's wider Circular Economy Action Plan and the resource recovery objectives of the Swachh Bharat Mission (CPCB, 2023) [44]. India still faces various challenges in e-waste management, including limited formal recycling capacity, low public awareness, and dominance of the informal sector. In practice, the 2022 framework not only integrates digital governance, producer accountability, and circular economy principles but also offers a model for other emerging economies. It marks a significant policy evolution so that they can achieve sustainable and ethically responsible e-waste management systems.

3.6 African approaches (e.g., Rwanda, Ghana) and others

Africa's approaches provide valuable insight into how developing countries are creating EPR-driven frameworks within their resource and infrastructure limits. It is especially visible in Rwanda, Ghana, and a few other emerging economies. Africa's e-waste management is undergoing significant change because of domestic consumption and the inflow of discarded electronics from developed countries at the same time. Though the continent produces a smaller share of global e-waste compared to Asia or Europe, informal recycling is widespread and has caused environmental damage as well as serious health impacts [1]. Therefore, several African countries have started including EPR in their national waste policies. This marks a shift from donor-dependent collection projects to formal, producer-funded systems that align with the Basel Convention and its Ban Amendment, which restricts e-waste imports into Africa.

Rwanda stands out as a continental leader in developing a centralized and policy-driven e-waste system. For example, Rwanda's 2018 National E-Waste Management

Policy and the 2021–2030 National E-Waste Management Strategy embed circular economy principles into the country's broader green-growth and climate-resilience plans [45]. They connect e-waste management to long-term goals for sustainable development and climate-resilient growth. The Bugesera E-Waste Dismantling and Recycling Facility was launched in 2017 through a public–private partnership under the Rwanda Green Fund (FONERWA). Since its start, this facility has become a national landmark: it processes thousands of tons of e-waste each year, offering a practical model of how African countries can deal with electronic waste safely and responsibly. Alongside recycling, the setup supports environmental safety, promotes new "green" jobs, and helps formalize what had largely been an informal recycling sector. The system is overseen by the Rwanda Utilities Regulatory Authority (RURA), which ensures that operations follow environmental standards, encourages inclusion (including gender inclusion), and works toward turning informal waste practices into formal, regulated recycling pathways.

Ghana has broken new ground in Africa with the adoption of the Hazardous and Electronic Waste Control and Management Act, 2016 (Act 917) [46]. It is one of the continent's first laws to put EPR into practice. Under this law, Ghana introduced an eco-levy on imported electronics. That levy helps fund the proper collection and recycling of waste electronics. The country's Environmental Protection Agency, Ghana (EPA-Ghana), oversees an e-waste fund, which finances certified recycling centers and supports clean-up or rehabilitation efforts in former informal e-waste hubs like Agbogbloshie. Ghana's EPR system has boosted occupational safety, improved chemical-waste management, and helped integrate previously informal workers into formal recycling chains.

Other African countries are also following similar paths. For example, Nigeria introduced EPR guidelines in 2020 under the National Environmental Standards and Regulations Enforcement Agency (NESREA), bringing producer responsibility to both electronics and packaging sectors [47]. Meanwhile, EPR regulations passed in South Africa in 2021 under the national waste law require producers of ICT equipment and lighting to manage the end-of-life of their products. Kenya has also taken steps with a draft EPR regulation (2022), which aims to establish producer-funded recycling frameworks, inspired by successful models in other African nations [3]. Taken together, these initiatives reflect Africa's growing regulatory maturity and show a genuine commitment to ethical, circular, e-waste governance. Table 2 provides an overview of the national regulations and EPR schemes.

Table 2: National regulations and EPR schemes.

Country/region	Policy name	Year introduced	Key features	Outcomes	Limitations	References
European Union	WEEE Directive (Waste Electrical and Electronic Equipment)	2003 (recast 2012)	Mandatory EPR; collection, recycling, and recovery targets; producer registration; harmonized reporting	Improved collection rates; clearer producer obligations; expanded recycling infrastructure	Persistent cross-border shipment issues; uneven enforcement among member states; illegal exports	[26]
United States (state-level laws)	Various State E-Waste Acts (e.g., California Electronics Waste Recycling Act)	2003 onward (state dependent)	Mix of EPR and advanced recycling fees; bans on landfill disposal; manufacturer-financed recycling in some states	Increased consumer awareness; established recycling programs in majority of states with laws	No federal law; inconsistent standards; major states (e.g., Texas) lack strong enforcement; patchy national coverage	[29–33]
Japan	Home Appliance Recycling Law (HARL)	2001	Consumer co-payment for recycling; strict producer take-back; high-quality recycling standards; designated collection points	High recycling rates; strong manufacturer participation; advanced dismantling technologies	Cost burden on consumers; limited scope (covers major appliances but not all electronics)	[4, 34–37]
China	Regulation on Management of the Recycling and Disposal of WEEE products	2011	National EPR fund financed by producer fees; licensed recyclers; crackdown on informal sector; list of covered products	Expansion of certified recyclers; improved monitoring; formalization of collection channels	Informal sector remains influential; enforcement challenges; limited transparency in EPR fund allocation	[40, 41]
India	E-waste (Management)	2011 (updated 2016, 2022)	Mandatory EPR; collection and recycling targets; producer	Growth of formal recycling sector; clearer EPR	Weak enforcement; informal sector dominance; target	[43, 44]

	Rules (with major amendments)		responsibility organizations (PROs); digital tracking	obligations; nationwide rules	fulfillment inconsistencies; limited auditing	
Ghana	Hazardous and Electronic Waste Control and Management Act (Act 917)	2016	Environmental handling fees; oversight of e-waste imports; infrastructure development funded through levy; formalization efforts (e.g., at Agbogbloshie)	Better tracking of imports; funding mechanisms for safer recycling; initial formalization steps	Continued informal recycling hazards; limited enforcement capacity; reliance on second-hand imports	[46]

4 Ethical considerations in e-waste management

4.1 Environmental justice (Global North to South dumping)

"Global North to South dumping" refers to the export of hazardous or unwanted e-waste and other environmentally harmful materials from high-income, industrialized nations (the *Global North*, e.g., the United States, EU, Japan, Canada) to low- and middle-income countries (the Global South, e.g., Ghana, Nigeria, China, and India). Global data reflect that e-waste generation is increasing rapidly, while only a small proportion is formally collected and recycled. A significant share is moved across borders, occasionally labeled as used or refurbished products, but often ends up as discarded waste in countries with inadequate waste management infrastructure [2]. The majority of WEEE generated and collected within the EU is exported to countries in the Global South for treatment and processing [11, 14]. The negative impacts on health and the environment from discarded electronics led to making environmental justice a central concern in e-waste governance. The marginalized and low-income countries of the Global South are facing disproportionate distribution of e-waste, resulting in health hazards and pollution. Because these recipient nations frequently lack proper waste infrastructure, environmental regulations, or worker protections, the result is a pattern of environmental injustice. There is frequent emission of toxic chemicals produced from burning, acid leaching, and dismantling e-waste, and the burden is never shared according to consumption. Moreover, there is also economic inequity, as the richer nations exteriorize the cost of waste disposal. Social inequity is another factor that cannot be ignored since vulnerable people, mostly the poor migrants and children, bear the burden of exposure. This dynamic reflects the historical resemblance of waste colonialism, a notion developed in the 1980s to denote the displacement of environmental damage from industrialized countries to the poorer countries of the world. This is evident in the Agbogbloshie scrapyard in Ghana and Alaba Market in Nigeria, which have received an enormous amount of e-waste discarded from Europe and North America [45–47]. The 2019 UNEP report indicates that, around 7–20% of e-waste gathered in the OECD countries ends up in developing countries through informal procedures [10]. Nonetheless, studies identify Ghana, Nigeria, and Tanzania as major African recipients of e-waste originating from the EU, UK, and other industrialized regions. Among these, Nigeria stands out as the largest importer of used EEE on the continent, receiving an estimated 60,000–71,000 tons each year through the ports of Lagos in 2015–2016, with roughly 77% of imports coming from the EU [2]. Several ports including Durban (South Africa), Bizerte (Tunisia), and Lagos (Nigeria), serve as significant gateways for used or end-of-life electronics into Africa. The transfer of e-waste from wealthier countries of the Global North to poorer countries in the Global South raises issues of ethical concern on key conceptual pillars of environmental justice. First, it creates issues of distributive justice [13, 14], because the people who suffer from the pollution, risk exposures, and unsafe working conditions created by the wastes are mostly in the poorer countries, who did not even have the benefits of those electronics. So, the harms are unfairly distributed among the rich and poor countries. Second, it challenges procedural justice, as the decisions are mostly made by companies and governments in the exporting countries. Communities in the receiving countries usually have little or no say in whether these e-waste shipments arrive or how they are handled. This process is leaving them excluded from the regulatory and policy processes that directly impact their lives. Finally, e-waste raises important questions of intergenerational justice, since hazardous materials from e-waste can persist in soil, water, and air for many years, impacting the health and environment of future generations. Overall, the ethical implications of North-South e-waste transfers go far beyond technical or financial concerns, raising critical questions of justice, human rights, and accountability. Although the Basel and Bamako Conventions prohibit the import of hazardous waste, illegal e-waste shipments still persist.

Exporters often conceal these shipments under the guise of "second-hand goods," through mislabeling and underreporting tactics. This ongoing inflow of e-waste exposes loopholes in global regulatory systems and reflects the disparities in environmental justice affecting the Global South, as e-waste flows from the Global North. Embedding environmental justice in e-waste recycling directly intersects with several human rights and Sustainable Development Goals (SDGs) such as the right to a healthy environment, safe working conditions, and protection of children, while advancing progress toward SDGs including SDG 3 (Good Health and Well-being), SDG 8 (Decent Work and Economic Growth), SDG 10 (Reduced Inequalities), and SDG 12 (Responsible Consumption and Production) [49].

4.2 Worker safety: child labor, hazardous exposures

Worker safety is one of the most pressing ethical concerns in global e-waste management, particularly within the vast informal recycling sectors of the Global South. According to the World Health Organization (2021), e-waste pollution poses a rising global health threat, with children and pregnant women identified as the most vulnerable groups [50]. In many countries of the Global South, e-waste recycling relies on vulnerable workers, often including young children. They are engaged in dismantling, burning, and acid-leaching of end-of-life electronics through unsafe practices to separate valuable metals such as copper and gold. The informal waste recycling sector employs about 18 million children and young people and 13 million women, with e-waste being a subsector of this work [51]. If not safely handled, hazardous waste can significantly cause cognitive and neurological issues in children, along with the risk for birth defects and cancer. Around 40,000 people live and work near Agbogbloshie in Ghana, Africa's largest and one of the biggest dumpsites in the world [51]. Moreover, some children as young as five years old work in informal e-waste recycling sites in Ghana. Beyond the working children, exposure to e-waste hazards affects all children as they consume more relative to their size and have slower toxin metabolism and closer contact with contaminated surfaces. In 2010, approximately 201,600 people across Ghana depended on e-waste recycling for their livelihood despite the risk of adverse birth outcomes, cognitive and behavioral dysfunctions, and a maximized chance of chronic diseases later in life. In another part of the Global South, Moradabad, known as India's "brass city," serves as a major informal e-waste hub where thousands of workers, including women and children, dismantle discarded e-waste and extract metals. These activities release toxic chemicals and gases, which are a reason for cancer and other diseases, whereas exposure to high mercury levels in soil, water, and aquatic life leads to chronic poisoning [52]. Guiyu, a town in Guangdong Province, China, was long regarded as the "e-waste capital of the world" because of its extensive informal e-waste recycling process from the 1990s through around 2015. Government statistics indicate that during its peak, over 80% of families and more than 100,000 people of Guiyu were involved in processing 150–300 million tons of e-waste annually. The process of pyrometallurgy, hydrometallurgy, and salt brine delamination contaminated the local waterways with As, Cr, Li, Mo, Sb and Se, Ag, Be, Cd, Co, Cu, Ni, Pb, and Zn. Studies in Guiyu also found extremely high PCB and lead contamination in the air, where exposure from open burning and polluted environment caused notably elevated toxin levels in e-waste workers and children. There is also evidence of high PCB levels in human milk and blood, with significantly higher rates of stillbirths, low birth weight, and reduced newborn health indicators compared to control areas [11, 14]. Despite varying geographic and socioeconomic contexts, all these cases collectively reveal the root ethical challenges embedded in global e-waste management. Such practices infringe upon basic human rights and international labor protections, defying ILO Conventions 138 and 182 on child labor and impeding global commitments under SDG 3 (Good Health and Well-Being) and SDG 8 (Decent Work and Economic Growth) [49].

4.3 Equity: fair distribution of economic value

The idea of equity in e-waste management is not only limited to protecting the environment but also about making sure that there is fair and inclusive access to economic benefits found in EEE and secondary raw materials. The secondary raw materials from e-waste refer to gold, copper, and other earth metals, which are vital for producing new technologies and sustaining the circular economy. Studies show that, even with just 17.4% of e-waste being recycled, there is a possibility to recover $10 billion worth of valuable materials and 4 million tons of reusable resources. The selected raw materials in e-waste were worth about $57 billion in 2019, representing about 25 million tons in total [2]. Moreover, e-waste contained approximately 31 billion kilograms of metals, including valuable materials like gold, copper, cobalt, and iron in 2022, which were valued at around US$91 billion. Despite their value, a global loss of about US$37 billion in 2022 is evident due to poor and informal e-waste management, mainly from health and environmental impacts [53]. However, the value gained from these resources is not distributed equally. Informal waste collectors extract valuable materials from discarded equipment to have a source of income, although earnings are highly variable among workers. On the other hand, countries of the Global North often restrict fair access to discarded electronics for the Global South. Some reports also indicate that European producers often export used branded electronic and electrical equipment labeled as "new" to Africa, as it is more profitable for them, although the equipment is near end-of-life, suggesting they were essentially e-waste. A global snapshot of the ethical issues regarding e-waste management is presented in Table 3.

Table 3: Ethical issues in e-waste management.

Ethical issue	Example case	Implications	Policy response
Worker safety	Informal recyclers in Agbogbloshie (Ghana) dismantling electronics without protective gear	Exposure to toxic metals (lead, mercury), respiratory illness, burns, long-term neurological damage	Ghana's Act 917 (2016) aims to regulate e-waste processing; some training and PPE programs supported by NGOs, though enforcement remains weak [46]
Child labor	Children engaged in e-waste picking and dismantling in Delhi and Moradabad (India)	Health risks, loss of education, and intergenerational poverty	India's Child Labour (Prohibition and Regulation) Act prohibits hazardous child labor; limited enforcement in informal recycling hubs [43]
Waste dumping/ illegal transboundary movement	Illegal shipments of used electronics from the EU and the United States to West Africa labeled as "second-hand goods"	Environmental contamination, burden of disposal shifted to poorer countries, violation of international norms	Basel Convention restricts hazardous waste exports; EU Waste Shipment Regulation strengthens inspections; persistent loopholes [18]
Inequitable trade/ environmental injustice	Global North exporting low-value or nonrepairable electronics to Global South	Disproportionate pollution and health harm borne by low-income communities; reinforce global inequalities	Basel Ban Amendment (2019) prohibits hazardous waste exports from OECD to non-OECD countries; uneven global adoption and compliance [18, 24]

These imported EEE have a very short life span and quickly end up in landfills or being burned in informal sites to recover materials, posing serious risks to both worker health and the environment. Informal recyclers in the Global North do not have access to a formal channel of selling, which could provide them with fair payment for their work and recovered materials. In contrast, countries in the Global North can easily recover secondary materials by using modern technologies, and they have access to formal market channels through which they get the required profit. All these mechanisms further strengthen global inequalities. So, it is necessary to establish an inclusive and formal recycling system, fair trade mechanisms, and transparent value chains to ensure equitable and fair distribution of value gained from EEE and secondary raw materials. Ethical policy frameworks should ensure the participation of developing regions in a meaningful way by supporting technology transfer, strengthening local capacities, and helping informal workers to become part of the formal system [54]. Moreover, international cooperation is also needed so that the developing countries are no longer only considered as passive sites of material extraction but active beneficiaries of the global circular economy. Integrating equity into e-waste governance thus reflects the principles of distributive justice and reinforces the realization of fundamental human rights, including the right to a healthy environment, fair working conditions, and social and economic inclusion. It also contributes to achieving Sustainable Development Goals, particularly SDG 8 (Decent Work and Economic Growth), SDG 10 (Reduced Inequalities), and SDG 12 (Responsible Consumption and Production) [49].

5 Case studies of policy and ethics in action

5.1 Europe: formal and regulated recycling

Differences in governance structure, technological and economic capacity, and social priorities affect the implementation of e-waste management policies and ethical frameworks. While developed countries tend to rely on formal recycling systems supported by advanced technologies and strict environmental standards, many developing economies depend on different approaches that shape not only the efficiency of recovering secondary raw materials but also determine how risks and benefits are shared among workers, communities, and national economies. By comparing case studies from Europe, Ghana, India, and Japan, this section explores the policy frameworks and enforcement and also their influence on labor conditions, environmental protection, public health, and the equitable distribution of economic value within the global e-waste recycling chain. The discussion highlights the importance of creating integrated approaches that bring together economic opportunity with environmental responsibility and social justice to achieve sustainable e-waste management.

European countries mostly operate institutionalized and highly formalized e-waste management systems under frameworks such as the EU's WEEE Directive. These regulations are designed to maximize recovery of secondary raw materials, minimize the environmental degradation caused by the recycling processes, and ensure occupational safety standards for workers in the e-waste recycling field. The ethical dimensions in Europe exhibit high compliance standards to protect labor rights and environmental health, aligning with commitments to human rights and the Sustainable Development Goals (SDGs). It is officially stated that the WEEE directive is directed toward safeguarding both the environment and human health by promoting sustainable ways of production and consumption. The directive focuses on achieving this by taking initiatives on reducing WEEE generation, encouraging reuse and recycling of EEE, and promoting optimized resource utilization through the extraction of secondary raw materials [11, 14]. As the recycling process is formalized, it has certain socioeconomic impacts, such as offering employment in safe working conditions and fair wages, covered by the EU Labor Rights framework. This framework contributes to the implementation of several principles in the European Pillar of Social Rights by providing the laws of adequate minimum wages and more transparent and predictable working conditions which respect workers' health, safety, and dignity. However, the capital-intensive system of these operations can potentially limit employment opportunities for less skilled workers. According to a report by Trinomics (2018) for the European Commission, circular economy policies generally possess a positive overall impact on employment in the EU, but the benefits are modest and primarily favor medium and high-skilled workers [55]. However, in general, these formal systems can achieve high recovery rates of valuable materials while promoting a circular economy by maintaining low health and environmental contamination, demonstrating a model of ethical and sustainable e-waste governance.

5.2 Ghana and India: informal recycling hubs

Agbogbloshie in Ghana and Indian cities such as Delhi, Mumbai, and Bangalore serve as major informal recycling hubs for e-waste. Large volumes of both locally generated and imported e-waste are processed using unsafe and the most basic techniques. Workers, even children, are frequently exposed to toxic and hazardous substances, leading to violations of labor rights and human rights. Local communities are bound to bear the disproportionate burden of environmental pollution and health impacts, which reflect serious concerns of environmental justice. Though informal recycling with limited alternatives provides livelihood for marginalized and low-income populations, the unsafe working conditions do not even ensure even and consistent wages. Moreover, the economic benefits rarely reach beyond small-scale local operators. On the other hand, significant environmental degradation is usual due to unregulated informal recycling of e-waste as the air, water, and soil are continuously getting contaminated by open burning, acid leaching, and unregulated disposal. So, there is always an ethical tension between poverty-driven survival and environmental sustainability.

5.3 Japan: advanced producer responsibility model

Japan provides a strong example of an effective EPR system. Under this framework, manufacturers are responsible for collecting, recycling, and safely disposing of their products once they reach the end of their life cycle. This policy requires companies to internalize the environmental costs of their manufactured products, and thus minimizes the risks associated with informal recycling. This protects informal workers and supports the country's shift toward a circular economy. Japan's EPR system has also helped create formal and well-regulated recycling jobs that rely on advanced technology. It enhances worker safety, encourages innovation in how materials are recovered, and contributes to sustainable economic growth. Overall, Japan's model achieves high recycling efficiency, reduces the amount of waste sent to landfills, and enables the recovery of important secondary raw materials. It shows how strong environmental policies can align ethical responsibility with both ecological protection and economic progress.

6 Governance challenges

6.1 An overview

Strong and effective governance is needed to form the foundation of sustainable e-waste management. Though a growing number of national and international regulations now address the e-waste issue, the weakness of governance continues to be the most significant barrier to achieving responsible and ethical recycling, along with environmental justice. The global chain through which e-waste moves is definitely a complex issue, and it requires transparent and well-coordinated frameworks. But the regulatory systems are often fragmented and poorly enforced in reality due to limitations of institutional capacity, corruption, and inadequate financing. As we have discussed above, many developing nations do not have a legal framework, and many others are based on external models that do not fit with the local economy and social conditions. Moreover, the enforcement agencies commonly experience overlapping responsibilities and inadequate expertise, and poor coordination among stakeholders. As a result, loopholes in regulations enable the illegal transboundary shipment of hazardous waste and also make way for unregulated and informal recycling industries without proper worker and environmental protection. Furthermore, the absence of reliable data and effective monitoring systems limits enforcement and hinders alignment with international standards. This ongoing deficiency in governance not only weakens environmental protection but also exacerbates global inequalities, as e-waste transfers remain heavily skewed toward the developing nations. Addressing this issue requires not only technical solutions but also responsible governance that emphasizes inclusion, accountability, and transparency. The following discussion looks at the main governance issues that affect sustainable e-waste management globally.

6.2 Weak enforcement and institutional gaps

Many nations, especially in the Global South, have adopted e-waste regulations which align with international frameworks such as the Basel Convention and EPR models. Yet, enforcement often falls short due to limited institutional capacity, inadequate oversight, and insufficient funding. Most of the e-waste in India is handled outside formal, regulated systems, as formal infrastructure is insufficient and enforcement is weak. There are relatively low formal recycling facilities in India, and many of those that exist mainly handle dismantling rather than full recycling processes. In Ghana, the implementation of the 2016 Hazardous and Electronic Waste Control and Management Act (Act 917) is weak because low stakeholder awareness and poor publicity undermine compliance [46]. As a result, despite the presence of regulations, most e-waste still ends up in informal channels, affecting both communities and the environment.

6.3 Corruption and informal trade

The effectiveness of e-waste regulatory frameworks is continually being weakened by corruption and illicit trade routes. There are loopholes in customs enforcement that enable e-waste to be entered under misleading labels of "donations," "second-hand electronics," or "used electronics," even though hazardous waste imports are banned. This phenomenon has been observed between the EU and West African countries, including Nigeria and Ghana. These practices contravene international agreements, such as the Basel and Bamako Conventions, and unfairly impose environmental and health risks on vulnerable populations.

6.4 Fragmented legal and institutional frameworks

E-waste management often involves multiple government bodies, including the environment, trade, industry ministries, and different local authorities. As a result, Fragmentation and inconsistency often arise in e-waste governance. The absence of a centralized coordinating agency in many ASEAN countries and African regions has led to overlapping mandates, redundant efforts, and weak accountability. ASEAN lacks the authority to establish region-wide legal frameworks for environmental governance, unlike the EU. Although e-waste has become an increasing concern across the region, there is still no unified policy comparable to Europe's WEEE and RoHS Directives. Many countries in ASEAN still lack clear laws or guidelines for managing e-waste. As a result, most efforts are small and have little impact. Thailand and the Philippines have spent years trying to create legal frameworks, but progress has been slow. Even though an e-waste law exists in Vietnam, missing implementation details has made it ineffective [3, 11]. Similarly, weak organizational structures are one of the existing reasons for poor management of waste in Africa. Although the WEEE Directive provides harmonized legislation, which is based on the EPR model as a regulatory tool forcing manufacturers and importers to take responsibility for their products throughout their lifecycles, there are noticeable gaps across the EU member states due to differences in collection practices and reporting systems, leading to cross-border e-waste trafficking. These practices cause e-manufacturers and e-waste handling operators to face challenges in their transition to more sustainable operations. Various studies examined the effectiveness of WEEE systems across European countries and noticed significant data gaps concerning the quantities of EEE placed on the market and the levels of environmental fees applied. Similarly, considerable differences are observed in how WEEE systems are organized even among European countries. The studies emphasized the importance of greater harmonization in national WEEE legislation and system implementation. Moreover, differences in whether environmental fees are visible along the reverse supply chain are not limited to Europe but occur globally. Therefore, it is necessary to address these structural gaps with clear rules and better coordination with centralized monitoring.

6.5 Lack of data and transparency

Many countries still do not have dependable data on how much e-waste is generated, collected, or recycled. This happens because informal recycling is widespread, many operators are unregistered, and manufacturers often underreport their outputs. Only about 22.3% of global e-waste was formally documented as collected and recycled in 2022 [54]. Moreover, recycling rates in some regions, especially in parts of Africa, have dropped below 1%, because the vast majority of discarded electronics remain unaccounted for. This lack of reliable data seriously undermines efforts to design effective policy and international cooperation.

6.6 Policy failures and lessons learned

Policy failures occur when the ambitious legislation does not match the corresponding investment, awareness, and infrastructure. Several countries in Africa and South Asia have adopted EPR laws modeled on Europe, but they lack the technical systems and enforcement mechanisms for effective implementation. Conversely, Japan's successful experience with EPR demonstrates that comprehensive monitoring, financial incentives, and producer accountability are critical for achieving compliance. This contrast highlights the importance of tailoring global policy frameworks to local contexts rather than just applying a "one-size-fits-all" regulatory approach.

7 Opportunities for integrated policy frameworks

7.1 Multistakeholder governance model

As the worldwide flow of e-waste is becoming more complex every day, policies need to be designed in a well-coordinated approach that involves multiple stakeholders. Fragmented governance and uneven distribution of responsibilities have already weakened the effective management of the e-waste chain for a long time. So, integrated frameworks can be a solution to this problem, as they can bring together governments, producers, consumers, recyclers, and international organizations. In this way the loopholes in regulatory systems can be closed and it can also improve the sense of accountability. Integrated systems can also be a strong effort toward balancing environmental and health protection with social and economic benefits. Such systems support the Sustainable Development Goals (SDGs), especially SDG 12 on Responsible Consumption and Production and SDG 17 on Partnership for the Goals.

Sustainable e-waste management requires collaboration among many different actors. These include policymakers, manufacturers, recyclers, informal sector workers, academics, and civil society organizations. OECD (2019) stresses that expanding circular economy policies will require engaging different stakeholders, including governments and the private sector, although it does not specifically measure the impact of such partnerships on collection performance [56]. According to the Basel Convention's PACE initiative, collaboration between public authorities and private actors can facilitate the adoption of environmentally sound management of end-of-life computing equipment by bringing together governments, manufacturers, recyclers, and civil society actors [18]. In addition, multistakeholder platforms, such as the Global E-waste Statistics Partnership (GESP) and the StEP Initiative (Solving the E-waste Problem), help share data, build capacity, and create coordinated policies across countries. The global e-waste statistics partnership provides for strengthening government capabilities, promoting coordinated approaches to data collection, and supporting policymakers with evidence for making e-waste laws consistent with international standards and the SDGs [48]. The StEP Initiative (Solving the E-waste Problem) similarly works across sectors, though specific statements on transboundary policy harmonization are not clearly documented. At the national level, integrated e-waste management frameworks need to incorporate the informal sector. In countries like India, Ghana, and Nigeria, informal recyclers are responsible for a significant portion of e-waste flows, so an integrated system is much needed there. Informal e-waste recyclers pose significant health and environmental risks, highlighting the need to establish scientifically grounded, safe, and environmentally sound management systems and strengthen regulatory frameworks [4, 14]. Enhancing access to formal recycling systems and integrating informal workers into structured programs can reduce environmental harm and create economic opportunities for local communities. Overall, a transparent multistakeholder approach makes governance work fairly and recognizes the contributions of marginalized workers.

7.2 Incentive mechanisms for producers

As the EPR principle gives producers both the financial and operational responsibilities for managing products at the end of their life, it is considered a central pillar of any effective e-waste management strategy. This approach ensures that producers pay their fair share of waste management costs and motivates them to design and produce electrical and electronic products in more sustainable ways. The EU WEEE Directive (Directive 2012/19/EU) remains one of the most detailed global frameworks for applying EPR to electronics. Producers are obliged to organize and finance the collection, recycling, and recovery of e-waste in the WEEE system. The directive also sets ambitious targets for collection rate and material recovery efficiency. However, EPR models focused only on compliance can limit innovation.

This happens when producers see them only as extra regulatory costs. To solve this problem, governments and policymakers are encouraging new incentive-based approaches. These approaches reward companies for designing sustainable products and acting responsibly. Some of these approaches include eco-design rules, variable fee systems, and deposit-return systems (DRS). The Ecodesign for Sustainable Products Regulation (ESPR) under the EU's Circular Economy Action Plan (2020) goes beyond traditional efficiency standards. It also sets requirements for durability, reparability, and recyclability. Its goal is to encourage a shift toward products that last longer and are easier to recycle. Moreover, it also aims to include environmental considerations during the design phase [57]. Another new policy tool is the deposit-return system (DRS), which has traditionally been used for beverage packaging but is now being considered for small electronics and batteries. In this system, consumers pay a small deposit when they buy a device and the return of devices at the end of their use can be incentivized through a deposit-refund system, allowing them to be harvested for parts or refurbished [55]. This encourages people to return devices, reduces illegal disposal, and improves material recovery and traceability. DRS models can improve compliance monitoring if combined with digital tracking technologies such as a blockchain-based traceability system. Moreover, they can ensure full transparency in waste flows. A blockchain-based system for tracking mobile phones through their end-of-life is described in a study, which shows that it enables transparent reverse logistics [58].

When the system of circular economy begins at the very start of a product's lifecycle, it results in smart product designs and production processes. As a result, it helps to save resources by offering tools and incentives, reduces inefficient waste management, and creates new business opportunities along with overseas trade in the EU internal market [55]. So, these incentive mechanisms work together to support a closed-loop production system by linking producer incentives to sustainability goals, where design, production, consumption, and recycling are all interconnected. Altogether, governments can promote the shift to a circular economy, and this can create green jobs and encourage a fairer distribution of value across global supply chains. In essence, effective EPR implementation supported by innovative policy tools connects environmental responsibility with industrial competitiveness. It also helps advance several Sustainable Development Goals, including SDG 8 (Decent Work and Economic Growth), SDG 12 (Responsible Consumption and Production), and SDG 13 (Climate Action).

7.3 Digital tracking and transparency

Digitalization offers new ways to improve compliance and accountability in e-waste management. New technologies like blockchain, Internet of Things tagging, and digital product passports (DPPs) make it possible to track electronic products in real time. This type of tracking covers the entire lifecycle, from production to end-of-life. The European Commission's Circular Economy Action Plan (2020) highlights DPPs as an important tool for traceability and transparency. The DPP is more than just basic product data with a QR code, where manufacturers register their products and input relevant details for authorities and consumers. The DPP collects detailed information from every stage of the value chain. Physical data carriers such as smart tags are attached to the product to provide access to dynamic data throughout its life stored in the cloud or database. These tags have access to real-time data about how the product has been used. All actors, such as manufacturers, suppliers, retailers, and recyclers, share and receive information. Such information helps decide what to do when the product reaches its end of life. Moreover, it can show whether the product should be reused, repaired, remanufactured, or recycled [59]. This exchange of data helps ensure that resources are used efficiently. It also supports more sustainable and transparent production processes. To put it simply, the DPP is a set of data that provides key information about a product, including its details on materials, components, and chemical substances. It also shows how a product can be repaired, the availability of spare parts, and how it should be disposed of safely. This information is collected throughout the product's entire life cycle. So, overall, it helps improve product design, manufacturing, use, repairability, recycling, reuse, disposal, and environmental impact [60]. This can be seen in pilot projects in Japan and South Korea, which have shown that digital tracking systems can reduce illegal dumping and improve material recovery efficiency. As an example, the Korea Environment Corporation manages an online waste information and management system called Allbaro to monitor hazardous waste. It uses radio-frequency identification (RFID) technology to track waste shipments. The system also allows businesses to prepare official transfer documents online. Currently, Allbaro covers the majority of waste produced by companies. The OECD (2019) report observes that Allbaro's data should be integrated with material-flow accounts to enhance monitoring of circular economy policies [24]. In contrast, a project named MATICS (Material Tracing Integrated Circuit System) in Japan used IC tags and RFID to track medical and industrial waste. The consortium launched a verification program to test the system where IC tags were attached to waste bales and containers. These tags helped trace waste movements in the Tokyo and Ibaraki Prefecture areas [48]. These initiatives illustrate how the integration of digital technologies can improve waste traceability, strengthen regulatory oversight, and enhance data transparency. These systems also help countries meet their international reporting duties under the Basel Convention. Moreover, they support compliance monitoring for cross-border e-waste shipments. Using digital traceability tools in national systems makes data more reliable. It also helps build consumer trust and corporate accountability, which are essential for ethical e-waste management.

8 Policy recommendations for sustainable e-waste management

Significant environmental, financial, and moral issues have been brought about by the growing amount of electronic garbage and the varied capacity for recycling around the world. Significant policy gaps still exist, especially in the areas of enforcement, inclusiveness, and international cooperation, despite the fact that many countries have implemented legal frameworks to lessen the effects of e-waste. A focused strategy for creating an equitable, secure, and sustainable global e-waste system is outlined in the following policy suggestions.

- Strengthening global treaties and international governance:
  Global e-waste regulation is still disjointed, with important agreements like the Basel Convention finding it difficult to keep up with the continuously evolving waste flows and emerging types of electronic goods. Closing loopholes that allow unlawful hazardous waste transfers requires strengthening these international accords. This entails amending treaty language to match contemporary electronics and recycling technologies, standardizing definitions of e-waste, enhancing monitoring systems, and increasing compliance reporting. Stronger enforcement would stop the export of hazardous waste, improve international accountability, and advance a more unified and well-coordinated system of global governance.

- Supporting developing countries through technology transfer and capacity-building:
  Despite frequently missing contemporary recycling facilities or sufficient financing for environmental control, developing nations suffer a disproportionate share of the global e-waste load. Richer countries and international organizations must offer ongoing support through financial aid, professional training programs, and technology transfer in order to rectify this imbalance. Developing nations can properly handle both domestic and imported e-waste by investing in ecologically friendly recycling facilities, digital tracking systems, and regulatory capability. In addition to lessening harm to the environment and human health, this kind of assistance also improves resource recovery, encourages green industries, and fosters long-term economic resilience.
- Formalizing the informal sector through training and certification:
  A large portion of the world's e-waste collection and recycling industry is dominated by unregulated, unprotected workers. Effective policy should concentrate on integrating these workers into formal systems rather than relocating them. Formalization requires training programs, skill certification, health and safety safeguards, and collaborations with PROs. Increasing collection volumes, improving worker well-being, raising recycling standards, and ensuring a more equitable distribution of economic gains along the e-waste value chain are all benefits of integrating informal recyclers into regulated channels.
- Including ethical clauses in international trade policies:
  Despite the well-established environmental inequities associated with e-waste flows, ethical considerations are still generally absent from global electronics commerce and waste legislation. Including ethical provisions in trade agreements can strengthen exporting nations' accountability and assist in stopping exploitative dumping activities. Prohibitions on the dumping of hazardous waste, human rights protections for recycling workers, transparent equipment quality monitoring, and legally obligatory accountability mechanisms are some examples of such clauses. Global e-waste flows are prevented from sustaining inequity, environmental damage, or abuses of fundamental rights by incorporating ethics into trade frameworks.

A comprehensive path toward a more sustainable and equitable global e-waste system is provided by the four recommendations put forth: formalizing the informal sector, strengthening international treaties, assisting developing nations through technological cooperation, and incorporating ethical standards into trade policies. Together, they put human dignity, economic justice, and environmental preservation first. The international community may get closer to developing a circular, morally sound model for handling the rising volume of electronic trash by pursuing these initiatives.

9 Conclusion

A significant waste management issue has been brought about by the quick expansion of EEE. In order to address the effects of e-waste on the environment, society, and economy, policies and ethics are essential. Global frameworks that offer direction for managing hazardous waste include the Basel, Stockholm, and Bamako Conventions. However, when national capacity is low, enforcement gaps and uneven implementation highlight the limitations of international agreements. The EU WEEE Directive, China's Circular Economy initiatives, India's E-waste Rules, and new African policies are examples of national regulations and EPR programs that demonstrate the potential of legal frameworks as well as the difficulties of informal recycling, insufficient data, and resource limitations. Additionally, ethical issues are quite important. It is impossible to overlook issues like child labor, worker safety, environmental justice, and equitable access to secondary raw materials. Human rights and the Sustainable Development Goals, which emphasize the moral necessity of taking action and safeguarding vulnerable groups, are related to these challenges. Case studies from Europe, Asia, and Africa demonstrate variations in the use of the circular economy, regulatory efficacy, and socioeconomic effects. Stronger compliance, improved resource recovery, and safer working conditions are all benefits of formalized, regulated systems. Communities are nevertheless at danger for environmental damage and health problems due to unregulated or informal systems. Integrated policy frameworks are crucial for the future. Results can be enhanced by multistakeholder cooperation, producer incentives, digital compliance tracking, and formalizing the unorganized sector. The gap between law and practice can be closed by bolstering international treaties, encouraging technology transfer to poor nations, and incorporating ethical provisions in trade and environmental regulations. To summarize, international collaboration led by ethics and policy is necessary for sustainable e-waste management. The social and ecological costs of e-waste will keep rising in the absence of concerted measures that put justice, openness, and environmental protection first. E-waste can be transformed into a chance for social justice, economic growth, and sustainability by bolstering enforcement, encouraging equity, and endorsing circular economy concepts.

References

[1] Baldé CP, Forti V, Gray V, Kuehr R, Stegmann P, The global e-waste monitor 2017: Quantities, flows and resources. United Nations University (UNU), International Telecommunication Union (ITU) & International Solid Waste Association (ISWA), Bonn/Geneva/Vienna, 2017, 1–109.

[2] Forti V, Balde CP, Kuehr R, Bel G, The Global E-waste Monitor 2020: Quantities, flows and the circular economy potential. United Nations University (UNU), International Telecommunication Union (ITU) & International Solid Waste Association (ISWA), Bonn/Geneva/Rotterdam, 2020, 120.

[3] Islam MK, Khatun MS, Mourshed M, An in-depth analysis and review of management strategies for E-waste in the south Asian region: A way forward towards waste to energy conversion and sustainability. Heliyon, 2024, 10.
[4] Masud MH, Akram W, Ahmed A, Ananno AA, Mourshed M, Hasan M, Joardder MUH, Towards the effective E-waste management in Bangladesh: A review. Environ Sci Pollut Res, 2019, 26, 1250–1276.
[5] Breivik K, Armitage JM, Wania F, Jones KC, Tracking the global generation and exports of e-waste. Do existing estimates add up?. Environ Sci Technol, 2014, 48, 8735–8743.
[6] Song Q, Li J, Zeng X, Minimizing the increasing solid waste through zero waste strategy. J Clean Prod 2015, 104, 199–210.
[7] Mourshed M, Kamal M, Masuk NI, Chowdhury SA, Masud MH, Anaerobic digestion process of biomass, In Rahimpour MR, Ed, Encyclopedia of Renewable Energy, Sustainability and the Environment. First Edition, Elsevier, Oxford, 2024, 771–782.
[8] Khatun MS, Islam MK, Sultana A, Mourshed M, Application of plastics for production of fuel, In: Environmental Hazards of Plastic Wastes. Elsevier, 2025, 79–96.
[9] Grant K, Goldizen FC, Sly PD, Brune M-N, Neira M, Van den Berg M, Norman RE, Health consequences of exposure to e-waste: A systematic review. Lancet Glob Health, 2013, 1, e350–e361.
[10] Masud MH, Mourshed M, Hossain MS, Ahmed NU, Dabnichki P, Generation of waste: Problem to possible solution in developing and underdeveloped nations, In: Waste management and resource recycling in the developing world. Elsevier, 2023, 21–59.
[11] Masud MH, Mourshed M, Mahjabeen M, Ananno AA, Dabnichki P, Global electronic waste management: Current status and way forward Paradigm Shift in E-waste Management. 2022, 9–47.
[12] Ongondo FO, Williams ID, Cherrett TJ, How are WEEE doing? A global review of the management of electrical and electronic wastes. Waste Manag, 2011, 31, 714–730.
[13] Kibria MG, Masuk NI, Safayet R, Nguyen HQ, Mourshed M, Plastic waste: Challenges and opportunities to mitigate pollution and effective management. Int J Environ Res, 2023, 17(20), 1–37.
[14] Mourshed M, Khatun S, Islam K, Masuk NI, Masud MH, Generation, composition, collection, and treatment of E-Waste Management of Electronic Waste: Resource Recovery. Technol Regul, 2024, 39–72.
[15] Mourshed M, Masud MH, Rashid F, Joardder MUH, Towards the effective plastic waste management in Bangladesh: A review. Environ Sci Pollut Res, 2017, 24, 27021–27046.
[16] Widmer R, Oswald-Krapf H, Sinha-Khetriwal D, Schnellmann M, Böni H, Global perspectives on e-waste. Environ Impact Assess Rev, 2005, 25, 436–458.
[17] Peiry KK, Ed, The Basel Convention on the control of transboundary movements of hazardous wastes and their disposal: The Basel Convention at a glance, In Proceedings of the ASIL Annual Meeting. Cambridge University Press, 2013.
[18] Convention B, Amendment to the Basel Convention on the Control of Transboundary Movements of Hazardous Wastes and their Disposal. Basel Convention, Geneva, 2019.
[19] Silva Filho CR, Velis CA, United Nations' plastic pollution treaty pathway puts waste and resources management sector at the centre of massive change. SAGE Publications Sage UK, London, England, 2022, 487–489.
[20] United Nations Treaty Collection [Internet]. 2022 [cited August 12, 2025]. Available from: https://treaties.un.org/.
[21] Pollutants PO, Stockholm Convention on persistent organic pollutants. Geneva, 2011.
[22] Union A Bamako Convention on the Ban of the Import to Africa and the Control of Transboundary Movement and Management of Hazardous Wastes within Africa 1991.
[23] Webster-Main A, Keeping Africa out of the global backyard: A comparative study of the Basel and Bamako conventions Environs: Envtl L &. Pol'y J, 2002, 26, 65.
[24] OECD, Waste management and the circular economy in selected OECD countries: Evidence from environmental performance reviews. OECD; 2019.

[25] EUR-Lex. Access to European Union Law [Internet]. European Union, 2001.
[26] Union E Waste from Electrical and Electronic Equipment (WEEE). 2012.
[27] Waste statistics – electrical and electronic equipment [Internet]. 2025. Available from: https://ec.europa.eu/eurostat/statistics-explained/index.php?title=Waste_statistics_-_electrical_and_electronic_equipment.
[28] Taylor B Europe's WEEE regulations might need a rethink. Recycling today. 2025.
[29] CalRecycle, Electronic Waste Recycling Act of 2003. State of California, USA, 2003.
[30] CalRecycle, SB 1215 Covered Battery-Embedded Products. State of California, USA, 2003.
[31] E-Cycle WA information for manufacturers [Internet]. Department of Ecology, WA. 2025 [cited August 12, 2025]. Available from: https://ecology.wa.gov/regulations-permits/guidance-technical-assistance/electronics-ecycle-guidance-and-reports/manufacturers.
[32] Quality ODoE, Modernizing Oregon E-Cycles: What Will House Bill 3220 (2023) Change for Manufacturers?. Oregon Department of Environmental Quality, USA, 2025.
[33] Department of Energy & Environmental Protection C. Electronics Manufacturer Requirements for Compliance with Connecticut's Electronics Recycling Law. Department of Energy & Environmental Protection, USA, Connecticut, 2025.
[34] Ministry of Economy TaI. Act on Recycling of Specified Home Appliances. Ministry of Economy, Trade and Industry, Japan, Japan, 2001.
[35] Japan for Sustainability. Revised Law Requires Recycling of Additional Home Appliances. Japan 2009.
[36] Ministry of Economy TaI, Current State of Enforcement of the Act on Recycling of Specified Kinds of Home Appliances (Collection Results) and Results of Recycling Efforts by Home Appliance Manufacturers. FY2021 Results, Ministry of Economy, Trade and Industry, Japan, Japan, 2022.
[37] Translation JL., Act on Promotion of Recycling of Small Waste Electrical and Electronic Equipment Japanese Law Translation. Tokyo, Japan, 2013, 1–8.
[38] Terazano A, Recycling Small Home Appliances and Metals in Japan. NIESCMW Online Magazine, 2013.
[39] Circular Economy Promotion Law [Internet]. Ministry of Ecology and Environment the People's Republic of China. 2008 [cited August 12, 2025]. Available from: https://english.mee.gov.cn/Resources/laws/envir_elatedlaws/201712/t20171212_427823.shtml?utm_source=chatgpt.com.
[40] IEA, Plan for the Implementation of the Extended Producer Responsibility System. People's Republic of China; 2016.
[41] China SCotPsRo, The Regulations for the Administration of the Recovery and Disposal of Waste Electric and Electronic Products. People's Republic of China. State Council of the People's Republic of China, 2009.
[42] Stern N, Xie C, China's new growth story: Linking the 14th Five-Year Plan with the 2060 carbon neutrality pledge. Chin Econ Bus Stud, 2023, 21, 5–25.
[43] Bhaskar K, Turaga RMR, India's e-waste rules and their impact on e-waste management practices: A case study. J Indus Ecol, 2018, 22, 930–942.
[44] Board CPC, Annual Report. India: Ministry of Environment, Forest and Climate Change. Govt. of India, 2022.
[45] Twagirayezu G, Uwimana A, Kui H, Birame CS, Irumva O, Nizeyimana JC, Cheng H, Towards a sustainable and green approach of electrical and electronic waste management in Rwanda: A critical review. Environ Sci Pollut Res, 2023, 30, 77959–77980.
[46] Bimpong FAK, Asibey MO, Inkoom DKB. Ghana's recently introduced e-waste regulatory policy: A hope for a better e-waste sector?. Waste Manag Res, 2024, 42, 1031–1041.
[47] Nwankwoala HO, Okujagu D, A review of wetlands and coastal resources of the Niger Delta: Potentials, challenges and prospects. Environ Ecosyst Sci, 2014, 5, 37–46.
[48] Japan for Sustainability, Consortium develops integrated circuit tag waste tracking system. Japan, 2009.

[49] DPI UN. The Sustainable Development Goals. Stylus Publishing, LLC, 2017.
[50] Organization WH, Children and digital dumpsites: E-waste exposure and child health. World Health Organization, 2021.
[51] Lebbie TS, Moyebi OD, Asante KA, Fobil J, Brune-Drisse MN, Suk WA, et al, E-waste in Africa: A serious threat to the health of children. Int J Environ Res Public Health, 2021, 18, 8488.
[52] Islam MR, Ruponti SA, Rakib MA, Nguyen HQ, Mourshed M, Current scenario and challenges of plastic pollution in Bangladesh: A focus on farmlands and terrestrial ecosystems. Front Environ Sci Eng, 2023, 17, 66.
[53] Owen C, Scougall B, Environmental Injustice: How Informal E-waste Recycling Impacts Human Rights. Norton Rose Fulbright, Australia, 2025.
[54] The Global E-waste Monitor 2024–Electronic Waste Rising Five Times Faster than Documented E-waste Recycling: UN The global E-waste Monitor 2024-Electronic Waste Rising Five Times Faster than Documented E-waste Recycling: UN, 2024.
[55] Katarina Svatikova JV, Williams R, Smit T, Moerenhout J, The effect of the circular economy on the labour market in Europe. DG Environment (ENV) (European Commission), Belgium, 2018.
[56] OECD, Global outlook on financing for sustainable development 2019: time to face the challenge. OECD. 2019.
[57] EU, Ecodesign for Sustainable Products Regulation. European Commission, Belgium, 2024.
[58] Dasaklis TK, Casino F, Patsakis C, Eds, A traceability and auditing framework for electronic equipment reverse logistics based on blockchain: The case of mobile phones, In: 2020 11th International Conference on Information, Intelligence, Systems and Applications (IISA). 2020, 15–17 July 2020.
[59] Hakola L, Abedi F, Nordman S, Smolander M, Paltakari J, Smart tags as enablers for digital product passports in circular electronics value chains. Circ Econ Sustain, 2025, 5, 2033–2055.
[60] Adisorn T, Tholen L, Götz T, Towards a digital product passport fit for contributing to a circular economy. Energies, 2021, 14, 2289.

Riti Thapar Kapoor*, Mohd Rafatullah, Aseem Vashisht, Surendra Gotherwal

Strategies for electronic waste management: advanced treatment and recovery technologies toward environmental restoration

Abstract: Electronic waste (e-waste) is the fastest-growing waste due to the huge demand for various electronic appliances, which are disposed of by their users because of their short lifespan. The improper disposal of e-waste is life-threatening because it includes several hazardous contaminants and toxic metals that may leach into groundwater, soil, and disperse into the air and show adverse impact on the environment and health of human beings. The present book chapter provides an in-depth understanding of different aspects of e-wastes, their constituents, and the negative effects on aquatic and terrestrial ecosystems due to their informal disposal. It also focuses on the sustainable approaches toward its effective management. The chapter advocates for awareness among different stakeholders, transformation of the informal sector, extended producer responsibility, accountability for e-waste recycling procedures to achieve sustainable development goals and protect the environment for future generations.

Keywords: Electronic waste, recovery technologies, recycling, sustainability, waste management

---

Acknowledgments: The authors extend their appreciation to Amity University Uttar Pradesh, Noida, Uttar Pradesh, India, Department of Mathematics and Natural Sciences, Prince Mohammad Bin Fahd University, Al-Khobar, Saudi Arabia. University Centre for Research and Development, Chandigarh University, Mohali, Punjab, India, and the Ministry of Electronics and Information Technology, Electronics Niketan, Lodhi Road, New Delhi, for providing all necessary facilities and valuable support.

***Corresponding author: Riti Thapar Kapoor,** Centre for Plant and Environmental Biotechnology, Amity Institute of Biotechnology, Amity University Uttar Pradesh, Noida 201 313, Uttar Pradesh, India. e-mail: rkapoor@amity.edu
Mohd Rafatullah, Department of Mathematics and Natural Sciences, Prince Mohammad Bin Fahd University, P.O. Box 1664, Al-Khobar 31952, Saudi Arabia
Aseem Vashisht, University Centre for Research and Development, Chandigarh University, Mohali, 140413 Punjab, India
Surendra Gotherwal, Ministry of Electronics and Information Technology, Electronics Niketan, Lodhi Road, New Delhi 110 003

 | https://doi.org/10.1515/9783112248072-009

1 Introduction

The increase in population, urbanization, industrialization, advancements in science and technology, expansion of the digital infrastructure, and global demand have led to the production of a wide range of electronic equipment. Development of electronic equipment or devices is a blooming sector that promotes the global economy. However, every electronic equipment becomes electronic waste at the end. Electronic waste, often referred to as "e-waste," denotes electronic devices or equipment that have been discarded and are no longer needed or functional, or any electronic or electrical device that has reached its end of life, whether it is in good working order or not [1]. According to the European Union, e-waste is the waste from electrical and electronic equipment (WEEE) that includes all the components of electronic equipment, its subassemblies, which are part of the product at the time it is discarded. It includes damaged, discarded, outdated, and malfunctioning electrical equipment. According to a recent study, only 9% of the global economy is circular (reused or recycled into products), whereas the rest 91% follows a linear model in which they use and throw [2]. As per the reports of the United Nations, each person generated approximately 7.6 kg per capita of E-waste in the year 2021, thus resulting in around 57.4 MMT across the globe [3]. As per the Global E-Waste Monitor, 62 million tons of e-waste were produced in 2022, and the generation of e-waste will rise to 82 million tons in 2030 [4]. E-waste is a rapidly expanding category of solid waste generating significant environmental concern globally [5]. E-wastes can be generated from different sources, such as domestic, business circles, or industries. The advent of novel technologies in the market replaces old electronic devices frequently, thus increasing e-waste. Due to the continuous upgrading of the existing electronic equipment, manufacturing units replace the old electronic gadgets with new devices with better features but this increases e-waste in the environment. The composition of e-waste includes metals (60%), plastics (15%), screens (12%), metal-plastic mix (5%), pollutants (3%), circuit boards (2%), cables (2%), and other components (1%) [6]. Indeed, electronic appliances make human life more comfortable, but their high rate of production, short life span, mishandling, and informal management practices generate problems for the environment and entire society. The e-waste generation by different countries is given in Figure 1.

2 Sources of electronic wastes

The e-waste generation has increased due to the growing demand of different stakeholders in society. The e-waste can be classified based on the type of equipment, its constituents, and the materials used. The e-waste domain comprises a variety of electronic and electrical equipments covering domestic appliances (washing machines, re-

Figure 1: Electronic-wastes generation by different countries (reprinted from Sandwal et al. [5]).

frigerators, televisions, etc.), information and communication technology devices (radio, mobile/cell phones, computer monitors, network system hardware, and small satellite system), lighting equipment (tungsten bulbs and compact fluorescent lamp), monitoring devices (CCTV cameras), and automated dispensers, and vending machines, etc. [7]. These devices function by using batteries, cable wires, printed circuit boards, cathode ray tubes, semiconductors, capacitors, etc. The production of electric and electronic devices is predicted to increase in the future because of the shift toward electric vehicles to reduce environmental pollution (Figure 2). Furthermore, developing nations target to promote the use of energy-efficient lights such as LED or CFLs to meet energy requirements, thus again reinforcing the enhancement in the production of e-wastes [8].

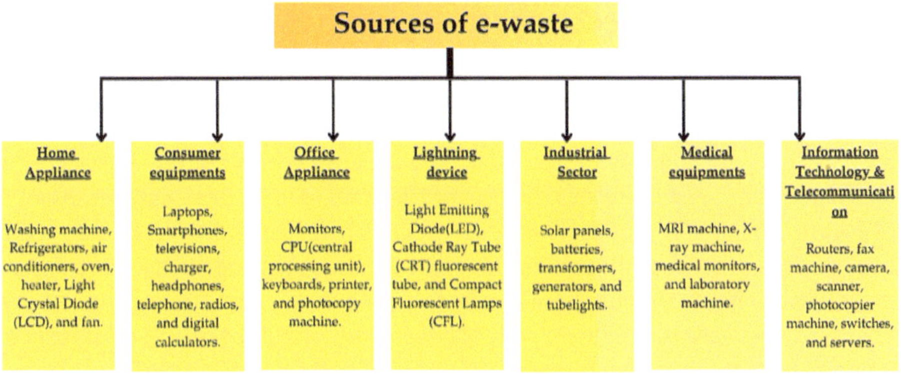

Figure 2: Various sources of electronic waste (reprinted from Sandwal et al. [5]).

3 Composition of electronic wastes

Electronic waste is a heterogeneous mix of materials, including metals, plastics, glass, and hazardous materials [9]. The composition of e-waste depends on the type of electronic devices, manufacturing year, and the materials used for production. Li and Achal [10] reported that a used and discarded computer is made up of plastics (23%), metals (44%), glass (15%), and electronic components (17%). The composition of e-waste can be divided into the following categories:

A major part of e-waste is composed of both valuable and toxic metals. Circuit boards and connectors use expensive metals like gold, silver, platinum, and palladium. Other metals such as copper, tin, and aluminum are used for casings, wiring, and soldering. However, CRT monitors, batteries, and circuit boards are composed of toxic heavy metals like lead, cadmium, mercury, etc. Plastics or polymers are used in electronic devices as casings, insulation, and circuit board laminates. Cables are made from polyvinyl chloride, whereas in keyboards and casings, acrylonitrile butadiene styrene is found. Polypropylene is applied in structural components. Glass is found in display screens, silicon is the main component of integrated circuits, semiconductor chips, and microprocessors. The capacitors, resistors, and insulating parts contain ceramic materials.

Different components present in the e-wastes contain toxic substances, and without proper treatment, they may pose environment and health risks. The brominated flame retardants are found in plastic casings and circuit boards; however, after burning, they may release toxic compounds. The capacitors and transformers contain polychlorinated biphenyls, whereas in refrigerators and air conditioners, chlorofluorocarbons are used, which can deplete the ozone layer. The lithium-ion batteries are found in laptops, smartphones, and power banks. In circuit boards, capacitors and resistors are found, which contain rare earth metals.

4 Hazardous effects of informal e-waste recycling

The huge amount of e-wastes has been generated due to growing demand of electronic devices. The informal practices employed in many developing and underdeveloped nations pose both environmental and public health concerns. If e-wastes are recycled properly, precious metals can be extracted and utilized. The improper disposal of e-wastes in landfills produces harmful gases and toxic metals in the soil, which can enter surface or groundwater via leaching and diffusion processes and reach different areas. The informal recycling techniques, such as burning plastic casings to remove precious metals, melting in open pots, and dissolving circuit boards in acids, expose informal workers to toxic substances [11]. Burning of circuit boards and wires generates fumes, which are inhaled by informal workers during occupational exposure, who are employed as recyclers without using any protective equipment. It has been reported that in developing and underdeveloped nations, women and teenage children form around 40% of the entire labor force involved in the gathering, disassembling, and informal e-waste recycling activities.

When these hazardous substances are ingested into the human body, they are stored in the fatty tissues and can cause health problems for residents who live near informal e-waste communities [11]. The informal e-waste recycling areas trigger pollution in nearby communities, leading to adverse consequences on the health of the overall population. Zhang et al. [12] found that dumpsites where e-wastes are informally processed have shown high concentrations of particulate matter and heavy metals in the atmosphere; these concentrations threatened the environment and health of the people residing near those sites. All electronic devices have a carbon footprint, which may lead to global warming. Recent studies have shown that one laptop can potentially emit about 10 tons of CO_2 and show an adverse impact on the environment [10].

5 Impact of e-waste on the environment

The mishandling and indiscrete disposal of e-wastes into landfills and nonregulated areas contaminate the entire ecosystem and pose health risks to living organisms. Recycling of e-wastes is executed by informal workers in residential complexes, inadequate infrastructure, and in unregulated commercial spaces. In such cases, e-waste is processed by hazardous processes, including open burning, manual diassembling, smelting, pulverizing, acid dissolving, salt leaching, and mercury amalgamation, which release hazardous pollutants into the ecosystem and cause degradation in the qualities of air, water, and soil, and pollutants bioaugmentation [13].

Liu et al. [11] reported elevated levels of heavy metals and halogenated compounds in informal e-waste-recycling areas, which may lead to direct or indirect exposure to air, water, and soil around the e-wastes recycling regions. Electronic-wastes are nondegradable, and when not disposed of properly in landfills, they release toxic chemicals and gases, ultimately degrading air quality and compromising soil and water quality, which adversely affect public health [14]. Harmful gases like dioxins, furan, and particulate matter are released due to the dismantling of e-wastes, which enter the food chain through the aquatic and terrestrial ecosystems.

Most of the pollutants released from e-wastes are persistent pollutants, nonbiodegradable, and have hazardous impacts when exposed to biological systems [15]. Due to prolonged transport of leached contaminants through soil, the soil strata in the proximity of the e-waste processing site are permeable; groundwater contamination is most likely to occur due to the infiltration of persistent contaminants [16]. The 95% of the total e-waste recycling is executed by informal sectors in developing nations, making it susceptible to the improper disposal of e-waste components in the environment. The precious metals are recovered by local recyclers without following proper guidelines, thus open burning releases particulate matter and other volatile organic compounds into the atmosphere, which affects the environment and population. Due to easy separation, burning of cable wires is recommended for recovery of copper and other elements; however, controlled operations are needed to check the dissemination of undesired contaminants.

Shen et al. [17] reported 36 types of polychlorinated biphenyls, and 16 types of polyaromatic hydrocarbons from soils near an informal e-waste-recycling site in Guiyu, China, and high concentrations of polybrominated diphenyl ethers and polybrominated biphenyls were found in e-waste dumping sites. The surface of the soil of a typical informal e-waste-recycling site in Brazil revealed contamination with copper, zinc, and lead [18]. The rice samples taken from Zhejiang province, an e-waste-processing town in Eastern China, reported 2–4 times higher concentrations of lead and cadmium in rice as compared to the permissible concentration of these elements [19].

6 Negative effects on air quality

The informal disposal practices used for e-waste management, such as shredding, dismantling, release toxic chemicals, dust particles, dioxins, and fumes, which lead to environmental pollution and negatively affect animals, birds, human beings, and other living organisms (Figure 3). Precious metals can be recovered from e-wastes with the use of acids, chemical agents, and desoldering tools; however, this process releases fumes and dust if recycling is not conducted properly and affects the health of the ecosystem. The incineration of e-wastes releases chlorinated and brominated gases, furan, dioxins, CO_2, and N_2O, which lead to global warming and deplete the ozone layer. Informal recycling practices can also release high concentrations of lead into the environment, which are absorbed by plants, animals, and human beings through inhalation and ingestion and show adverse impacts [20].

7 Negative effects on soil

Different living organisms reside in the soil as they assist in food production and maintain microbial communities. Due to improper handling, illegal dumping, and disposal in landfills, e-wastes generate toxins and flame-retardant agents mostly present in the plastics. Toxic substances pollute the soil and leach via soil particles, affecting the growth of plants cultivated in nearby areas. The heavy metals can be discharged into water bodies through soil erosion and seepage of the soil [22]. Crops planted in

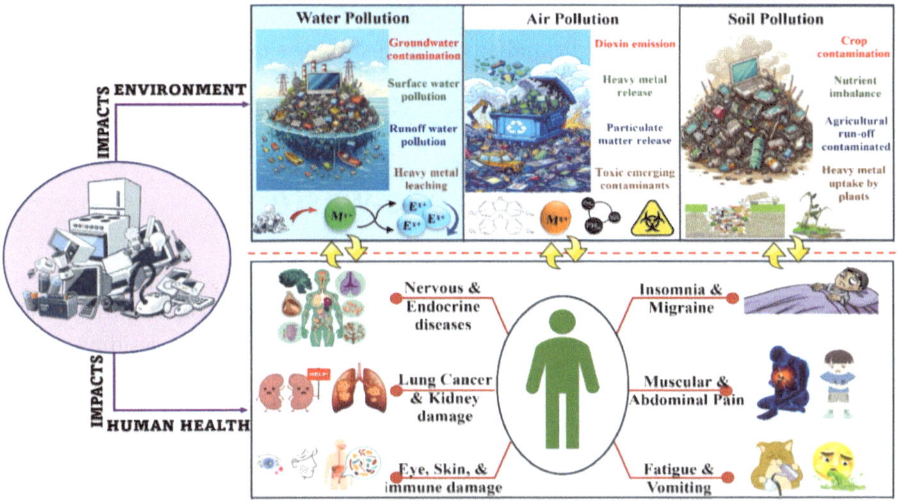

Figure 3: Effect of e-waste dumping on the environment and health of human beings (reprinted from Srivastava et al. [21]).

contaminated soil become vulnerable to the absorption of toxic metals such as arsenic and cadmium, and reflect a reduction in agricultural productivity by changing the soil composition and pH level [23]. Polybrominated diphenyl ether and polychlorinated biphenyls can remain in soil for a prolonged time and negatively affect microbes, adversely affect earthworm populations, cause loss of biodiversity and vegetation on the surface, and enter human beings through the food chain. If animals consume contaminated plants containing copper and lead, it can lead to stunted growth and health disorders. Some plant species, such as *Allium cepa*, *Zea mays*, and *Oryza sativa*, are adversely affected by heavy metals released from e-wastes in the soil [24]. Burning of e-wastes on the soil surface negatively affects soil microbes and physico-chemical properties of soil, ultimately hampering crop yield and growth [25].

8 Negative impact on water quality

The hazardous toxic metals such as arsenic, lead, mercury, cadmium, and lithium are released from e-wastes, and seep into the soil and pollute the groundwater [10]. Mercury and lead are released from cathode ray tube decomposition and enter the groundwater. The toxic substances travel miles from recycling sites through the soil and are released into water bodies, like rivers, reservoirs, ponds, wetlands, and lakes, making the water unfit for drinking and other uses, thus creating a challenge to find clean water [26]. Due to acidification and accumulation of toxic contaminants, freshwater and marine organisms die and adversely disrupt the biodiversity of the aquatic ecosystem.

9 Adverse health impacts of e-waste

The e-waste management is not strictly monitored or audited by competent authorities, specifically in developing nations, showing negative impacts on air, soil, and water bodies, and ultimately the health of human beings. Population residing near e-waste treatment areas has been reported to suffer from various disorders, such as asthma, neurological and cardiovascular disorders, adverse effects on the respiratory and immune system, and cancer. Many e-waste workers have reported various health issues ranging from dizziness, weakness, stress, headache, shortness of breath, and chest pain [27]. Adverse effects on the liver, male reproductive organs, and sperm quality, and high blood glucose levels have been reported due to exposure to informal e-waste recycling (Table 1).

Amoabeng et al. [28] reported a link between exposure to informal e-waste-recycling practices and adverse birth outcomes, such as miscarriage, premature birth, short gestational periods, lower birth weight, altered neurodevelopment, and adverse learning and behavioral outcomes with respect to children. Prolonged exposure to hazardous chemicals released through e-wastes has shown adverse effects in different organs of workers, such as diseases in the endocrine system, neurological system, reproductive anatomy, and bones [29]. Children and women are actively involved in collecting, dismantling, as well as burning of e-wastes. Toxic metals, such as mercury and lead, are released from these practices and can cross the placenta; contaminants from breast milk disrupt neuron development in the central nervous system in the fetal stage and in infants [30].

Long-term exposure to Pb from e-waste disposal sites can lead to brain tumors and abnormal brain growth in young children [31]. Chemicals leaching out from e-waste can affect unborn children if pregnant women are exposed to contamination, leading to lymphocyte formation in the umbilical cord and epigenome [32]. Xu et al. [33] reported that heavy metals, if present in high concentrations, can cause irregular methylation of 79 genes in DNA, which are required for various biological processes, such as ion binding, cell adhesion, embryonic morphogenesis, and signaling pathways. Most of the chemical compounds present in e-wastes are carcinogenic; thus, many toxic substances have been found in the blood of informal workers [19].

Table 1: Impact on health due to the contaminants present in e-wastes.

Contaminants	Components of e-wastes as source	Impact on health
Arsenic (As)	Semiconductors and integrated circuits	Skin-related problems and cancer
Cadmium (Cd)	Batteries, solar cells, stabilizers, power bank, and laptop	Defects in respiratory and skeletal systems, pain, fever, lung diseases, and kidney and reproductive organs damage
Barium (Ba)	CRT, CFL, optical devices, and transducers	Heart diseases, digestive system disorders, and paralysis
Copper (Cu)	Cables, wires, and printed circuit boards	Stomach discomfort, nausea, and liver damage
Nickel (Ni)	Batteries and CRT	Skin diseases, lung infection, behavioral abnormalities, and tumor
Antimony (Sb)	Plastics and fire retardants	Bronchitis and lung diseases
Lead (Pb)	Television, CRT computer monitor, lamp, circuit boards, and battery	Anemia, neurological disorders, and kidney and reproductive system damage
Lithium (Li)	Batteries, heart pacemaker, and toys	Fatigue, nausea, and irritation of eyes
Chromium (Cr)	Floppy CD, used as anticorrosive coating on metal bodies of electronic devices	Lung cancer, eye irritation, muscular contraction, cancer, and affects endocrine gland functions
Mercury (Hg)	LCD screens of televisions and monitors, printed circuit board, computer screen, and CFLs	Brain damage, neurological disorders, and anemia
Polycyclic aromatic hydrocarbons	Semiconductors and bio-imaging devices	Immune-toxicogenic, carcinogenic, teratogenic, kidney and liver damage, and jaundice
New flame retardants	Electronic devices and printed circuit boards	Neurological, endocrine and reproductive organ damage, and cancer
Per or polyfluoroalkyl substances	Semiconductors and water-resistant coatings in electronic devices	Cancer, liver and immune system damage, and low birth weight
Polybrominated diphenyl ethers	Plastic coatings and flame-retardant wires	Thyroid problems, cancer, and neurological and endocrine diseases
Plastics and polyvinyl chloride	Cable and computer bodies	Dioxins are produced during combustion, birth defects, and cancer
Dioxin	Smelting and soldering of electronic devices having polymer body	Fetal abnormalities, slow growth, and affects reproductive and immune systems

10 Management of e-wastes

Waste management strategies are essential for management of a huge amount of e-wastes. E-wastes contain both recyclable and reusable materials, providing environmental and economic advantages. One metric ton of mobile phones contains approximately 350 g of gold, 150 kg of copper, and 100 g of palladium [34]. Plastics contain 20–25% total e-waste, which can be segregated mechanically and reutilized; however, the presence of pollutants decreases recovery efficacy. The refurbished laptops and smartphones can be resold at a lower price, thus offer circular economy model and decreasing extraction of materials and saving energy. The recovery by shredding and hydrometallurgical extraction of components and application of reusables are essential for management of e-wastes. The combination of different strategies can enhance recovery of resources, economic and environmental benefits, and sustainability for the e-waste management system.

11 Traditional e-waste processing methods

The conventional methods aim to regain valuable resources and eliminate toxic components.

11.1 Collection and sorting

The first step in the processing of e-wastes is the collection of electronic devices from homes, companies, industries, and disposal facilities. The sorting can be performed manually by the employees disassembling gadgets to segregate reusable and waste items, like batteries and cathode-ray tubes. There is a need to maintain a central registry in order to collect data and regular monitoring activity [35].

11.2 Mechanical processing

It is the main step in e-waste recycling, which includes reduction in size and separation of materials. This procedure is used to shred or pulverize metals and minerals found in e-wastes. Based on the physical properties like size, weight, shape, density, electrical or magnetic properties, materials can be segregated into different output fractions. Gravity separation, eddy current separation, and magnetic separation are commonly used sorting techniques. After that, aluminum fractions are sent to aluminum smelters, copper and magnetic fractions are sent to a copper smelter or steel facility for additional processing. However, some fraction of plastics and nonrecyclable wastes are also produced.

11.3 Landfilling

The landfilling is a widely applied method for e-waste disposal, specifically in developing nations. The toxic compounds present in different constituents of e-wastes pollute soil and groundwater via leaching of heavy metals and hazardous elements and cause long-term adverse impacts to the environment. By improving landfill technology with plastic or clay barrier layers, leachates can be prevented from entering into the soil. It may be of three types: horizontal barriers that are useful in checking the downward movement of heavy metals, vertical barriers that can check the polluted water flow in deep aquifers. Liner barriers are used to inhibit movement of leachates, but they should be impermeable, flexible, with more tensile strength. However, land filling is not successful because of the leaking of landfills, chemical reactions, uncontrolled fires, etc., and is not suitable for volatile and persistent pollutants. Recently, landfill wastes have been valorized via anaerobic digestion for biogas production [36].

11.4 Metallurgical treatment

The metallurgical methods extract precious metals from e-wastes via different chemical and thermal procedures.

11.4.1 Pyrometallurgy

The pyrometallurgical process utilizes gas-phase operations at high temperature, sintering, melting, incineration, and smelting in a blast or plasma arc furnace. The refractory and metal oxides produce a slag phase when degraded leftovers are burned in a furnace or molten bath to eliminate polymers. This method can be used for any kind of e-waste, but its controlling methods are challenging. Due to smelting, some metals are lost in the slag as refining procedures are not selective. Dioxin production is a major problem related to pyrometallurgy if off-gas treatment or a filter system is not used.

11.4.2 Hydrometallurgy

The hydrometallurgical process involves leaching of solid materials with acid or caustic reactions. The ion exchange, solvent extraction, adsorption, and precipitation were used for isolation and to concentrate the precious metals. Then, solutions are subjected to electrorefining, chemical reduction, and crystallization for metal recovery. Despite the well-established methods such as cementation, solvent extraction, and precipitation, there are several hurdles to overcome for recovering metals from the leached e-waste solution [37].

12 Advanced treatment and recovery methods

Thermochemical and biochemical technologies are being used for processing and valorization of e-wastes.

12.1 Thermochemical treatment

Thermochemical methods like incineration, pyrolysis, and gasification are used for processing of e-wastes and recovery of valuable resources.

12.1.1 Incineration

Incineration is a controlled procedure in which e-wastes are combusted in incinerators at very high temperatures. This process reduces a huge volume of e-wastes, but releases hazardous gases such as furans and dioxins. After the separation of different components, copper foil is recovered from motherboards via the pit burning procedure. After the ash is rinsed away, the impure copper undergoes further recycling. Defective condensers and chips are also incinerated to obtain metals. Burning of printed circuit boards at 840–850 °C for 3 min under 6.2% oxygen decreases the material by up to 21% and produces byproducts, and after leaching, 99% of copper and 97% of silver can be recovered.

12.1.2 Pyrolysis

Thermochemical breakdown of organic material at high temperature in the absence of oxygen is known as pyrolysis. After pyrolysis, organic material produces carbon-rich solid residue along with liquid and gaseous products. Jadhao et al. [38] adopted an integrated approach for metal recovery from e-wastes through pyrolysis at 500 °C for 30 min, followed by ultrasonication for 30 min, and reported 90% efficiency for metal fraction recovery, along with 18% gas yield and 5% oil yield, with an energy potential of 34 MJ/kg. Mir and Dhawan [39] reported pyrolysis (300–700 °C) followed by leaching for metal recovery from integrated circuits installed in PCBs, showed 97% copper, 89% iron, 65% nickel, 57% lead, and 33% silver recovery.

12.1.3 Gasification

In the gasification process, partial oxidation of e-wastes produces char and syngas. Gasification and its slag can be used for the recovery of metals from e-wastes. The co-gasification of e-wastes and organic wastes is beneficial as it can provide biofuels and rare earth metals from e-wastes. Khaobang et al. [40] conducted e-waste pyrolysis and decoupling biomass gasification and reported metal recovery 84% and 90% from PCB char and PCB ash, respectively. The co-gasification of biomass and e-waste is an energy-efficient process for the generation of syngas and recovery of precious metals.

12.1.4 Advanced thermal plasma technology

Thermal plasma technology can be used to extract metals from wastes because of its advantages, such as high temperature assisting slag metals separation, increase in reaction kinetics with less reaction time. Samal and Blanco [41] reported the application of a thermal plasma arc system in treating e-wastes for metal recovery from PCBs. The plasma processing of printed circuit boards shows 76% metal recovery with the release of carbon monoxide. Szałatkiewicz [42] used a plasmatron plasma reactor equipped with three arc plasmatrons for processing 36 kg of waste PCBs at 1,750 °C and attained 76% recovery of precious metals; however, it requires excessive energy.

12.2 Biochemical Methods

The biological methods can break down the e-wastes and metal ores by using microorganisms. Metals from e-wastes or ores are transformed into soluble salts in liquid media through biological methods (Table 2). Bacteria such as *Acidithiobacillus ferrooxidans*, *Leptospiral ferroxidase*, and *Acidithiobacillus thiooxidans* remove metals from trash [43]. A bioleaching study on waste PCBs by using *Acidithiobacillus ferrooxidans* showed that bacterial culture not only increases the recovery of copper but also prevents the development of inhibitory byproducts such as libethenite, which reduces metal recovery [44]. The tenfold copper recovery was observed in a week via the bioleaching process. Mixed culture has shown high bioleaching capacity. Ilyas et al. [45] reported that *Sulfobacillus thermosulfidooxidans* consortia with *Thermoplasma acidophilum* and *Sulfobacillus acidophilus* were cultured at 45 ° C and used in e-wastes for metal recovery analysis through bioleaching. More than 75% recovery of copper, aluminum, nickel, and zinc was recorded with *Sulfobacillus thermosulfidooxidans* and *Thermoplasma acidophilum*.

12.2.1 Biometallurgical methods

Biometallurgy is a promising recycling method which uses microbes to recover metals. Microbes utilize metal species for structural or catalytic purposes, either by binding metal ions to the cell surface or transporting them intracellularly. These interactions take place via sorption, redox reactions, and sulfide precipitation. The commonly used microbes are bacteria (*Chromobacterium* violaceum and *Pseudomonas* species), *Acidithiobacillus* species, and various fungi like *Aspergillus niger.* Both hydrometallurgy and biometallurgy rely on leaching; the latter uses chemicals released by microbes as compared to the synthetic chemicals (Table 3). This process facilitates the formation of metal complexes or precipitates, which are then recovered from the culture broth for refining. Bioleaching and biosorption are the main types of biometallurgical methods.

Bioleaching serves as a biological method for recovering precious metals from sulfide ores. By facilitating the oxidation of sulfide into sulfate, bacteria release metals, making them accessible for extraction from the solution. Microbial leaching has two potential mechanisms; in the direct mechanism, oxidation of minerals occurs, whereas the indirect mechanism involves microbes which convert ferrous to ferric ions, then ferric ions act as oxidizing agents for minerals. However, bioleaching is a slow process and difficult to control its secondary reactions, and precipitation of lead, tin, and some other metals can also be obtained [46]. Biosorption is a passive mode of metal sorption mechanism that uses dead biomass of microbes, also known as biosorbents, and these can bind and accumulate metals [47]. The biosorption process can be either a physical or chemical process onto the cell wall. These are eco-friendly, feasible, no sludge generation, efficient process in detoxifying effluents and recovery of metals.

Table 2: Metal recovery from e-wastes by using various procedures.

Methods	Leaching agent used	Recovered metal	References
Hydrometallurgy	Thiourea	Au 98% and Ag 100%	[48]
Hydrometallurgy	Nitric acid	Cu 99%	[49]
Hydrometallurgy + biosorption	*Ammonium thiosulfate* and *Lactobacillus acidophilus*	Au 90%	[50]
Bioleaching	*Penicillium simplicissimum*	Cu 90% and Ni 89%	[51]
Bioleaching + direct current	*Acidithiobacillus*	Cu 97%	[52]
Bioleaching	*Frankia* species DDNSF-01 and *Frankia casuarinae* DDNSF-02	Au 75% and Cu 94%	[53]

12.3 Application of airborne nanoparticles for recycling of e-wastes

The wide range of airborne natural particles, which are rich in natural minerals, can be used as interfaces to remove airborne pollutants mainly produced from e-waste deposits and e-waste industrial processes [54]. The use of magnetite can remove contaminants efficiently, which are emitted from e-waste materials [55].

12.4 Application of natural minerals for treatment of e-waste

The natural minerals, such as clay minerals and montmorillonite nano- and micro-traps, can be used to remove and recycle the e-waste pollutants [56]. The production of montmorillonite-iron oxide nano-trap can remove mercury ions from water in a few seconds without any energy investment. Kaolin is a natural clay mineral which can easily remove harmful substances. Kaolin has a layered structure, which can be used efficiently as a natural gaseous adsorbent for gaseous trace metals, such as mercury.

12.5 Bioreactors and nanotechnology for treatment of e-wastes

In recent years, hybrid technologies, such as integrated bioreactor combined with recyclable iron oxide nano/micro-particles adsorption interfaces, have been developed to remove CO_2 and undesired air pollutants by using natural particles [57]. Hybrid technologies with novel features remove hazardous substances; however, maintaining the efficiency of the processes is needed for e-waste management.

Table 3: Different metal extraction technologies from e-wastes, advantages, and limitations.

Extraction methods	Advantages	Limitations	References
Mechanical separation	Low environmental impact if sorted properly	Labor-intensive process	[58]
Hydrometallurgy	High efficiency, moderate environmental impact	Wastewater management is required	[21]
Pyrometallurgy	High efficiency	High emissions, energy-intensive process	[21]

Table 3 (continued)

Extraction methods	Advantages	Limitations	References
Bioleaching	Eco-friendly process, less energy needed	Slow phenomenon, controlled conditions are needed	[59]
Supercritical fluid extraction	No toxic chemicals, more efficiency	High cost, complex procedure	[60]
Plasma arc recycling	Fast recovery of metals, less emission	Energy-intensive, expensive setup	[61]
Ionic liquid extraction	High selectivity, reusable solvents	High cost, limited scalability	[62]
Electrodialysis	Fast metal separation, less waste production	Energy-intensive process, requirement of pure reagents	[63]

12.6 Innovative technologies for e-waste recycling

Artificial intelligence (AI) is a promising technology which offers a wide range of applications for e-waste reduction. Now, AI is being used in the generation of green designs by many companies at a global scale, as AI can form better prototypes as compared to the earlier models. AI can produce novel electronic designs with easy disassembly and by identifying the constituents suitable for recycling or reutilization. Additionally, AI can be integrated directly into the product by using sensors and algorithms; then the device can predict the remaining lifespan of its components, and this information can be used to repair or maintain the products before they fail completely [64]. AI can analyze the trends of lifespan of electronic devices and can instruct the company to design products with easy recycling procedure. If certain locations accumulate waste faster than others, AI can assist in frequent collection, while reducing unnecessary pickups at less volume sites [65].

The rCubeBot mobile robot system has been generated for identification and sorting of specific e-wastes. If automatic robots are integrated into municipal waste vehicles, they can easily identify e-wastes for reuse, and by applying a convolutional neural network, this system achieves 96% precision in detecting electronic components through real-time image processing. Integrating robotic automation into e-waste collection infrastructure shows economical, scalable, and efficient, which lessens manual sorting and promotes sustainability for management of e-wastes.

AI can enhance the efficiency of segregation of reusable and discarded e-wastes by maintaining safety. Automated sorting powered by AI, using advanced sensors and cameras, can identify and categorize devices upon arrival, and streamlining the initial sorting phase [66]. AI can further assist in the separation of precious metals, and hazardous materials can be identified and eliminated to reduce health risks. Furthermore, by monitoring plant operations, AI can optimize the recycling process. There are several challenges, which can hinder the implementation of AI technologies for e-waste management. However, AI technologies are very expensive. Advanced components, like industrial robots, precision sensors, and high-performance computing systems, are expensive, and companies do not want to invest in these AI tools [67]. The integration of AI system in recycling facilities is not a simple process, but it requires space, substantial upgrades, high energy demand, and skilled workers for the operation and maintenance of the system. Adapting AI technology is an expensive endeavor but a long-term investment requires proper planning and restructuring of current system [4]. Further, standardized framework and guidelines are needed for the successful implementation of AI system in e-waste management. The security concerns are the main challenges in the execution of AI tools for e-waste management. The insufficient safety measures may lead to unauthorized access, data breaches, disrupt operations, or expose sensitive information; thus, cybersecurity measures and regulatory frameworks are needed for protection before its implementation at a large scale. AI tools have the potential to transform the e-waste management system at the global level in a more efficient, sustainable, and effective manner.

13 Recommendations

Several recommendations have been made to solve the problems related to e-waste management as follows:

1. The collaborations among government organizations, private firms, nongovernmental organizations through public–private partnerships are needed for developing sustainable infrastructure for safe e-waste management, which helps consumers for safe disposal of e-wastes.
2. The collaborations among national governments, international organizations, and investors are needed to implement extended producer responsibility (EPR) and takeback schemes for the reduction of e-wastes, specifically in developing nations.
3. All nations should follow the guidelines or best practices recommended by the United Nations and the World Health Organization for the management of e-wastes. This will allow stakeholders to understand their responsibilities toward the e-waste management system.
4. The companies should provide complete information about the composition of products, life span, repair options, refurbishment, etc., to the consumers. Awareness education is needed for all stakeholders of society about the adverse impacts on the environment and human health due to the mishandling or improper dumping of e-wastes.

5. The government should promote manufacturers to develop green products, and they must be accountable for the environmental impact of the products. Applying EPR will prompt manufacturers to develop green devices with responsibility and proper recycling of e-waste. However, financial and technological assistance is needed to establish a sustainable system for e-waste management.

14 Conclusion

Electronic waste is one of the most serious concerns for the environment and public health. Management of e-wastes is a challenging task due to complex composition, inadequate infrastructure, and informal recycling. However, it generates opportunities through technological innovations for developing green products and resource recovery for promoting the circular economy. Thus, conventional e-waste disposal methods should be replaced with advanced technologies to ensure the recovery of valuable resources with minimum or zero impact on the ecosystem. However, technology alone can not solve the problem. The stringent regulations and strong policy frameworks are needed for transformation. The government should encourage and provide incentives for developing green products, which can be easily reused, repaired, and recycled. International support is needed to check e-waste export to developing or underdeveloped nations without proper management infrastructure. Thus, efforts are needed to develop effective collection methods, safe recycling practices, eco-friendly disposal methods, and proper implementation of e-waste management guidelines. Then, we can shift toward a sustainable circular economy for managing e-waste by preserving valuable resources for future generations.

References

[1] Ramprasad C, Gwenzi W, Chaukura N, Azelee NIW, Rajapaksha AU, Naushad M, Rangabhashiyam S, Strategies and options for the sustainable recovery of rare earth elements from electrical and electronic waste. Chem Eng J, 2022, 442, 135992.
[2] Ali S, Shirazi F, The paradigm of circular economy and an effective electronic waste management. Sustainability, 2023, 15, 1998.
[3] Murthy V, Ramakrishna S, A review on global e-waste management: Urban mining towards a sustainable future and circular economy. Sustainability, 2022, 14, 647.
[4] Quinto S, Law N, Fletcher C, Le J, Antony Jose S, Menezes PL, Exploring the e-waste crisis: Strategies for sustainable recycling and circular economy integration. Recycling, 2025, 10, 72.
[5] Sandwal SK, Jakhar R, Styszko K, E-waste challenges in India: Environmental and human health impacts. Appl Sci, 2025, 15, 4350.
[6] Andeobu L, Wibowo S, Grandhi S, A systematic review of e-waste generation and environmental management of Asia Pacific countries. Int J Environ Res Public Health, 2021, 18, 9051.

[7] Ouro-Salim O, Urban mining of e-waste management globally: Literature review. Clean Waste Syst, 2024, 9, 100162.
[8] Huang Y, Hsiang EL, Deng MY, Wu ST, Mini-LED, micro-LED and OLED displays: Present status and future perspectives. Light Sci Appl, 2020, 9, 105.
[9] Parvez SM, Jahan F, Brune MN, Gorman JF, Rahman MJ, Carpenter D, Islam Z, Rahman M, Aich N, Knibbs LD, Sly PD, Health consequences of exposure to e-waste: An updated systematic review. Lancet Planet Health, 2021, 5, 905–920.
[10] Li W, Achal V, Environmental and health impacts due to e-waste disposal in China – A review. Sci Total Environ, 2020, 139, 745.
[11] Li J, Liang Y, Li R, Huang L, Li J, Kang H, Sustainable management for electronic waste: A review. J Clean Prod, 2020, 258, 120894.
[12] Zhang B, Huo X, Xu L, Cheng Z, Cong X, Lu X, Xu X, Elevated lead levels from e-waste exposure are linked to decreased olfactory memory in children. Environ Pollut, 2017, 231, 1112–1121.
[13] Beula D, Sureshkumar M, A review on the toxic e-waste killing health and environment – today's global scenario. Mater Today Proc, 2021, 47, 2168–2174.
[14] Kaushik PR, Herat S, Current state of e-waste management in India. Int J Environ Waste Manag, 2020, 25, 322.
[15] Lu C, Qiao Z, Lin K, Hu S, Fu M, Peng C, Zhou B, Zhou S, Zhang W, Ling S, Distribution characteristics and risks assessment of brominated flame retardants in surface soil from both a legacy and a new e-waste dismantling site. J Clean Prod, 2022, 373, 133970.
[16] Chakraborty P, Selvaraj S, Nakamura M, Prithiviraj B, Cincinelli A, Bang JJ, PCBs and PCDD/Fs in soil from informal e-waste recycling sites and open dumpsites in India: Levels, congener profiles and health risk assessment. Sci Total Environ, 2018, 621, 930–938.
[17] Shen C, Chen Y, Huang S, Wang Z, Yu C, Qiao M, Xu Y, Setty K, Zhang J, Zhu Y, Lin Q, Dioxin-like compounds in agricultural soils near e-waste recycling sites from Taizhou area, China: Chemical and bioanalytical characterization. Environ Int, 2009, 35, 50–55.
[18] Fujimori T, Takigami H, Pollution distribution of heavy metals in surface soil at an informal electronic-waste recycling site. Environ Geochem Health, 2014, 36, 159–168.
[19] Fu J, Zhang H, Zhang A, Jiang G, E-waste recycling in China: A challenging field. Environ Sci Technol, 2018, 52, 6727–6728.
[20] Khanna R, Saini R, Park M, Ellamparuthy G, Biswal S, Mukherjee P, Factors influencing the release of potentially toxic elements (PTEs) during thermal processing of electronic waste. Waste Manag, 2020, 105, 414–424.
[21] Srivastava AN, Sikarwar VS, Bisen D, Fathi J, Maslani A, Lopez Nino BN, Barmavatu P, Kaviti AK, Pohorelý M, Buryi M, E-waste unplugged: Reviewing impacts, valorization strategies and regulatory frontiers for efficient e-waste management. Processes, 2025, 13, 2014.
[22] Briffa J, Sinagra E, Blundell R, Heavy metal pollution in the environment and their toxicological effects on humans. Heliyon, 2020, 6, e04691.
[23] Cui J, Luo C, Tang CW, Chan T, Li X, Speciation and leaching of trace metal contaminants from e-waste contaminated soils. J Hazard Mater, 2017, 329, 150–158.
[24] Ankit N, Saha L, Kumar V, Tiwari J, Sweta N, Rawat S, Singh J, Bauddh K, Electronic waste and their leachates impact on human health and environment: Global ecological threat and management. Environ Technol Innov, 2021, 24, 102049.
[25] Moeckel C, Breivik K, Nost TH, Sankoh A, Jones KC, Sweetman A, Soil pollution at a major West African e-waste recycling site: Contamination pathways and implications for potential mitigation strategies. Environ Int, 2020, 137, 105563.
[26] Aziz KHH, Mustafa FS, Omer KM, Hama S, Hamarawf RF, Rahman KO, Heavy metal pollution in the aquatic environment: Efficient and low-cost removal approaches to eliminate their toxicity: A review. RSC Adv, 2023, 13, 17595–17610.

[27] Huo X, Wu Y, Xu L, Zeng X, Qin Q, Xu X, Maternal urinary metabolites of PAHs and its association with adverse birth outcomes in an intensive e-waste recycling area. Environ Pollut, 2019, 245, 453–461.
[28] Amoabeng Nti AA, Arko-Mensah J, Botwe PK, Dwomoh D, Kwarteng L, Takyi SA, Acquah AA, Tettey P, Basu N, Batterman S, Robins TG, Fobil JN, Effect of particulate matter exposure on respiratory health of e-waste workers at Agbogbloshie, Accra, Ghana. Int J Environ Res Public Health, 2020, 17, 3042.
[29] Sonny JN, Samali VM, Bosco A, Boniface A, Judith A, Abdallah IN, Joseph O, Cosmas O, The impact of electronic-electrical waste on human health and environment: A systematic literature review. J Eng Technol Res, 2023, 15, 1–16.
[30] Rautela R, Arya S, Vishwakarma S, Lee J, Kim K, Kumar S, E-waste management and its effects on the environment and human health. Sci Total Environ, 2021, 773, 145623.
[31] Cai H, Xu X, Zhang Y, Cong X, Lu X, Huo X, Elevated lead levels from e-waste exposure are linked to sensory integration difficulties in preschool children. Neurotoxicology, 2019, 71, 150–158.
[32] Zeng X, Xu X, Qin Q, Ye K, Wu W, Huo X, Heavy metal exposure has adverse effects on the growth and development of preschool children. Environ Geochem Health, 2019, 41, 309–321.
[33] Xu X, Liao W, Lin Y, Dai Y, Shi Z, Huo X, Blood concentrations of lead, cadmium, mercury and their association with biomarkers of DNA oxidative damage in preschool children living in an e-waste recycling area. Environ Geochem Health, 2018, 40, 1481–1494.
[34] Jaiswal P, Verma S, Paul AR, Haque N, Bruckard W, Economic valuation of metal recovery from mobile phones in India. Minerals, 2025, 15, 259.
[35] Kumar A, Sharma D, Gaur D, Liu Y, Sustainable waste electrical and electronic equipment management guide in emerging economies context: A structural model approach. J Clean Prod, 2022, 336, 130391.
[36] Swami S, Suthar S, Singh R, Thakur AK, Gupta LR, Sikarwar VS, Integration of anaerobic digestion with artificial intelligence to optimise biogas plant operation. Environ Dev Sustain, 2025, 27, 9773–9803.
[37] Wang Z, Huang G, An C, Recyclability and recovery of carbon from waste printed circuit boards within a circular economy perspective: A review. J Environ Manag, 2025, 375, 124266.
[38] Jadhao PR, Panda R, Pant KK, Nigam KD, Integrated approach for metallic fraction recovery and generation of valuable products from electronic waste. Ind Eng Chem Res, 2023, 62, 19372–19380.
[39] Mir S, Dhawan N, Investigation of pyrolysis followed by leaching for metal recovery from integrated circuits. Jom, 2023, 75, 3467–3479.
[40] Khaobang C, Kathongthung N, Phitsuwan P, Sitthichirachat P, Wibowo H, Areeprasert C, In-situ adsorptive pyrolysis of e-waste using coal and rice husk fly ash as alternative adsorbents for energy and precious metal recovery by solvent extraction. J Anal Appl Pyrolysis, 2024, 179, 106465.
[41] Samal S, Blanco I, An overview of thermal plasma arc systems for treatment of various wastes in recovery of metals. Materials, 2022, 15, 683.
[42] Szałatkiewicz J, Metals recovery from artificial ore in case of printed circuit boards, using plasmatron plasma reactor. Materials, 2016, 9, 683.
[43] Gulliani S, Volpe M, Messineo A, Volpe R, Recovery of metals and valuable chemicals from waste electric and electronic materials: A critical review of existing technologies. RSC Sustain, 2023, 1, 1085–1108.
[44] An J, Characteristics of metals leached from waste printed circuit boards using *Acidithiobacillus ferrooxidans*. Minerals, 2021, 11, 224.
[45] Ilyas S, Lee JC, Chi RA, Bioleaching of metals from electronic scrap and its potential for commercial exploitation. Hydrometallurgy, 2013, 131, 138–143.
[46] Baniasadi M, Vakilchap F, Bahaloo-Horeh N, Mousavi SM, Farnaud S, Advances in bioleaching as a sustainable method for metal recovery from e-waste: A review. J Ind Eng Chem, 2019, 76, 75–90.

[47] Chatterjee A, Abraham J, Efficient management of e-wastes. Int J Environ Sci Technol, 2016, 14, 211–222.
[48] Lee H, Molstad E, Mishra B, Recovery of gold and silver from secondary sources of electronic waste processing by thiourea leaching. JOM, 2018, 70, 1616–1621.
[49] Rao MD, Singh KK, Morrison CA, Love JB, Recycling copper and gold from e-waste by a two-stage leaching and solvent extraction process. Sep Purif Technol, 2021, 263, 118400.
[50] Sheel A, Pant D, Recovery of gold from electronic waste using chemical assisted microbial biosorption (hybrid) technique. Bioresour Technol, 2018, 247, 1189–1192.
[51] Arshadi M, Nili S, Yaghmaei S, Ni and Cu recovery by bioleaching from the printed circuit boards of mobile phones in non-conventional medium. J Environ Manag, 2019, 250, 109502.
[52] Wei X, Liu D, Huang W, Huang W, Lei Z, Simultaneously enhanced Cu bioleaching from e-wastes and recovered Cu ions by direct current electric field in a bioelectrical reactor. Bioresour Technol, 2020, 298, 122566.
[53] Marappa N, Ramachandran L, Dharumadurai D, Nooruddin T, Recovery of gold and other precious metal resources from environmental polluted e-waste printed circuit board by bioleaching Frankia. Int J Environ Res, 2020, 14, 165–176.
[54] Ganguly M, Tao Y, Lee B, Ariya PA, Natural kaolin: Sustainable technology for the instantaneous and energy-neutral recycling of anthropogenic mercury emissions. ChemSusChem, 2019, 13, 165–172.
[55] Eltouny N, Ariya PA, Competing reactions of selected atmospheric gases on Fe_3O_4 nanoparticles surfaces. Phys Chem Chem Phys, 2014, 16, 23056–23066.
[56] Ganguly M, Dib S, Ariya PA, Fast, cost-effective and energy efficient mercury removal-recycling technology. Sci Rep, 2018, 8, 1–10.
[57] Rocha AA, Wilde C, Hu Z, Nepotchatykh O, Nazarenko Y, Ariya PA, Development of a hybrid photo-bioreactor and nanoparticle adsorbent system for the removal of CO_2, and selected organic and metal co-pollutants. J Environ Sci, 2017, 57, 41–53.
[58] Nandan A, Suresh AC, Saole P, Jeevanasai SA, Chandrasekaran R, Meili L, WanAzelee NI, Selvasembian R, An integrated approach for electronic waste management – overview of sources of generation, toxicological effects, assessment, governance, and mitigation approaches. Sustainability, 2023, 15, 16946.
[59] Zhang S, Yu Y, Dechlorination behavior on the recovery of useful resources from WEEE by the steam gasification in the molten carbonates. Procedia Environ Sci, 2016, 31, 903–910.
[60] Nguyen TH, Won S, Ha MG, Nguyen DD, Kang HY, Bioleaching for environmental remediation of toxic metals and metalloids: A review on soils, sediments, and mine tailings. Chemosphere, 2021, 282, 131108.
[61] Yang Y, Hu B, Bio-based chemicals from biorefining: Lipid and wax conversion and utilization, In: Advances in Biorefineries. Elsevier, Amsterdam, The Netherlands, 2014, 693–720.
[62] Changming D, Chao S, Gong X, Ting W, Xiange W, Plasma methods for metals recovery from metal-containing waste. Waste Manag, 2018, 77, 373–387.
[63] Al Hassan MK, Alfarsi A, Nasser MS, Hussein IA, Khan I, Ionic liquids and NADES for removal of organic pollutants in wastewater: A comprehensive review. J Mol Liq, 2023, 391, 123163.
[64] Prentice C, Dominique Lopes S, Wang X, Emotional intelligence or artificial intelligence – an employee perspective. J Hosp Mark Manag, 2020, 29, 377–403.
[65] Puntoni S, Reczek RW, Giesler M, Botti S, Consumers and artificial intelligence: An experiential perspective. J Mark, 2021, 85, 131–151.
[66] Shreyas Madhav A, Rajaraman R, Harini S, Kiliroor CC, Application of artificial intelligence to enhance collection of e-waste: A potential solution for household WEEE collection and segregation in India. Waste Manag Res J Sustain Circ Econ, 2022, 40, 1047–1053.
[67] Chun H, Hur J, Kim D, Son NS, Cross-border vertical integration and technology in factory Asia: Evidence from Korea. Dev Econ, 2020, 58, 99–133.

Sonal Thukral, Apoorva Jain*, Arjun Mittal*

Factors affecting producers' behavior toward e-waste management: a relational analysis using total interpretive structural modeling and MICMAC analysis

Abstract: According to the Central Pollution Control Board (CPCB) of India, the growth rate of e-waste was 7.71 lakh tons in 2018–2019 and 10.14 lakh tons in 2019–2020, which is a 31% increase. Thus, the increasing rate of e-waste in India demands greater attention from various stakeholders in determining their responsibility toward e-waste management. Despite the primary responsibility assigned to producers under extended producer responsibility, significant efforts are still required to motivate them to meet these obligations. Although producers may intend to fulfil their e-waste management responsibilities, these intentions often do not translate into concrete actions. To explore this intention-behavior gap, the study draws on these specific triggers that can motivate individuals to convert their intentions into actual behavior from the Fogg Behavior Model (FBM). Drawing on FBM, the present study identifies the key factors that affect producers' behavior toward e-waste management. Further, the study presents its hierarchical linkages by using TISM. These factors were identified through in-depth interviews of the experts, and a hierarchy-based model was developed for them by applying TISM. These factors were also analyzed based on their driving power and dependence using cross-impact matrix multiplication applied to the classification analysis. The results indicate that the factors affecting producers' behavior toward e-waste management are spread across five levels, wherein Government laws and regulations drive the entire system and play a very important role in e-waste management. Appropriate government laws and regulations create awareness and education among producers. Such educated producers ultimately educate consumers and shape their attitude toward e-waste management.

Keywords: Awareness, e-waste, management, Fogg Behavior Model, producer's behavior, TISM, MICMAC Analysis

Acknowledgments: The work presented is largely an outcome of the Major Research Project sponsored by the Indian Council of Social Science Research (ICSSR), awarded to the author. However, the responsibility for the facts stated, opinions expressed, and the conclusions drawn is entirely that of the authors.

***Corresponding author: Apoorva Jain**, Delhi School of Management, Delhi Technological University, Delhi 110042, India, e-mail: apoorva.jain@dtu.ac.in
***Corresponding author: Arjun Mittal**, Department of Commerce, Hans Raj College, University of Delhi, Delhi 110007, India, e-mail: arjunmittal@ymail.com
Sonal Thukral, Faculty of Commerce and Business, Delhi School of Economics, University of Delhi, Delhi 110007, India

 | https://doi.org/10.1515/9783112248072-010

1 Introduction

According to the United Nations' Global E-Waste Monitor Report 2024, global e-waste generation has surged by a record 62 million tons – an 82% increase – between 2010 and 2022. It is projected to reach 82 million tons by 2030, representing an additional 32% rise [1]. As electronic waste continues to escalate worldwide, developing countries such as India and China are struggling with this challenge with mounting urgency. India, in particular, has reported the highest levels of e-waste generation, experiencing a staggering growth rate of 163% from 2010 to 2022 [2]. This alarming increase in e-waste in India underscores the need for greater attention from various stakeholders regarding their responsibilities in e-waste management. Although the e-waste (Management) Rules 2016 emphasize the concept of extended producer responsibility (EPR), India currently has a limited number of formal recyclers, and their processing capacities are insufficient [3]. Moreover, there is a lack of information about the waste management strategies employed by these authorized recyclers [4], suggesting that the role of producers in managing e-waste needs an investigation in the Indian context. While several studies have examined e-waste management in India [4–7], there remains a limited focus on the responsibilities of producers. Consequently, it is essential to examine the factors influencing producers' behavior toward e-waste management in India.

Despite the primary responsibility assigned to producers under EPR, significant efforts are still required to motivate them to meet these obligations. Although producers may intend to fulfill their e-waste management responsibilities, these intentions often do not translate into concrete actions. To explore this intention–behavior gap, we turn to behavioral models. However, many of these models fail to adequately explain how intentions translate into actual behavior. While numerous studies have sought to bridge the intention-behavior gap [8, 9], challenges remain in understanding the connection between intention and action, particularly in the context of e-waste management.

Research indicates that specific triggers can motivate individuals to convert their intentions into actual behavior. This study draws on these triggers from the Fogg Behavior Model (FBM), a behavioral design theory developed by Fogg [10]. A trigger is something that prompts immediate action, and the FBM identifies three key factors – motivation, ability, and triggers – that govern human behavior. While many studies have examined motivation and ability, the FBM stands out by emphasizing the importance of triggers. Fogg categorizes triggers into three types: sparks, facilitators, and signals. A spark ignites a user's motivation, fostering the self-drive necessary for engaging in desired behavior. A facilitator provides clear guidance on how to achieve the target behavior, while a signal serves as a reminder or cue to encourage action. Fogg argues that even with motivation and ability, behavior change will not occur without triggers; thus, triggers are critical interventions that prompt individuals to engage in target behaviors.

The study begins by identifying the factors that influence producers' behavior in fulfilling their e-waste management responsibilities. This is achieved through in-depth interviews with electronic product producers, after which these factors are classified into Fogg's trigger categories. Interviews were conducted with six experts, including directors, managers, vice presidents, and CEOs from companies with annual revenues ranging from USD 1 billion to 5 billion. Existing research [11] indicates that even a small sample size can enhance the researcher's understanding of both the participants and the phenomena being studied.

The study proceeds further by employing total interpretive structural modeling (TISM) to examine the multilayered hierarchical relationships between these triggers and their influence on producers' behaviors concerning e-waste management. Some triggers may drive others, ultimately eliciting the desired behaviors. Thus, these triggers are crucial for translating producers' intentions regarding e-waste management into action, narrowing the intention-behavior gap. TISM is widely used for modeling conceptual frameworks across various management and technical fields, helping to clarify the "what," "why," and "how" of theory building by transforming poorly articulated models into clear representations [12]. Third, the study utilizes Matrice d'Impacts Croisés Multiplication Appliquée à un Classement (MICMAC) to identify the driver-dependent connections among the triggers.

Through in-depth interviews, the study finds that producers' behavior regarding e-waste management results from a complex interplay of various triggers. These include regulatory requirements (facilitator trigger), economic incentives (signal triggers), educational and awareness programs for both producers and consumers (spark triggers), training programs for producers (signal triggers), and the availability of infrastructure (signal triggers). The study further maps the multilayered linkages between these triggers using TISM and subsequently applies MICMAC to analyze them in terms of their driving-dependence power.

Findings also indicate that the regulatory environment underpins the entire e-waste management system for producers, demonstrating high dependence according to MICMAC. Regulatory factors shape producers' educational and awareness programs and stimulate training within their organizations. These linkage clusters, as identified by MICMAC, drive producers to educate consumers, incentivize responsible behavior, and enhance their own infrastructural capabilities – factors that exhibit high dependence in the MICMAC analysis.

The study opens avenues for future research by proposing that triggers (sparks, facilitators, and signals) be integrated into established behavioral models, offering practical insights to help translate producers' intentions regarding e-waste management into actionable outcomes.

2 Review of literature

Research on e-waste management has gained momentum in recent years but has yet to become rich considering the different aspects affecting it. Studies indicate that there are many factors which lead to the rise of e-waste, such as advancements in technology, production of a variety of electrical and electronic products, and continuous upgrading of them [13], short life cycle of EEE (electronic and electrical equipment), low recycling [14], consumer demands, etc. Besides these, e-waste statistics have jumped significantly due to the huge export of e-waste into developing countries from developed ones. The United Nations Environmental Programme (UNEP) suggests that only 10% of the e-waste produced in the world today is managed in developed countries; the rest 90% is sent to developing countries in different parts of the world. The developing countries become the dumping ground for e-waste by developed countries in the name of donation so that they can keep up with the latest technologies in the developed countries [15]. In India, under current regulations, e-waste can be imported only for recycling and reuse, but like other countries, we too lack adequate infrastructure and recycling equipment for dealing with such a huge amount of e-waste [16]. Industrialization and consumerization are proliferating, and this dumping process is expected to continue in the future [17]. One of the possible reasons the literature indicates for such mismanagement is the weak regulations in developing countries. Another factor that contributes to the ineffective working of the regulatory body is the presence of informal handling of e-waste.

Annually, India generates approximately 1.85 million tons of e-waste and stands among the top five countries in the world in the production of e-waste. More than 90% of the e-waste in India is managed by the informal sector, which negatively affects the well-being of the people and results in massive environmental pollution [18]. Although since February 2020, EPR has been provided to 1,500 producers, there are only 34 registered producer responsibility organizations. Thus, due to ineffective implementation of EPR, a large chunk of e-waste is pumped into the informal sector. According to a study by Assocham, formal recyclers recycled only 1.5% of the total e-waste generated. It is observed that presently, companies are not even taking back 10% of their products, nor are they undertaking any steps to make their customers aware of such schemes.

Informal recycling practices are quite prevalent in developing countries, which result in informal handling and extraction of the resources from e-waste, and expose the workers to hazardous content like lead, cadmium, and chromium, etc. [19]. The majority of the toxic and hazardous materials are released into the environment from improper disposal and handling of e-waste [20]. These organic and other toxic pollutants can bio-accumulate and bio-magnify through the food web [20] and have serious health impacts for the whole ecosystem [21]. Therefore, here the role of government rules and regulations regarding e-waste management becomes critical as it acts as a regulatory body for the legal and illegal activities performed in a nation.

Producers have an important role to play in e-waste management. In accordance with the E-Waste (Management and Handling) Rules from 2011, India, like several other advanced and developing countries, has adopted an EPR strategy for managing electronic trash (e-waste). In accordance with these regulations, producers are now in charge of establishing e-waste collection facilities as well as funding and putting up a system for ecologically friendly e-waste handling [16]. Notwithstanding current laws, there continues to be a knowledge and awareness gap among the people regarding the problems associated with the processing and treatment of e-waste. And not only consumers but producers' awareness plays a big role in e-waste management. Training provided to producers will not only make them aware of e-waste management but will also help to disseminate the information to consumers via producers.

In spite of the important role of producers in e-waste management, there hasn't been much research carried out from producer's perspective. In order to fill this gap, the present study attempts to understand the factors affecting E-Waste management from the producer's perspective. The TISM model is employed to understand the hierarchical relationship among the factors that affect the producers' behavior toward e-waste management.

3 Research methodology

The present study employs TISM to identify the key factors that affect producers' behavior toward e-waste management and present their hierarchical linkages. TISM is widely used for modeling conceptual frameworks in different arenas of management and technical fields [22]. This concept is developed by answering "what," "how," and "why" questions. For understanding and answering "what," the unstructured interviews were conducted in order to extract the factors affecting producers' behavior toward e-waste management. Similarly, "how" and "why" are used for developing the hierarchical relation and interpreting the links among the factors [12, 23]. The factors identified through in-depth interviews were verified by the experts, and the factors that have been confirmed are utilized to build the e-waste management model.

3.1 Total Interpretive Structural Modeling (TISM)

Warfield [24] defined interpretive structural modeling (ISM) as "it is a well-known approach for recognizing associations among specific factors which pertain to the problem or study at hand." ISM helps in establishing inter-relationships among various factors. The relationship between the identified factors is interpreted on the basis of the expert opinions; this method is Interpretive. Further, a structured approach is employed to extract outcomes from complex problems; this method is called Structured.

In the ISM method, the final outcome is modeled in hierarchical and graphical format [25]. A recent extension of ISM is known as TISM, which is an upgraded qualitative modeling technique. ISM has several limitations, which lead to the development of TISM [12]. These limitations are:

- Only those individuals who are familiar with ISM methodology are able to use it and deduce information from it.
- The implementation benefit of the ISM methodology can be easily reaped only when computer facilities are available.
- ISM methodology fails to provide the same interpretations of the model for different individuals since the interpretation of links is somewhat feeble.
- ISM methodology answers only "what" and "how" in theory building; it is unable to answer the "why" part in theory building since it stays quiet on the causality of linkages.
- The transitive relationships in the diagraph diagram are not considered in ISM.

Thus, ISM may misinterpret the entire decision-making process since the interpretation for direct linkages is comparatively weak in this technique. Consequently, the TISM approach is employed to account for the limitations of the ISM approach [26]. By employing interpretive matrix, it helps in overcoming the key drawback of ISM, i.e., poor interpretation of the links. TISM aids in depicting the entire structure and interpretation for direct as well as significant transitive relationships by a diagraph model. Thus, complicated systems are portrayed in a simple manner through the TISM approach. It provides a structural model which helps in determining how the different elements are connected and why they are supposed to connect in that way. Consequently, it facilitates answering "what," "why," and "how" in theory building by transforming feebly articulated rational models into straightforward and unambiguous models [12, 26].

The current study identifies the barriers and enablers of effective e-waste management by producers and employs the TISM approach to delineate the inter-relationship among them. These barriers and enablers were identified through in-depth interviews of the experts, and a hierarchy-based model was developed for them by applying TISM. These barriers and enablers were also analyzed based on their driving power and dependence using cross-impact matrix multiplication applied to the classification (MICMAC) analysis. The step-wise process of TISM has been adopted from [12] and is presented in Figure 1.

Step I. Identify and define elements

The first step in TISM is to identify and define the elements for which inter-relationships will be modeled. Such elements can be identified either through the review of existing literature or using any idea generation method as a small group exercise.

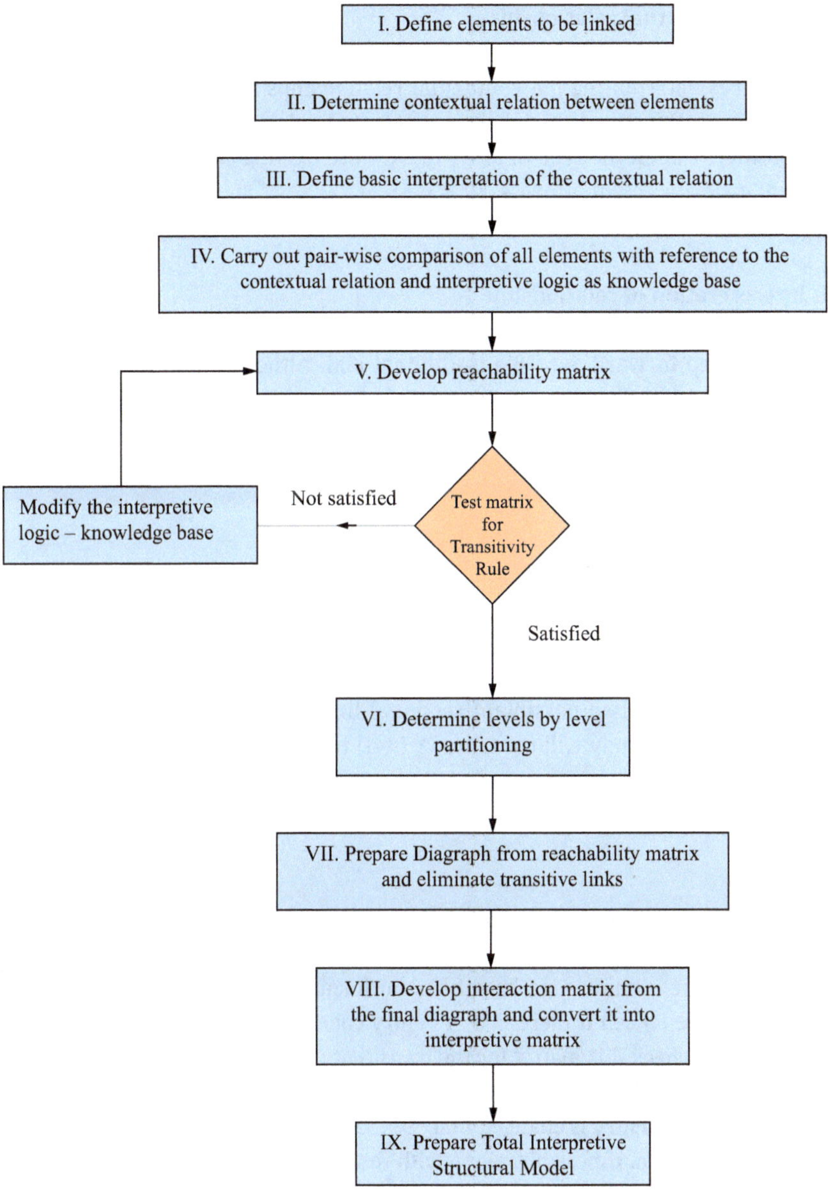

Figure 1: Process of total interpretive structural modeling (TISM).

Step II. Define contextual relationships

This step involves determining the contextual relationships between the elements in order to develop a structural model. The contextual relationship between the elements is contingent upon the type of structure we are dealing with, such as attribute enhancement, priority, intent, process, etc.

Step III. Interpretation of relationship

This is the first step forward over the traditional ISM. Although contextual relationships are captured in traditional ISM, it remains silent on how the relationship actually works. Consequently, such clarification is enhanced in TISM by mentioning in what ways one element will influence another.

Step IV. Interpretive logic of pair-wise comparison

Each element is compared individually with all the other elements in a pair-wise comparison. For n identified elements, the total number of pair-wise comparisons will be $n\ X\ (n\text{-}1)$ since there are two possible directional links i–j or j–i. Consequently, the knowledge base for the study will contain $n\ X\ (n–1)$ number of rows. Expert opinions are recorded for each pair-wise comparison by entry code "Y" for Yes and "N" for No, and if it is "Y," it is further interpreted.

Step V. Reachability matrix and transitivity test

The interpretive logic – knowledge base helps in forming the reachability matrix by making entry 1 in the i–j cell if there is a "Y" entry code, otherwise 0 if there is an "N" entry code. We then check this matrix for the transitivity rule and update it until the full transitivity is achieved. The knowledge base is also updated whenever there is a new transitive link. The "No" entry is changed to a "Yes" entry, and "Transitive" is entered in the interpretation column. If possible, along with the "Transitive" entry, the meaningful explanation of the transitive relationship is also given.

Step VI. Level partition on reachability matrix

The level partition is employed in order to know the placement of elements level-wise. We determine the reachability set and antecedent set for each element with the help of the final reachability matrix. For any particular element, the reachability set consists of the element itself and all other elements that it will influence/enhance. On the other hand, for any particular element, the antecedent set will consist of the element itself and all other elements that will influence/enhance it. Further, for all these elements, the intersection of these two sets is determined.

Those elements achieve the top level in the TISM hierarchy for which the reachability set and intersection set are the same. These top-level elements would not reach any other elements above their own level. The next level elements are obtained by carrying out the same procedure. This iteration is repeated till all the levels of elements are determined.

Step VII. Developing diagraph

In accordance with the relationships shown in the reachability matrix, the directed links are drawn, and elements are arranged graphically. In the diagraph, only those transitive relationships are retained whose interpretation is crucial.

Step VIII. Interaction matrix

A binary interaction matrix is developed from the final digraph by depicting all the interactions as "1" in the respective cells. Further, each cell with "1" is interpreted with respective relevant interpretations taken from the interpretive logic-knowledge base to form the interpretive matrix.

Step IX. Total interpretive structural model

Finally, with the help of the digraph and interpretive matrix, TISM is developed for the identified elements. The nodes in the digraph are substituted by the interpretation of the elements placed in boxes. The information furnished by the interpretive matrix is depicted by the side of the respective links in the structural model. Consequently, it leads to the development of the Total Interpretive Structural Model in terms of interpreting nodes as well as links.

4 Findings

4.1 Application of the TISM model for factors affecting e-waste management of producers

As mentioned above, the present study employs a TISM-based hierarchical model to determine the contextual relationship between the factors affecting producers' behavior toward e-waste management. The step-wise procedure is presented below:

Step I. Identify and define elements

The present study identifies the factors affecting e-waste management by producers. Eight factors have been identified by conducting in-depth interviews of experts from the industry. Finding the relevant area's stakeholders is done through subjective sampling. A total of six experts were chosen from organizations that manufacture electrical and electronic products. The experts included directors, managers, vice presidents, and CEOs. The factors identified through in-depth interviews were verified by the experts, and the factors that have been confirmed are utilized to build the e-waste management model. The following are the factors which have been identified through in-depth interviews:
- F1: Educating Producers – Spreading information about e-waste management among producers.
- F2: Educating Consumers – Consumers' education toward e-waste management rules and formal disposal mechanisms.
- F3: Producers' Awareness – Producers' awareness about government laws and policies and recycling and disposal processes of e-waste is expected to influence the producers' e-waste management.
- F4: Government Laws and Regulations – The current government laws and policies regarding e-waste management are expected to influence the producers' e-waste management.
- F5: Training Organization – Imparting training will influence organizations' e-waste management practices.
- F6: Infrastructure – Present technology and infrastructure with respect to e-waste management are also expected to influence producers' access to these and their e-waste management.
- F7: Consumers' Attitude – Consumers' concern toward proper disposal of e-waste.
- F8: Incentivizing Consumers – Giving incentives to consumers to encourage them to practice proper e-waste management.

Step II. Define contextual relationships

Contextual relationships between the factors were identified on the basis of the expert's opinion. However, the interpretation gives a better understanding of the relationship. It assists by giving a detailed road map of the connections. A group of experts rationally defined these links.

Step III. Interpretation of relationship

The interpretation of relationships between the factors was captured through experts' opinions. The experts were asked if "Factor 1 (F1) will influence/enhance Factor 2 (F2)" or not. Further, if the contextual relationship appears to be yes, "In what way Factor 1 (F1) will influence or enhance Factor 2 (F2)" was also asked.

Step IV. Interpretive logic of pair-wise comparison

In order to undertake the pair-wise comparison of the identified factors, an "interpretive logic-knowledge base" is developed. The relationship between the factors has been answered in terms of Yes ("Y") or No ("N"). We followed with the work of [27] wherein the response was taken "Y" if more than 70% of the responses (direct) were Yes, otherwise "N."

Step V. Reachability matrix and transitivity test

From the above "interpretive logic-knowledge base," the reachability matrix is developed by making the entry 1 for "Y" code and 0 for "N" code. Further, the reachability matrix was examined for the transitivity rule, i.e., if F2 influences F3 and F3 influences F4, then F2 relates to F4. Again, we followed the work of [27] to determine the threshold for transitivity links. The link was considered a significant transitive link if captured responses were more than 50%. The final reachability matrix with the transitivity rule is presented in Table 1.

Step VI. Level partition on reachability matrix

Level partitioning is undertaken till the level of each factor is determined. The reachability set, antecedent set, and their interaction set are determined by following the original ISM/TISM methodology. Level partitioning helps in categorizing and systematically dividing these factors affecting e-waste management into different levels. The

Table 1: Final reachability matrix with transitive links.

	F1	F2	F3	F4	F5	F6	F7	F8	Driving Power
F1	1	1	1	0	1	1	1*	0	6
F2	0	1	0	0	0	0	1	0	2
F3	1	1*	1	0	0	0	1	0	4
F4	1*	0	1	1	0	0	1*	0	4
F5	0	0	0	0	1	1	1*	1	4
F6	0	0	0	0	0	1	1	0	2
F7	0	0	0	0	0	0	1	0	1
F8	0	0	0	0	0	0	1	1	2
Dependence	3	3	3	1	2	3	8	2	

Note: 1* entries depict the transitivity relationship.

highest level is attained by the factors having the same reachability and antecedent sets, and the remaining levels are determined after we iteratively remove previous factors. The level partitioning of these factors is presented in Table 2.

Table 2: Hierarchical partitioning.

Variables	Reachability set	Antecedent set	Intersection set	Level
Iteration 1				
1	1,2,3,5,6,7	1,3,4	1,3	
2	2,7	1,2,3	2	
3	1,2,3,7	1,3,4	1,3	
4	1,3,4,7	4	4	
5	5,6,7,8	1,5	5	
6	6,7	1,5,6	6	
7	7	1,2,3,4,5,6,7,8	7	I
8	7,8	5,8	8	
Iteration 2				
1	1,2,3,5,6	1,3,4	1,3	
2	2	1,2,3	2	II
3	1,2,3	1,3,4	1,3	
4	1,3,4	4	4	
5	5,6,8	1,5	5	
6	6	1,5,6	6	II
8	8	5,8	8	II
Iteration 3				
1	1,3,5	1,3,4	1,3	
3	1,3	1,3,4	1,3	III

Table 2 (continued)

Variables	Reachability set	Antecedent set	Intersection set	Level
4	1,3,4	4	4	
5	5	1,5	5	III
Iteration 4				
1	1	1,4	1	IV
4	1,4	4	4	
Iteration 5				
4	4	4	4	V

Step VII. Developing digraph

After level partitioning, all the factors numbered from "F1" to "F8" are arranged in hierarchical order in the form of a digraph. A digraph is a visual depiction of variables on an equal footing. Lower-level components make up the top of the digraph, whereas higher-level factors make up the base. The right interpretation of the relationships between the factors as depicted in the digraph is used to create the overall interpretative structural model. In this diagram, factors are represented by nodes, and arrows are used to indicate the direction of linkages between the nodes. Important transitive linkages are shown by dotted lines, while direct links are shown by direct lines. The final digraph with transitivity links is shown in Figure 2.

Step VIII. Interpretive matrix

All the interactions in the digraph are translated by 1 to form a binary interaction matrix. The binary interaction matrix and interpretive matrix are presented in Tables 3 and 4, respectively.

Step IX. Total interpretive structural model

The final TISM-based model for the factors affecting the e-waste management of producers is developed on the basis of the information obtained from the interpretive matrix and digraph. The nodes in the digraph are substituted by the interpretation of the inhibitors placed in boxes. A final TISM-based model is presented in Figure 3.

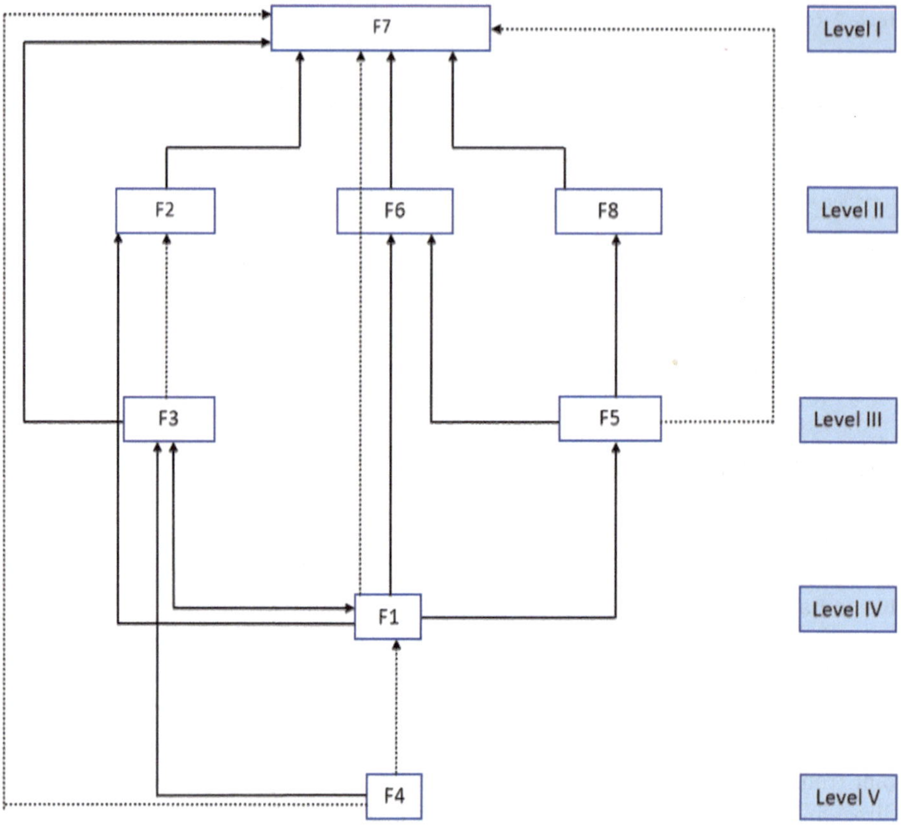

Figure 2: Diagraph with significant transitive links.

Table 3: Binary interaction matrix.

	F1	F2	F3	F4	F5	F6	F7	F8
F1	–	1	1	0	1	1	1*	0
F2	0	–	0	0	0	0	1	0
F3	1	1*	–	0	0	0	1	0
F4	1*	0	1	–	0	0	1*	0
F5	0	0	0	0	–	1	1*	1
F6	0	0	0	0	0	–	1	0
F7	0	0	0	0	0	0	–	0
F8	0	0	0	0	0	0	1	–

Note: 1* entries depict a significant transitivity relationship.

Table 4: Interpretive matrix.

	F1	F2	F3	F4	F5	F6	F7	F8
F1	–	Educated producers will also educate consumers about effective e-waste management practices	Educated producers will be more aware of the harmful effects of e-waste	–	Educated producers will be able to train the organization about e-waste management	Educated producers will be able to develop better infrastructure for e-waste management	Educated producers will be able to shape consumers' attitudes toward effective e-waste management	–
F2	–	–	–	–	–	–	If consumers are educated then it will shape their attitude toward effective e-waste management	–
F3	Producers' awareness about e-waste management will educate them about effective e-waste management practices	Producers' awareness about e-waste management rules will help in educating consumers also	–	–	–	–	Producers' awareness about e-waste will also affect consumers' attitudes toward e-waste management	–
F4	Appropriate government laws will educate producers about effective e-waste management practices	–	If government laws are in place, they will create greater awareness among producers about effective e-waste management practices	–	–	–	Appropriate government laws will develop a positive consumer attitude toward e-waste management	–

(continued)

Table 4 (continued)

	F1	F2	F3	F4	F5	F6	F7	F8
F5	–	–	–	–	–	A trained organization will develop better infrastructure for effective e-waste management	If producers are trained about effective e-waste management practices, they will also influence the attitude of consumers to participate in such practices	Trained Producers know how to incentivize customers for effective e-waste management
F6	–	–	–	–	–	–	If appropriate infrastructure is in place for e-waste disposal, it will develop a positive consumer attitude to participate in e-waste management practices	–
F7	–	–	–	–	–	–	–	–
F8	–	–	–	–	–	–	If consumers are incentivized for participating in e-waste management practices, it will develop a positive attitude in consumers toward e-waste management	–

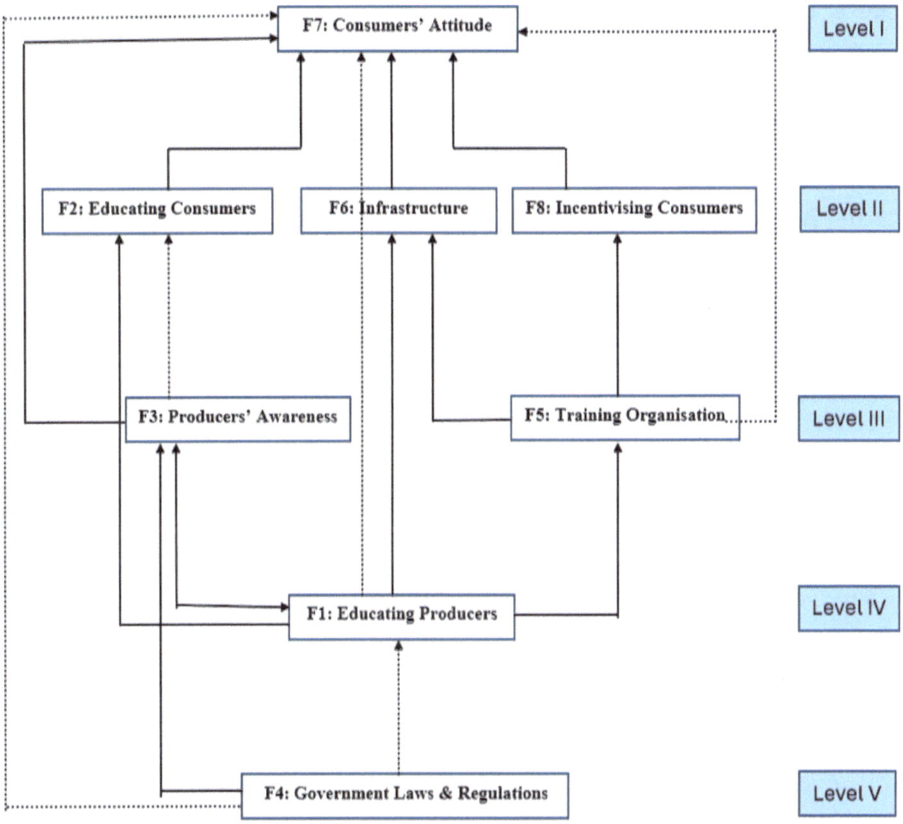

Figure 3: TISM-based model for factors affecting e-waste management of producers.

4.2 Classification of factors: MICMAC analysis

The dependence and driving power of the factors are analyzed by performing MICMAC analysis. MICMAC analysis helps in identifying the factors that drive the entire system. For each factor, the driving power and dependence are represented by "1" in the rows and columns, respectively. The analysis represents the factors into four clusters, which are:

Cluster I: autonomous factors

These factors represent weak dependence as well as driving power. These factors have hardly any links and are comparatively detached from the system. There is no factor in this cluster.

Cluster II: dependent factors

The factors in this cluster represent weak driving power but strong dependence. There are four factors in this cluster, i.e., F2, F6, F7, and F8.

Cluster III: linkage factors

These factors represent strong driving power as well as dependence. Any action conducted on these factors will influence others too, thus they are unstable. These also take into consideration any feedback effect on themselves. There are three factors in this cluster, i.e., F1, F3, and F5.

Cluster IV: independent factors

These factors represent weak dependence but strong driving power. They are known as key factors since they have a very high driving power. There is one factor in this cluster, i.e., F4.

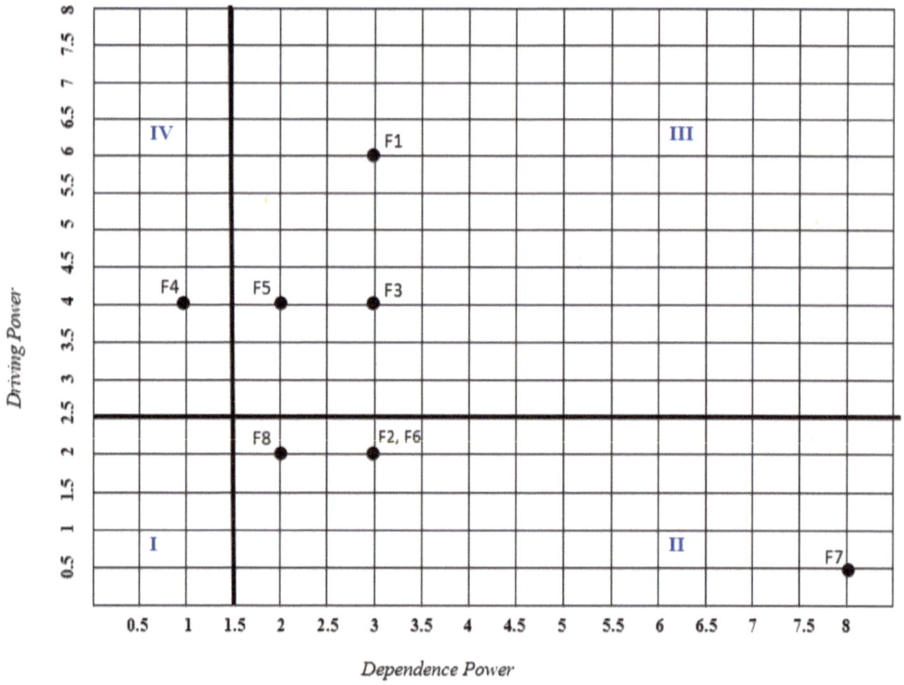

Figure 4: MICMAC analysis.

5 Discussion and implications

The study identifies the key factors that affect producers' behavior toward e-waste management and presents their hierarchical linkages by using TISM. Such analysis aids in the development of clear knowledge of the driving-dependent relationship among these factors. The TISM analysis indicates that these factors are spread across five levels, wherein, Government Laws and Regulations (F4) is at Level V; Educating Producers (F1) is at Level IV; Producers' Awareness (F3) and Training Organization (F5) are at Level III; Educating Consumers (F2), Infrastructure (F6), and Incentivizing Customers (F8) are at Level II; and Consumers' Attitude (F7) is at Level I.

Thus, the results indicate that Government Laws and Regulations (F4) drive the entire system and play a very important role in E-waste management from the producers' perspective. Appropriate Government Laws and Regulations create awareness among producers about effective e-waste disposal and thus occupy the bottom position in the hierarchical model. Consequently, the implementation of Government laws and regulations is the first step toward effective e-waste management in any nation. Educating Producers (F1) is at Level IV, which indicates that implementation of appropriate Government laws and regulations educates producers about effective e-waste disposal mechanisms. E-Waste (Management) Rules, 2016, introduced the concept of "Extended Producer Responsibility" (EPR), which made producers or manufacturers liable for the safe disposal of electronic goods. After the EPR policy's effect, the producer's role in managing e-waste has become critical. The E-Waste Rules inflict the producers with greater responsibility to manage e-waste. Thus, only educated producers will be able to take this responsibility of effective e-waste management, thus this factor occupies Level IV in the model.

Beyond this, the framework shows three paths for managing e-waste. One path shows that Government Laws and regulations lead to awareness in producers, which ultimately affects the consumer's attitude toward e-waste management. In literature, it is pointed out that the behavior of consumers is one of the most important factors for e-waste management. In order to effectively solve the present E-waste situation, worldwide experiences on consumers' e-waste disposal behavior and knowledge may be beneficial [28]. Another path identified in the model shows that educated producers train the organization and the employees, which leads to the development of appropriate infrastructure for effective disposal of e-waste. Appropriate e-waste disposal facilities will positively shape the consumer's attitude toward effective e-waste management. The third path shows that trained producers will incentivize consumers in order to encourage them to actively participate in e-waste disposal. Consequently, it will positively shape consumers' attitudes toward e-waste management. Due to a lack of incentives for consumers, it is observed that only 5–15% of consumers participate in formal e-waste management [16]. Convenience in recycling and economic benefits is considered an important factor in determining consumers' attitudes toward e-waste management. Considering recycling to be troublesome and time-consuming are the major reasons for the majority of consumers for not participating in e-waste recycling activities [29]. An inherent preference by consumers for the informal sector was being noticed due to the convenient services and better competitive prices being provided by the informal sector [30]. Thus, incentivizing consumers toward formal e-waste disposal is an important factor in effective e-waste management.

Different businesses have alternatives about how to manage waste in accordance with their practices owing to these routes. Due to the fact that they are on the same level, these two bifurcation paths have a similar impact on waste management. Any of the previously described pathways can be taken to advance to the next level.

Further, the study undertakes MICMAC analysis, which provides information about the relative importance and interdependencies among these factors. As presented in Figure 4, there are no autonomous factors in Cluster I. The autonomous factors are weak dependents and drivers, which do not represent any significant impact on e-waste management by producers. Consequently, it indicates that all the factors in the model are connected and play a significant role in affecting the e-waste management by producers.

Cluster II contains the following factors: Educating Consumers (F2), Infrastructure (F6), Consumers' Attitude (F7), and Incentivizing Consumers (F8). Among these factors, Educating Consumers (F2), Infrastructure (F6), and Incentivizing Consumers (F8) occupy the same level in the TISM hierarchy, i.e., Level II, while Consumers' Attitude (F7) occupies the top level in the TISM hierarchy, i.e., Level I. Factors in this cluster have high dependence but weak driving power. It is evident from the TISM model that these factors are highly dependent on other factors in the model since they occupy the top level in the hierarchical model. As mentioned above, proper e-waste infrastructure, consumers' education, and incentives for consumers are influenced by producers' awareness and education, as well as government laws and regulations with respect to e-waste management. Implementation of e-waste management laws and regulations by the government helps in educating producers and creating awareness about effective ways of disposing of e-waste. Further, these educated and informed producers also educate and incentivize consumers about effective mechanisms for e-waste management. Educated producers also participate in developing appropriate infrastructure for formal e-waste disposal by consumers. Finally, formal e-waste disposal infrastructure, consumers' education toward e-waste management, and incentivizing consumers to participate in formal e-waste disposal positively shape the attitude of consumers toward effective e-waste management. It is observed that consumers' behavior toward e-waste management is influenced by producers' behavior toward e-waste management as well as government rules and regulations for the appropriate disposal of e-waste. Consequently, it becomes imperative for the government to undertake appropriate initiatives and implement e-waste management rules and regulations for effective disposal.

Cluster III contains Educating Producers (F10), Producers' Awareness (F3), and Training Organization (F5). Factors in this cluster have high driving power as well as dependence. In the TISM model, it is evident that producers' education, producers' awareness, and organizational training are strongly dependent on government laws and regulations; and they have a strong effect on consumers' education, infrastructure, incentivizing consumers, and consumers' attitude.

Cluster IV contains Government Laws and Regulations (F4). Factors in this cluster have strong driving power and low dependence. As evident in the model, Government laws and regulations are not dependent on other factors; however, they are driving the entire system. Thus, government laws and regulations are the key factor and form the base of the TISM hierarchy. Consequently, as the first step, it is of utmost importance for any nation to implement government laws and regulations for the effective disposal of e-waste. Such laws and regulations shall clearly define producers' as well as consumers' responsibility in e-waste management. The government shall also organize several programs to make producers as well as consumers aware of the harmful effects of e-waste and their corresponding responsibility in e-waste management.

Further, the triggers employed in this investigation are informed by the FBM [10]. By elucidating the interconnections among these triggers – categorized as sparks, facilitators, and signals – the study opens avenues for future research, proposing the integration of these triggers into established behavioral frameworks, such as the theory of planned behavior [31]. This integration is anticipated to enhance understanding of how to convert producers' intentions into actionable outcomes. Current behavioral models often encounter challenges in bridging the intention-behavior gap; future researchers may address this gap by examining the moderating effects of these triggers on both intentions and the hierarchical relationships among various triggers, as delineated in the TISM model. Prior studies have suggested that the explanatory power of the Theory of Planned Behavior could be augmented through the incorporation of these triggers [32].

According to Fogg [10], three specific types of triggers are identified as effective in eliciting behavior among individuals with varying levels of motivation and ability. A facilitator serves as a nudge for individuals who are motivated but lack the necessary ability; a spark ignites motivation in those who possess the ability but lack the corresponding drive; and a signal acts as a reminder for individuals who are both motivated and capable. This study elucidates the hierarchical relationships among these triggers and, when contextualized within the FBM, offers practical insights for managers, policymakers, and businesses in formulating strategies to encourage producers to fulfill their e-waste management responsibilities.

For example, producers exhibiting high levels of self-motivation to manage e-waste are likely to implement educational and awareness programs aimed at both producers and consumers (F2, F3), alongside training initiatives (F5) within their organizations. Conversely, producers characterized by high motivation yet limited ability may require facilitation through government regulations (F4). On the other hand, producers equipped with the necessary infrastructure and training capabilities but lacking motivation can benefit from spark triggers, such as educational programs (F1), which can inspire them to further promote awareness (F3) and provide training (F5). For those who are both motivated and capable, reminders to incentivize consumers (F8) for the proper disposal of electronic goods are of paramount importance.

6 Conclusion and limitations

The study investigates the factors influencing producers' intentions to manage e-waste in India, a country facing significant challenges due to a rapid increase in electronic waste generation. Utilizing in-depth interviews and the FBM, the research identifies critical triggers – classified as sparks, facilitators, and signals – that can bridge the intention-behavior gap. Employing TISM and MICMAC, the study reveals hierarchical relationships among these triggers, revealing the foundational role of government regulations in driving producers to fulfill their e-waste management responsibility.

Through in-depth interviews, the study finds that producers' behavior regarding e-waste management results from a complex interplay of various triggers. These include regulatory requirements (facilitator trigger), economic incentives (signal triggers), educational and awareness programs for both producers and consumers (spark triggers), training programs for producers (signal triggers), and the availability of infrastructure (signal triggers). The study further maps the multilayered linkages between these triggers using TISM and subsequently applies MICMAC to analyze them in terms of their driving-dependence power.

Findings also indicate that the regulatory environment underpins the entire e-waste management system for producers, demonstrating high dependence according to MICMAC. The study contributes significantly to refining theoretical frameworks for understanding behavior by emphasizing the critical role of triggers in narrowing the intention–behavior gap, especially in the case of e-waste management behavior.

The study provides actionable insights for producers, governments, and other stakeholders, advocating for the integration of the triggers into existing behavioral frameworks to enhance compliance with e-waste responsibilities. Furthermore, this study underscores the relevance of the 2030 Agenda for Sustainable Development, particularly in the context of sustainable consumption and production, as outlined in SDG 12.

References

[1] United Nations Institute for Training and Research (UNITAR), & International Telecommunication Union (ITU). (2024). Global e-waste monitor 2024: Electronic waste rising five times faster than documented e-waste recycling. https://ewastemonitor.info/wp-content/uploads/2024/12/GEM_2024_EN_11_NOV-web.pdf

[2] United Nations Conference on Trade and Development, 2024, Digital Economy Report: Shaping an environmentally sustainable and inclusive digital future. United Nations Conference on Trade and Development, Geneva, 2024.

[3] Nivedha R, Sutha DA, The challenges of electronic waste (e-waste) management in India. Eur J Mol Clin Med, 2020, 7(03), 4583–4588.

[4] Sengupta D, Ilankoon IM, Kang KD, Chong MN, Circular economy and household e-waste management in India: Integration of formal and informal sectors. Miner Eng, 2022, 184, 107661.

[5] Arya S, Kumar S, E-waste in India at a glance: Current trends, regulations, challenges and management strategies. J Clean Prod, 2020, 271, 122707.

[6] Kumar S, Agarwal N, Anand SK, Rajak BK, E-waste management in India: A strategy for the attainment of SDGs 2030. Mater Today Proc. 2022, 60, 811–814.

[7] Debnath, B., Das, A., & Das, A. (2022). Towards a circular economy in e-waste management in India: Issues, challenges, and solutions. Circular Economy and Sustainability, 2, 523–543. https://doi.org/10.1007/s43615-021-00074-8

[8] Wall R, Devine-Wright P, Mill GA, Interactions between perceived behavioral control and personal-normative motives: Qualitative and quantitative evidence from a study of commuting-mode choice. J Mixed Methods Res, 2008, 2(1), 63–86.

[9] Kraft P, Rise J, Sutton S, Roysamb E, Perceived difficulty in the theory of planned behaviour: Perceived behavioural control or affective attitude? British J Soc Psychol, 2005, 44(3), 479–496.

[10] Fogg, B. J. (2009). A behavior model for persuasive design. In Proceedings of the 4th International Conference on Persuasive Technology (pp. 1–7). Association for Computing Machinery. https://doi.org/10.1145/1541948.1541999

[11] Eisenhardt KM, Graebner ME, Theory building from cases: Opportunities and challenges. Acad Manag J, 2007, 50(1), 25–32.

[12] Sushil, Interpreting the interpretive structural model. Glob J Flex Syst Manag, 2012, 13(2), 87–106.

[13] Wang B, Ren C, Dong X, Zhang B, Wang Z, Determinants shaping willingness towards on-line recycling behaviour: An empirical study of household e-waste recycling in China. Resour Conserv Recycl, 2019, 143, 218–225.

[14] Rabani B, Thakur B, Recycling potential of e-waste for Jammu city. Int J Progress Res Sci Eng, 2020, 1, 29–32.

[15] Albuquerque CA, Mello CHP, Paes VC, Balestrassi PP, Souza LB, Electronic junk: Best practice of recycling and production forecast case study in Brazil, In: Lecture Notes in Management and Industrial Engineering. Springer, Berlin/Heidelberg, Germany, 2018, 127–134.

[16] Bhaskar K, Turaga RMR, India's e-waste rules and their impact on e-waste management practices: A case study. J Ind Ecol, 2018, 22(4), 930–942.

[17] Armijo C, Puma A, Ojeda S, A set of indicators for waste management programs, In: 2nd International Conference on Environmental Engineering and Applications. LACSIT Press, Singapore, 2011.

[18] Henzler MP, Hemkhaus M, Frederik E, Sinha S, Mahes P, Mehra G, Creating successful formal-informal partnerships in the Indian e-waste sector. Practical Guidance Indian E-Waste Rules, 2018, https://www.adelphi.de/en/publication/creating-successful-formal-informal-partnerships-indian-e-waste-sector.

[19] Kaushik PR, Herat S, Current state of e-waste management in India. IJEWM, 2020, 25(3), 322–339.

[20] Borthakur A, Policy implications of e-waste in India: A review. IJEWM, 2016, 17(3–4), 301–317.
[21] Robinson BH, E-waste: An assessment of global production and environmental impacts. Sci Total Environ, 2009, 408(2), 183–191.
[22] Srivastava AK, Sushil, Modeling organizational and information systems for effective strategy execution. J Enterp Inf Manag, 2015, 28(4), 556–578.
[23] Dubey R, Gunasekaran A, Wamba SF, Bag S, Building theory of green supply chain management using total interpretive structural modeling (TISM). IFAC-PapersOnLine, 2015, 48(3), 1688–1694.
[24] Warfield, J. N. (1979). Development of an interpretive structural model and strategies for implementation based on a descriptive and prescriptive analysis of resources for environmental education/studies. In A sourcebook for the design of a regional environmental learning system: Conducting collective inquiry (Vol. IV). U.S. Department of Housing and Urban Development.
[25] Jharkharia S, Shankar R, IT enablement of supply chains: Modeling the enablers. Int J Product Perform Manag, 2004, 53(8), 700–712.
[26] Jena J, Sidharth S, Thakur LS, Kumar Pathak D, Pandey VC, Total interpretive structural modeling (TISM): Approach and application. J Adv Manag Res, 2017, 14(2), 162–181.
[27] Prasad UC, Suri RK, Modeling of continuity and change forces in private higher technical education using total interpretive structural modeling (TISM). Glob J Flex Syst Manag, 2011, 12(3), 31–39.
[28] Borthakur A, Govind M, Emerging trends in consumers' e-waste disposal behaviour and awareness: A worldwide overview with special focus on India. Resour Conserv Recycl, 2017, 117, 102–113.
[29] Wang Y, Luo C, Li J, Yin H, Li X, Zhang G, Characterization of PBDEs in soils and vegetations near an e-waste recycling site in South China. Environ Pollut, 2011, 159(10), 2443–2448.
[30] Awasthi AK, Li J, Management of electrical and electronic waste: A comparative evaluation of China and India. Renew Sustain Energy Rev, 2017, 76, 434–447.
[31] Ajzen I, The theory of planned behavior. Organ Behav Hum Decis Process, 1991, 50, 179–211.
[32] Wang Y, Kang H-K, The role of triggers in physical activity among college students: An extended model of the theory of planned behavior. Behav Sci, 2024, 14(4), 328.

John Tsado Mathew*, Jonathan Hussaini, Abel Inobeme, Elijah Yanda Shaba, Yakubu Azeh, Musah Monday, Tanko Ezekiel, Aliyu Mohammed Sakpe, Etsuyankpa Bini Muhammad, Ismaila Aliyu Oga, Mohammed Chado Isah, Adam Ibn Aliyu, Muhammad Aishetu Ibrahim, Jibrin Noah Akoji, Charles Oluwaseun Adetunji, Amos Mamman

Valorization strategies and regulatory frontiers for efficient e-waste management

Abstract: The escalating generation of electronic waste (e-waste) has spawned pressing environmental, economic, and public health issues. Valorization options, including metal recycling, plastic recycling, energy recovery, and the creation of circular economy systems, provide sustainable avenues for converting e-waste into valuable resources. In parallel, policy regimes are developing to implement extended producer responsibility, enhance collection infrastructure, and encourage eco-design of electrical and electronic equipment. Nevertheless, enforcement gaps, global regulatory inconsistencies, and informal recycling activities undermine achievements. Policy reinforcement, technology innovations, and harmonization of sustainable valorization frameworks are paramount in ensuring effective and safe e-waste management globally.

Keywords: E-waste, life cycle assessment, recycling, regulatory frameworks, valorization

***Corresponding author: John Tsado Mathew**, Department of Chemistry, Ibrahim Badamasi Babangida University, Lapai, Niger State, Nigeria, e-mail: johntsadom@gmail.com
Jonathan Hussaini, Yakubu Azeh, Musah Monday, Muhammad Aishetu Ibrahim, Amos Mamman, Department of Chemistry, Ibrahim Badamasi Babangida University, Lapai, Niger State, Nigeria
Abel Inobeme, Elijah Yanda Shaba, Department of Chemistry, Edo State University, Uzairue, PMB 04, Auchi, Edo State, Nigeria
Tanko Ezekiel, Adam Ibn Aliyu, Department of Chemistry, Federal University of Technology, Minna, Nigeria
Aliyu Mohammed Sakpe, Department of Basic Sciences, Niger State College of Agriculture, Mokwa, Nigeria
Etsuyankpa Bini Muhammad, Ismaila Aliyu Oga, Department of Chemistry, Federal University of Lafia, Nasarawa State, Nigeria
Mohammed Chado Isah, Department of Biology, Ibrahim Badamasi Babangida University, Lapai, Niger State, Nigeria
Jibrin Noah Akoji, Department of Chemistry, Baze University, Abuja, Nigeria
Charles Oluwaseun Adetunji, Applied Microbiology, Biotechnology and Nanotechnology Laboratory, Department of Microbiology, Edo State University, Uzairue, PMB 04, Auchi, Edo State, Nigeria

 | https://doi.org/10.1515/9783112248072-011

1 Introduction

Electronic waste, commonly referred to as e-waste, has become the fastest-growing category of waste in the modern world. As society becomes increasingly dependent on digital devices, the volume of discarded electronics continues to rise sharply. Global estimates indicate that approximately 53 million tons of e-waste were generated in 2021, and this figure is projected to reach 120 million tons by 2030. This growing burden is more than a landfill issue – it represents a significant environmental, economic, and social challenge that urgently demands innovative and sustainable solutions [1]. E-waste exhibits a dual character that makes its management both critical and complex. On one side, it contains highly toxic substances such as lead, mercury, cadmium, and brominated flame retardants, which can contaminate soil and groundwater when improperly handled, posing serious health and ecological risks. On the other side, e-waste is a valuable reservoir of recoverable materials including gold, silver, palladium, copper, aluminum, nickel, rare-earth elements like neodymium and dysprosium, and various reusable organic components. This combination of risks and opportunities highlights the need for strategies that address environmental hazards while unlocking economic value [2]. Several key factors drive the global push toward e-waste valorization, which focuses on maximizing resource recovery and minimizing waste. Resource scarcity is among the major drivers, as many of the minerals required for electronic manufacturing are scarce and geographically concentrated. Extracting them from e-waste reduces reliance on environmentally damaging primary mining. Economic incentives also play an important role, since recovered metals have high market value, encouraging investment in advanced recycling technologies. Increasing regulatory pressure from governments further compels industries to adopt cleaner waste-management practices. Rising public awareness is also significant, as consumers increasingly demand corporate responsibility and sustainability. At the same time, technological advancements such as robotics, artificial intelligence, and improved chemical processes are making material recovery more efficient and feasible [3]. The regulatory environment governing e-waste management is complex and varies significantly across regions. International conventions such as the Basel Convention of 1989 regulate the transboundary movement of hazardous waste, including e-waste. The Rotterdam Convention of 1998 introduced prior informed consent procedures for hazardous chemicals, while the Bamako Convention of 2000 bans the export of hazardous waste to African countries. National regulations differ widely; for instance, the European Union's WEEE Directive places responsibility on producers to fund collection and recycling, while the United States applies the Resource Conservation and Recovery Act (RCRA) alongside state-specific rules. In many developing countries, regulations exist but enforcement remains weak due to limited institutional capacity [4, 5].

Valorization strategies fall into several major categories. Material recovery techniques include mechanical methods such as shredding and magnetic separation, alongside chemical approaches like hydrometallurgy and pyrometallurgy, each offering trade-offs between cost, purity, and energy use. Advanced recycling technologies – such as direct recycling, improved disassembly guided by sensor-based systems, and the use of bio-based solvents or electrochemical processes – are gaining attention as low-toxicity alternatives. Circular economy models promote designing electronics for easy disassembly, repair, and refurbishment, enabling longer product lifespans and reducing waste generation. Emerging opportunities include the use of nanomaterials, innovative waste-to-resource pathways, and bioremediation techniques, all of which present promising new frontiers for e-waste utilization [6]. To keep pace with technological progress, regulatory systems must also evolve toward more flexible and adaptive frameworks. Approaches such as regulatory sandboxes allow innovators to test new recycling technologies under temporary, controlled conditions that inform future policy development. Public-private partnerships help align government incentives – such as subsidies or tax relief – with private sector expertise and capacity. Data-driven compliance systems that employ Internet of Things (IoT) sensors and blockchain can track e-waste from its source to final processing, improving transparency and discouraging illegal dumping. These innovations promote the rapid spread of best practices and strengthen the overall resilience of e-waste management systems [7, 8]. A holistic perspective is crucial for the long-term success of e-waste valorization. Beyond the technical solutions, the social and economic dimensions must be acknowledged. Recycling industries have the potential to generate employment, while community involvement can improve collection efficiency. Issues such as data security in decommissioned devices also require careful consideration. Effective management therefore depends on collaboration between manufacturers, recyclers, policymakers, researchers, and civil society organizations – all working together to align objectives and develop a shared vision for sustainability [9].

The broader roadmap for the future integrates technological, economic, and regulatory insights to shape a more sustainable e-waste system. This includes identifying scalable and efficient material recovery methods, highlighting advanced recycling technologies with reduced environmental impacts, and exploring circular economy models capable of transforming linear consumption patterns into closed-loop systems. It also involves analyzing innovative regulatory approaches and drawing lessons from global case studies to understand successful strategies across different contexts [3]. Addressing barriers such as inadequate funding, limited technological capacity, and governance gaps will be essential in building an effective and comprehensive management framework. Ultimately, the goal is to shift the perception of e-waste from a growing environmental burden to a valuable resource pool that supports sustainability, economic growth, and responsible environmental stewardship [10, 11].

2 E-waste composition and characterization

Electronic waste, commonly referred to as e-waste, encompasses discarded electrical and electronic equipment ranging from small devices like mobile phones to large installations such as data center servers. As global dependence on electronic technologies continues to rise, the composition of e-waste has become increasingly complex, creating significant challenges for recovery, recycling, and safe disposal. A clear understanding of what e-waste contains is essential for developing efficient circular-economy systems, preventing hazardous environmental releases, and unlocking the economic value embedded in discarded devices [12, 13]. In general, e-waste is composed of several major material categories, each contributing differently to its overall weight and environmental impact. Metals – especially precious metals such as gold, silver, palladium, and rhodium, along with base metals like copper, nickel, aluminum, and iron – make up a substantial portion. These may account for between 10% and 30% of a typical electronic device and often represent the highest economic value due to their recoverability and market demand. Plastics constitute another major fraction, typically between 20% and 40%, and include materials such as polypropylene, polyethylene terephthalate, polyvinyl chloride, and epoxy resins [9].

Many of these plastics contain flame retardants and pigments that complicate recycling and may introduce hazardous compounds. Glass and ceramics, including fused silica, borosilicate, and lead-based glass, add another 5–10%. Electrochemical components such as lithium-ion and alkaline batteries introduce electrolytes and solvents that can be flammable or toxic. Organic materials, fillers, and other miscellaneous components contribute smaller portions, while hazardous additives – such as mercury, cadmium, arsenic, lead, brominated flame retardants, and Perfluoroalkyl and Polyfluoroalkyl Substances (PFAS) – are present in trace quantities but pose serious risks. The exact composition varies depending on the device type and manufacturing era; for instance, modern smartphones contain high amounts of copper but minimal lead, whereas older cathode-ray tube televisions contain large volumes of lead-glass [14, 15]. Given the heterogeneous nature of e-waste, proper sampling and pretreatment are essential before conducting any analytical assessment. Industry-based sampling often relies on shredding or mechanical separation to isolate major fractions such as metals, plastics, and glass. Randomized sampling helps average out variability across diverse devices, and the use of certified reference materials ensures analytical accuracy. Pretreatment typically involves shredding, magnetic separation, and density-based sorting to reduce complexity; however, it must be performed carefully to avoid altering chemical species or introducing contamination [16]. Physical characterization techniques provide insight into the structural and morphological features of e-waste. Measurements of density and particle size help differentiate between plastics, metals, and glass. X-ray diffraction enables identification of crystalline materials, while scanning electron microscopy reveals micro-scale surface features, especially useful for studying batteries and semiconductor components. Thermogravimetric analysis further distinguishes between organic and inorganic fractions by monitoring weight loss during controlled heating [17, 18]. Chemical characterization focuses on determining the elemental and molecular composition of e-waste. For metals, inductively coupled plasma optical emission spectroscopy offers highly sensitive multi-element analysis and is commonly paired with acid digestion techniques. Atomic absorption spectroscopy provides lower-cost, targeted measurement of specific metals, while X-ray fluorescence allows nondestructive elemental mapping of bulk materials. Plastics and organic compounds are often identified using Fourier-transform infrared spectroscopy, which rapidly distinguishes polymer types [19].

Gas chromatography–mass spectrometry is used to detect volatile organics, flame retardants, and plasticizers, while pyrolysis-GC-MS helps analyze polymer structure by breaking plastics into identifiable monomers. Hazardous compounds such as PFAS or organophosphorus flame retardants are frequently assessed using high-performance liquid chromatography, and specialized X-ray methods like XANES can determine oxidation states of heavy metals. Electrochemical techniques, including anodic stripping voltammetry, detect trace metals in leachates produced during toxicity assessments [20, 21]. Advanced spectroscopic tools provide deeper insight into the micro- and nano-scale features of electronic materials. Raman spectroscopy helps identify crystalline carbon-based materials and can be applied at micro-scale resolutions. X-ray photoelectron spectroscopy reveals chemical bonding environments and oxidation states on device surfaces, while electron energy loss spectroscopy, often combined with transmission electron microscopy, enables characterization of nanoscale domains containing light elements [22].

Because modern electronics increasingly incorporate micro- and nano-fabricated components, specialized analytical instruments are necessary to capture these small but important fractions. Laser ablation ICP-MS allows highly localized sampling of precious metals on printed circuit boards, while nano-SIMS produces detailed isotopic maps of contaminants and rare-earth elements at sub-micrometer scales. These high-resolution methods generate crucial data for designing selective, high-efficiency recovery strategies [23]. Once the compositional data are obtained, they are integrated into broader analytical and decision-making frameworks. Mass-balance models quantify the total stock of valuable materials within a waste stream, providing insight into potential recovery yields. Life cycle assessments evaluate environmental impacts associated with different recycling technologies. Economic analyses combine material prices, processing costs, and expected recovery efficiencies to assess feasibility. Increasingly, machine learning methods are used to cluster compositional profiles, predict waste characteristics, and optimize sorting and collection pathways [24]. E-waste remains highly heterogeneous, with substantial variability even among devices of the same model. Hazardous additives such as brominated flame retardants can interfere with analytical detection and increase handling risks. Safety concerns are heightened due to the presence of toxic metals, volatile electrolytes, and legacy materials like lead-based glass. In addition, emerging technologies such as flexible displays and nanomaterials often lack standardized compositional data, limiting the accuracy of predictive models. However, significant opportunities exist. The development of standardized sampling protocols and reference materials would reduce laboratory-to-laboratory inconsistencies. Automation, through technologies like inline XRF or laser-ablation systems, can rapidly screen large waste volumes (Table 1) [25].

Table 1: Composition and characterization of e-waste.

E-waste component	Description/examples	Significance/issues	References
Metals	Includes ferrous metals (iron and steel), nonferrous metals (copper and aluminum), and precious metals (gold, silver, and palladium).	High economic value; can be recovered through mechanical, chemical, or biological processes; improper disposal can cause soil and water contamination.	[26]
Plastics	Covers thermoplastics, thermosetting plastics, and mixed polymer materials used in casings, circuit boards, and insulation.	Can be recycled into new products; may release toxic compounds if incinerated improperly.	[27, 28]
Glass	Mostly from display screens, CRTs, and LCD panels.	Can be reused in new displays or construction materials; contains lead in CRTs which poses environmental hazards.	[29]
Circuit boards	Printed circuit boards containing metals, plastics, and flame retardants.	Rich source of precious metals; challenging to recycle due to mixed materials.	[30]
Batteries	Lithium-ion, nickel-cadmium, lead-acid batteries from laptops, phones, and vehicles.	Contain heavy metals and toxic electrolytes; improper disposal causes environmental and health risks; can be valorized for metal recovery.	[10]
Cables and wires	Copper, aluminum, and insulation materials used for power and data transmission.	Valuable for metal recovery; improper disposal may release hazardous substances from insulation.	[2]
Other hazardous components	Mercury switches, fluorescent lamps, capacitors, and brominated flame retardants.	Toxic to human health and environment; require special handling and disposal methods.	[31]

3 Valorization strategies for e-waste

E-waste is no longer an environmental liability; it constitutes a dynamic resource base that can be converted into high-value products. Effective valorization requires a holistic approach that blends technology, design, policy, and market mechanisms. The following pragmatic roadmap illustrates the most promising strategies for transforming discarded electronic equipment into a competitive economic asset [9]. Before a device reaches the landfill, its design determines how readily it can be disassembled, refurbished, or recycled. Modern electronics benefit from modular architectures, standardized connectors, and surface-mount components that facilitate rapid disassembly. By embedding design-for-disassembly principles into the product life cycle, manufacturers reduce sorting costs and increase the quality of recovered materials. Incentives such as extended producer responsibility (EPR) schemes compel designers to adopt these practices, thereby turning the concept of "design for reuse" from a best practice into a regulatory requirement [32, 33]. Valorization begins with reliable collection. City-wide e-waste collection points, drop-off kiosks, and retailer take-back programs create high-volume streams that can be economically processed. Once collected, rapid-sorting technologies – optical sorting, X-ray fluorescence, and AI-driven robotic disassembly – enable the separation of valuable metals, plastics, and glass with minimal manual labor. The cleaner the feedstock, the higher the recovery rates and the lower the need for secondary processing steps [34]. Mechanical recycling remains the most cost-effective method for many e-waste streams, especially for high-density metals such as copper, gold, and palladium. The process typically involves shredding, magnetic separation, and density-based sorting. Recent advances in ultrasonic and acoustic separation allow for the recovery of finer metal particles, thereby raising the overall yield. Mechanical recycling is especially valuable for low-content items like small capacitors, where the recovery cost per gram is comparatively low [13, 35].

For components containing trace amounts of precious metals or rare-earth elements, purely mechanical methods are insufficient. Physical and chemical recovery techniques – such as hydrometallurgical leaching, bioleaching, and pyrometallurgical smelting – enable the selective extraction of high-value elements. For example, cyanidation can recover gold and silver from printed circuit boards, while ion-exchange processes can separate rare-earths from complex matrices. As the market price of these elements rises, the economic viability of such intensive processes improves, encouraging investment in specialized recovery facilities [36]. Recovered materials can feed into secondary manufacturing chains, creating new products that are as valuable – if not more so – than the original components. Recycled copper can be produced into high-conductivity wires for renewable energy infrastructure, while recycled plastics can be formulated into high-grade fibers for aerospace applications. Upcycling also encompasses refurbishing usable devices, turning them into "re-born" products that can be sold at a lower price point. This approach reduces waste and provides a sustainable supply of refurbished electronics to underserved markets [37].

To build trust in the circular economy, transparent supply chains are essential. Digital traceability platforms using blockchain technology can record every step of an e-waste item – from collection to final product – ensuring that recovered materials are genuinely recycled. This traceability provides proof of sustainability for end-users and can unlock premium pricing for "clean-label" products. Additionally, smart contracts can automatically trigger payments to responsible recyclers when certain recovery milestones are met, incentivizing compliance [38].

Governments can accelerate valorization through a mix of regulatory and incentive-based tools. EPR mandates compel manufacturers to take responsibility for end-of-life products, while feedstock taxes on virgin material encourage the use of recycled input. Grants and low-interest loans for recycling infrastructure development lower the barrier to entry for small and medium enterprises. Moreover, "green procurement" policies give recycled products a competitive advantage in public sector purchases, creating a stable demand anchor for recycled materials [39, 40]. PPPs can catalyze the development of integrated e-waste valorization ecosystems. Public entities bring regulatory oversight and infrastructure subsidies, while private firms contribute capital, technology, and market expertise. Successful examples include joint ventures that pool waste streams from multiple municipalities into a central recycling hub, thereby achieving economies of scale. Collaboration between universities and industry can also drive R&D into novel deconstruction and recovery technologies that are both economically viable and environmentally benign [41]. Even the most sophisticated valorization strategy fails without consumer participation. Educational campaigns that highlight the environmental and economic benefits of e-waste recycling can shift consumer behavior. Incentive schemes – such as deposit-return programs for electronic devices – encourage consumers to return old equipment for proper processing. Mobile apps that track personal recycling contributions and reward users with loyalty points or discounts further strengthen the social loop that sustains the circular economy [42, 43].

The e-waste valorization landscape is constantly evolving, driven by rapid advances in electronics design, materials science, and recycling technology. Continuous investment in process optimization – e.g., refining leach solutions to reduce acid consumption, deploying AI for predictive maintenance of recycling plants – ensures that the industry remains competitive. Moreover, cross-sector collaboration, such as applying e-waste recovery techniques to other industrial waste streams, expands the overall economic impact and drives systemic efficiencies [2]. Valorizing e-waste is a multilayered endeavor that demands coordination across the design, collection, processing, and market stages of the product life cycle. By integrating modular design, advanced sorting, mechanical and chemical recovery, secondary manufacturing, digital traceability, supportive policy, and consumer engagement, stakeholders can transform e-waste from a growing environmental challenge into a robust economic resource. The collective objective is clear: to build a circular ecosystem where electronic products are engineered, used, and reclaimed in a way that maximizes value while minimizing environmental footprints [2].

4 Economic and environmental assessment of valorization

Valorizing waste streams – whether agricultural residues, food industry by-products, or municipal solid waste – offers a compelling route to circular economy targets. This section evaluates the economic feasibility and environmental performance of a representative valorization pathway: converting lignocellulosic biomass into a high-value biochemicals platform (e.g., levulinic acid and 5-hydroxymethylfurfural). The assessment synthesizes capital and operating cost estimates, market dynamics, and lifecycle environmental metrics, thereby providing stakeholders with a balanced view of the trade-offs involved [44]. In the last decade, demand for bio-derived platform chemicals has surged, driven by strict petrochemical substitution policies in the EU and North America. Current market prices in 2024 are USD 1,200 per ton for levulinic acid and USD 1,500 per ton for 5-hydroxymethylfurfural. Assuming a 60% levulinic acid yield and a 30% 5-HMF yield from one ton of dry biomass, a single ton of feedstock can generate a revenue of USD 1,320 from levulinic acid and USD 450 from 5-HMF, totaling USD 1,770. These figures are conservative; co-products such as residual sugars and lignin-derived aromatics can increase net revenue by an additional 10–15% [45].

Operating expenses average USD 0.25 per kilogram of product for raw materials and utilities, which equals USD 250 per ton. When combined with product revenue, the gross margin per ton of biomass is roughly USD 1,520 [46, 47]. The life cycle assessment (LCA) of the valorization process was conducted in accordance with the ISO 14040 framework, which evaluates environmental burdens across all stages of a product's life. In this assessment, greenhouse gas (GHG) emissions were quantified from the point of feedstock cultivation through to the final disposal of residual wastes. The analysis covered key operational phases, including cultivation and transportation of the feedstock, pretreatment using enzymatic methods, catalytic conversion, product separation and purification, as well as waste handling activities [48]. The results of the assessment show that cultivation and transport contribute an estimated 50 kg CO_2-equivalent emissions per ton of product, while enzymatic pretreatment accounts for approximately 120 kg CO_2e. The catalytic conversion stage adds a further 80 kg CO_2e, and separation and purification processes contribute about 70 kg CO_2e. Waste handling is responsible for the remaining 10 kg CO_2e. Altogether, the full life cycle emissions amount to 330 kg CO_2-equivalent per ton of final product, illustrating the relative contribution of each operational phase to the overall environmental footprint [49]. Compared to conventional petrochemical routes (about 800 kg CO_2e per ton of levulinic acid), the bio-based process delivers a 59% reduction in emissions. This advantage is amplified when the feedstock is sourced from waste streams, effectively turning a negative-carbon activity into a positive one [50]. Beyond GHG savings, the process mitigates methane emissions by diverting anaerobic digestible organics from landfills. Using a typical landfill gas capture efficiency of 30%, the valorization pathway effectively sequesters methane that would otherwise emit 0.22 tons of CH_4 per ton of biomass, which equals 0.63 tons of CO_2e avoided. Combined with the direct CO_2e savings, total emissions for the process are reduced by 1.2 tons of CO_2e per ton of biomass processed [51].

Water usage is a critical factor; the plant consumes 5 L/kg of biomass for pretreatment and purification. Recirculation systems and membrane-based purification reduce fresh water demand to 2 L/kg. The residual lignin fraction can be sold as a low-grade carbon feedstock, such as for activated carbon production, or combusted to offset plant electricity needs, contributing to a closed-loop operation. Solid waste streams total less than 3% of the original feedstock mass, which can be incorporated into soil amendments or returned to the feedstock supply chain [52, 53]. The integrated facility generates several co-products that further enhance environmental benefits. Lignin-based phenolic resins, for instance, can replace petroleum-derived plastics, reducing the overall environmental burden of downstream applications. The surplus sugars from hydrolysis can be fermented to produce bio-ethanol or biobutanol, diversifying revenue streams and reducing single-product risk [54].

The economic analysis demonstrates that a 10-kW lignocellulosic platform chemical plant is financially viable, delivering attractive returns (IRR about 18%) within a moderate payback window of roughly 4 years. The environmental assessment underlines the process's ability to cut GHG emissions by up to 60% relative to conventional petrochemical routes, while also reducing waste and water footprints. Moreover, the co-product portfolio broadens market resilience and adds incremental value [55]. The valorization pathway presents a compelling case for investment from both a profitability and sustainability standpoint (Table 2). By converting otherwise discarded biomass into high-value chemicals, the project aligns with circular economy goals, regulatory incentives, and growing consumer demand for greener products. Continued optimization – through enzyme cost reductions, scale-up, and process integration – could further amplify both economic gains and environmental stewardship [56].

Table 2: Economic and environmental assessment of e-waste valorization strategies.

Assessment aspect	Description	Implications/significance	References
Economic feasibility	Evaluates the costs and benefits of e-waste valorization processes, including capital investment, operational costs, and potential revenue from recovered materials.	Determines whether a valorization strategy is financially viable and attractive for investors or governments. Cost-effective strategies encourage wider adoption and industrial scaling.	[3]

Table 2 (continued)

Assessment aspect	Description	Implications/significance	References
Environmental impact	Assesses the ecological footprint of valorization processes, including emissions, energy consumption, and the release of hazardous substances.	Helps identify environmentally friendly methods that minimize soil, air, and water pollution while recovering valuable resources. Supports sustainable waste management policies.	[57, 58]
Life cycle assessment (LCA)	A systematic approach to quantify environmental impacts throughout the entire life cycle of e-waste management, from collection to disposal or material recovery.	Provides a holistic view of the sustainability of valorization strategies, allowing optimization to reduce carbon footprint, resource depletion, and environmental hazards.	[59]
Material recovery efficiency	Measures the proportion of metals, plastics, and other materials successfully recovered from e-waste.	Higher recovery rates enhance economic returns, reduce demand for virgin resources, and support circular economy objectives.	[3]
Social and health considerations	Evaluates the impacts on human health, worker safety, and community well-being.	Ensures that valorization processes are safe for workers and communities, particularly in informal recycling sectors, and promotes responsible waste management practices.	[60]

5 Regulatory frameworks and policy frontiers

The velocity of technological and economic transformation has, in numerous contexts, eclipsed the cadence of policy innovation. Simultaneously, emerging risks – cybersecurity threats, climate-induced financial instability, and the ethical quandaries posed by nascent technologies – necessitate a reevaluation of the conventional regulatory toolkit. In this chapter, we provide an exhaustive survey of the contemporary regulatory landscape, scrutinizing the institutional architecture that sustains it, the dominant policy frontiers that command today's agenda, and the challenges that loom ahead. Our objective is to furnish a coherent framework for apprehending how public policy can keep abreast of, and shape, the swiftly evolving realms of finance, technology, and sustainability [61].

The traditional regulatory paradigm – characterized by sectoral statutes, supervisory agencies, and compliance regimes – was forged in an era when technological advancement was gradual. Today, that same paradigm is being stretched by fintech start-ups that circumvent bank intermediaries, data-driven investment platforms that operate transnationally, and climate-related physical risks that mandate novel risk-transfer mechanisms. Consequently, regulators confront a dual imperative: preserving the safeguards that have proven efficacious while cultivating new tools capable of addressing emerging challenges [62]. Regulators are experimenting with "sandbox" environments – controlled settings whereby firms can trial new products under regulatory supervision. The FCA's regulatory sandbox in the United Kingdom, for example, has become a model for balancing innovation with consumer protection. Concurrently, "regulatory technology" (RegTech) aids supervisors in monitoring compliance in real time, leveraging data analytics and machine learning to flag suspicious activity [62]. Digital assets and cloud-based services frequently transcend borders, generating jurisdictional puzzles. The European Union's Markets in Crypto-Assets (MiCA) regulation, still under consideration, aims to provide a single regulatory framework for digital assets across the EU. In contrast, the United States remains fragmented across federal and state regulations. An emerging trend is the shift toward voluntary, industry-led best-practice codes that can serve as a common denominator when formal harmonization stalls [63, 64]. Europe's "Ethics Guidelines for Trustworthy AI" emphasize transparency, accountability, and human oversight. The U.S. Consumer Financial Protection Bureau (CFPB) is exploring risk-based supervisory checkpoints for AI-driven credit decisions. Mandates for algorithmic transparency (e.g., the EU's proposed AI Act) require auditable decision logs. The principal challenge lies in reconciling innovation incentives with the imperative to prevent discriminatory outcomes and systemic risk [65]. Financial markets are increasingly exposed to climate risk – both physical (e.g., extreme weather) and transition (e.g., regulatory shifts). Policy initiatives comprise: A voluntary framework encouraging companies to disclose climate risk metrics. The EU's commitment to net-zero emissions by 2050 includes a "Climate Transition Taxation" on non-essential financial products, such as short-term debt that amplifies carbon emissions [66].

The dual mandate of protecting consumer data while enabling data-driven services generates a regulatory conundrum. The fine line between permissible data sharing and privacy infringement can become blurred, particularly when data is repurposed for AI training. Modeling climate risk is intrinsically uncertain. Adoption of disparate climate pathways (RCP 4.5 versus RCP 8.5) yields vastly divergent risk projections. Regulators must therefore balance prudence with flexibility, allowing firms to adjust to emerging scientific consensus [67].

Even the most comprehensive rules are only as effective as the supervisory capacity to enforce them. The rapid expansion of digital assets and decentralized finance (DeFi) presents novel challenges: many DeFi protocols operate, for all practical purposes, in a largely unregulated milieu, complicating law enforcement. Supplant one-size-fits-all rules with frameworks that adjust to the risk profile of each product or institution, thereby nurturing innovation while preserving robust safeguards [68].

Expand multilateral frameworks (e.g., Basel III, MiCA) and foster the creation of joint supervisory bodies for emerging digital assets. Deploy data analytics, AI, and blockchain for real-time monitoring and compliance. RegTech not only enhances enforcement but also diminishes compliance costs for firms. Commence with robust voluntary frameworks to build industry capacity, then phase in mandatory reporting as data quality and comparability improve [69]. Craft a clear, sector-agnostic standard for data collection, sharing, and protection that can be embedded into regulatory regimes. Leverage industry expertise to devise practical, enforceable rules, while safeguarding public interests. Employ scenario analysis and stress testing to anticipate future shocks. Regulators should periodically revise frameworks to reflect new scientific insights, market trends, and geopolitical developments [70].

Regulatory frameworks constitute the invisible scaffolding that sustains contemporary economies as they navigate an era of rapid technological evolution, heightened environmental risk, and shifting consumer expectations. The policy frontiers – data privacy, AI governance, climate regulation, fintech innovation, and ESG disclosure – are intricately interwoven, each amplifying the other. Policymakers face the twin challenge of crafting rules that are both forward-looking and grounded in proven risk management principles. The most successful regulatory landscapes will blend rigidity with flexibility, enforceability with innovation, and national priorities with global best practices. Only by doing so can regulators preserve financial stability, protect consumers, and foster an ecosystem wherein technology serves the public good rather than merely corporate profit [61].

6 Technological Innovations and Emerging Frontiers

The growing volume and complexity of electronic waste have created a pressing need for innovative technologies that enhance the efficiency, safety, and environmental sustainability of e-waste management. Traditional recycling methods, while effective in recovering metals and some materials, are often labor-intensive, energy-consuming, and limited in their ability to process heterogeneous e-waste streams. Emerging technological innovations are addressing these challenges by integrating automation, advanced material recovery techniques, digital tracking, and circular economy principles. These innovations not only improve material recovery but also minimize environmental impact and enhance economic viability, aligning closely with the principles highlighted in the economic and environmental assessments discussed in Table 3 [3, 71].

Automated dismantling systems represent one of the most significant advancements in e-waste processing. These systems utilize robotics and artificial intelligence to identify, segregate, and dismantle electronic devices with high precision. Unlike manual dismantling, which is labor-intensive and exposes workers to hazardous substances, automated systems efficiently separate components such as printed circuit boards, batteries, and displays while reducing contamination and minimizing occupational hazards. The integration of AI allows the system to adapt to different device types, enhancing recovery efficiency and ensuring that valuable materials, including precious and rare metals, are extracted effectively. By improving precision and throughput, automated dismantling aligns with economic objectives by reducing operational costs and increasing material recovery rates [72]. Advanced sorting and separation technologies complement automated dismantling by further refining the segregation of e-waste materials. Sensor-based sorting, optical recognition, eddy current separation, and density-based separation techniques enhance the accuracy of separating metals, plastics, and glass. These methods ensure that each material fraction is appropriately processed, reducing losses and contamination. For example, eddy current separation effectively extracts nonferrous metals such as aluminum and copper from shredded e-waste, while optical sensors can identify specific plastics for recycling. The application of these technologies improves overall material recovery efficiency, contributing to both economic feasibility and environmental sustainability as highlighted in Table 3 [73]. Bioleaching and microbial recovery have emerged as environmentally friendly alternatives for extracting metals from complex e-waste streams. Certain bacteria, fungi, and enzymes can solubilize metals, including copper, gold, and nickel, through natural metabolic processes. This method reduces the need for high-temperature processes or harsh chemicals, resulting in lower energy consumption and minimal toxic emissions. Although bioleaching is generally slower than chemical or physical methods, it is particularly effective for low-grade or complex waste that is otherwise challenging to process. The integration of biological techniques into a broader valorization strategy supports sustainable metal recovery and aligns with environmental assessment goals by minimizing ecological impact [74].

Hydrometallurgical methods and green chemistry approaches offer additional pathways for selective metal recovery. Utilizing aqueous solutions, organic acids, or eco-friendly solvents, these methods target specific metals with high efficiency. Compared to pyrometallurgical processes, hydrometallurgical techniques operate at lower temperatures and consume less energy, reducing greenhouse gas emissions and operational costs. Selective leaching and solvent extraction processes allow for the purification of valuable metals, enhancing the economic return of recycling operations and supporting circular economy objectives [75]. Thermal treatment technologies, including pyrolysis and gasification, provide solutions for recovering energy from non-recyclable fractions of e-waste, particularly plastics and organic materials. Pyrolysis decomposes plastic waste into fuels, syngas, and char under controlled temperatures, while gasification converts waste into a combustible gas mixture that can be used for electricity generation. These processes not only reduce the volume of waste destined for landfills but also generate energy, offering both environmental and economic benefits. Proper emissions control and residue management are essential to ensure that these thermal methods do not compromise environmental safety [76, 77].

The integration of digital technologies, such as the IoT and data analytics, is transforming e-waste management into a more transparent and traceable process. Digital tracking systems, including barcoding, RFID tags, and sensor networks, allow for real-time monitoring of e-waste from collection to processing. These systems enhance regulatory compliance, improve operational efficiency, and provide valuable data for economic and environmental assessments, as discussed in Table 3. By enabling traceability, digital technologies also prevent illegal dumping and ensure that hazardous components are managed according to established standards [78]. Emerging circular economy practices, including 3D printing and upcycling of recovered materials, represent a frontier in technological innovation. Recovered metals and plastics can be transformed into new components or products, reducing the demand for virgin resources and extending the life cycle of materials. This approach not only generates economic value but also promotes sustainability by integrating recovered materials directly into new manufacturing processes [79]. Technological innovations and emerging frontiers in e-waste management are redefining the possibilities of sustainable recycling and valorization. Automated dismantling, advanced sorting, bioleaching, hydrometallurgical recovery, thermal treatments, digital tracking, and circular economy practices collectively enhance material recovery efficiency, reduce environmental impact, and improve economic feasibility [80]. As highlighted in Table 3, integrating these technologies into a comprehensive e-waste management strategy ensures that discarded electronics are transformed from environmental liabilities into valuable resources. By fostering innovation, policy alignment, and sustainable practices, these emerging technologies pave the way for a resilient, efficient, and environmentally responsible approach to managing the growing global challenge of electronic waste [81].

Table 3: Technological innovations and emerging frontiers.

Technology/ Innovation	Description/Method	Significance/Advantages	References
Automated Dismantling Systems	Use of robotics and AI-driven machinery to disassemble electronic devices efficiently.	Improves precision, reduces labor costs, minimizes exposure to hazardous materials, and increases recovery efficiency.	[82]
Advanced Sorting and Separation	Techniques such as sensor-based sorting, optical recognition, eddy current separation, and density-based separation.	Enhances material segregation accuracy, maximizes recovery rates, and reduces contamination in recovered fractions.	[73]
Bioleaching and Microbial Recovery	Use of bacteria, fungi, or enzymes to extract metals from e-waste, including precious and rare earth metals.	Environmentally friendly, lower energy consumption, and capable of processing complex or low-grade materials.	[83]

Table 3 (continued)

Technology/ Innovation	Description/Method	Significance/Advantages	References
Hydrometallurgical and Green Chemistry Methods	Application of aqueous chemistry, organic acids, or eco-friendly solvents to recover metals.	High selectivity for valuable metals, reduced environmental impact, and potential for scaling up industrially.	[84]
Pyrolysis and Gasification	Thermal decomposition of plastic and organic e-waste fractions to produce fuel, syngas, or char.	Provides energy recovery, reduces landfill burden, and converts waste into useful energy resources.	[85]
IoT and Digital Tracking Systems	Integration of sensors, barcoding, and data analytics to monitor e-waste throughout the life cycle.	Improves traceability, ensures regulatory compliance, and enables efficient collection and processing.	[86]
3D Printing and Material Reuse	Upcycling recovered plastics or metals into components for new devices.	Promotes circular economy, reduces demand for virgin materials, and enhances sustainable product design.	[87, 88]

7 Stakeholder Engagement and Public Awareness

Stakeholder engagement and public awareness serve as essential drivers that transform project ideas into practical outcomes. Whether the initiative involves environmental protection, infrastructure development, public health promotion, or policy implementation, the participation of key actors – government institutions, communities, NGOs, private organizations, academic bodies, and the general public – is fundamental to both the legitimacy and long-term sustainability of the project. This section highlights the importance of stakeholder participation and public sensitization, outlines evidence-based engagement strategies, presents relevant examples, identifies common barriers, and provides practical recommendations for embedding these elements throughout the project life cycle [89, 90].

Stakeholder engagement strengthens the credibility and acceptance of a project. When stakeholders are consulted and included, they develop trust in the process, and the initiative becomes more resilient to political, social, or institutional resistance. Engagement also enriches project design by incorporating local knowledge, community experiences, and contextual insights that enhance relevance and reduce risks. In addition, stakeholders contribute valuable resources – financial, technical, or human support – that expand the project's capacity. Early consultation further minimizes conflict by identifying competing interests and establishing dialogue that respects diverse perspectives. By bringing different sectors together, engagement also stimulates innovation, allowing teams to explore alternative solutions and adapt effectively to emerging challenges [91].

Public awareness is equally crucial because it influences behavior and promotes collective responsibility. Awareness campaigns encourage individuals and communities to adopt practices such as environmental conservation, energy saving, or health-seeking behavior. An informed public also strengthens accountability, as citizens are better positioned to evaluate decisions and demand transparency. Effective communication fosters social cohesion by clarifying project benefits, addressing misinformation, and encouraging community participation. Moreover, public opinion holds significant political influence, shaping policy priorities, electoral decisions, and funding allocations [92]. Various strategies have proven effective in achieving meaningful engagement. Mapping and prioritizing stakeholders help identify their roles, interests, and influence, enabling targeted communication. Participatory workshops and co-design sessions create platforms for shared problem-solving and improve ownership of outcomes. Multi-channel communication – using radio, print media, social media, SMS, and in-person meetings – ensures inclusivity across different literacy levels and technological access. Youth-led initiatives inject creativity and mobilize digital communities. Feedback and monitoring systems – such as surveys, dashboards, and citizen-science tools – promote responsiveness and transparency. Capacity-building programs further strengthen local leadership and ensure sustainability [93]. Despite the benefits, stakeholder engagement often faces challenges such as power imbalances, digital exclusion, misinformation, and limited funding. These issues can be addressed through transparent governance structures, the use of low-tech communication channels, simplified messaging, community ambassadors, cross-sector funding, and phased engagement plans. It is also crucial to manage information overload by presenting key messages clearly and providing tools that support better decision-making [94, 95].

To ensure meaningful results, several recommendations should guide stakeholder engagement and public awareness efforts. Engagement must begin early and continue throughout the project. Governance structures should be designed to include diverse stakeholder representation. Communication should integrate both qualitative narratives and data-driven insights. Local champions who understand community dynamics can help translate technical information into relatable messages. Incentives – whether financial, educational, or skill-based – can encourage participation. Establishing a centralized communication hub helps streamline updates and feedback. Additionally, feedback loops should be formalized, and engagement metrics should be tracked to measure progress. Messaging must also respect cultural norms, languages, and values. Finally, exit strategies should be planned to ensure that local communities can maintain engagement efforts after the project's active phase ends [65]. Stakeholder engagement and public awareness are fundamental to building inclusive, effective, and resilient projects. When these components are woven into planning, design, and implementation, they empower communities, strengthen governance, and create long-lasting social and environmental benefits. Integrating these principles into future operational frameworks ensures that insights translate into tangible, measurable outcomes that serve societies for generations to come [96].

8 Challenges and Future Perspectives

Despite the growing recognition of the importance of effective e-waste management and the advancements in valorization strategies, technological innovations, and regulatory frameworks, significant challenges remain in implementing sustainable systems worldwide. Electronic waste presents complex problems due to its heterogeneous nature, rapid generation rate, and the presence of hazardous substances. These challenges are compounded by gaps in policy enforcement, insufficient infrastructure, economic constraints, and low public awareness, particularly in developing regions. Understanding these challenges is essential to identifying future perspectives and guiding research, policy development, and industrial practices toward more efficient and sustainable e-waste management [10]. Heterogeneity and complexity of e-waste are the primary challenges. Electronic devices contain a wide variety of materials, including metals, plastics, glass, and hazardous components such as mercury, cadmium, and brominated flame retardants. The composition of e-waste varies significantly between device types, brands, and models, making standardization of recycling and valorization processes difficult. Complex products, such as smartphones and laptops, integrate precious metals with plastics and other components in compact and layered designs. Extracting valuable materials efficiently from these complex matrices requires advanced technological interventions, including automated dismantling, bioleaching, and hydrometallurgical techniques. However, these technologies often require high initial investment, skilled labor, and controlled processing environments, limiting their adoption in regions with constrained resources [9]. Informal Recycling and Health Hazards continue to be a critical challenge in many developing countries. Informal e-waste recycling provides livelihoods for thousands of people but frequently operates outside regulatory frameworks, with little access to protective equipment or environmentally safe processing methods. Open burning, acid leaching, and unregulated dismantling expose workers and nearby communities to toxic metals and persistent organic pollutants, creating significant health risks. Integrating informal sectors into formal recycling systems, offering training, and providing financial incentives are essential strategies, but achieving widespread adoption requires careful policy design and robust stakeholder engagement [97]. Regulatory and Policy Gaps pose significant barriers to efficient e-waste management. Although international agreements, regional directives, and national laws provide frameworks for handling e-waste, enforcement is often weak. Illegal transboundary movement of e-waste persists due to lax monitoring, insufficient penalties, and fragmented legal systems. Many countries lack harmonized standards for collection, recycling, and disposal, which complicates compliance for manufacturers and recyclers. Even in regions with strong policies, the enforcement of EPR programs can be inconsistent, limiting the accountability of producers for end-of-life product management. Strengthening regulatory frameworks, improving monitoring mechanisms, and fostering cross-border cooperation are necessary to close these gaps [9].

Economic constraints and market fluctuations also impact the sustainability of e-waste management systems. Establishing formal recycling facilities with advanced technologies is capital-intensive. The profitability of material recovery depends on market demand and prices for recovered metals and plastics. Fluctuating commodity prices can render recycling economically unviable, especially for precious metals in low-grade e-waste streams. Additionally, energy costs, labor expenses, and transportation challenges can further increase operational costs. Combining multiple valorization strategies, such as integrating physical, chemical, and biological recovery methods, can improve material recovery and economic feasibility. Governments and financial institutions can support e-waste management through subsidies, tax incentives, and funding for research and infrastructure development to address economic constraints [10, 98]. Technological Limitations remain a concern despite recent innovations. Many emerging technologies, including automated dismantling, bioleaching, and pyrolysis, are still in the research or pilot stages and require optimization for large-scale deployment. Scale-up challenges include process efficiency, energy consumption, equipment durability, and the handling of mixed or contaminated waste streams. Moreover, the integration of digital tracking systems, IoT-based monitoring, and smart logistics for e-waste collection requires a robust information technology infrastructure, which may be lacking in low-resource settings. Continuous research, development, and collaboration between academia, industry, and governments are essential to overcome these technological limitations [99].

Public awareness and behavioral challenges hinder the effectiveness of e-waste management initiatives. Many consumers remain unaware of the environmental and health risks associated with improper disposal of electronics or the availability of recycling programs. Low participation in formal collection schemes and a preference for informal or illegal disposal channels limit the quantity and quality of e-waste entering sustainable management systems. Public awareness campaigns, community engagement programs, and educational initiatives targeting schools, universities, and local communities are crucial to fostering responsible behavior [42]. Looking forward, future perspectives in e-waste management revolve around improving sustainability, efficiency, and inclusivity. A major focus is the advancement of circular economy practices, which prioritize design for recyclability, repairability, and resource recovery. By embedding circular economy principles in manufacturing, producers can minimize the generation of complex and hazardous waste while maximizing material value. Policies promoting eco-design, standardized components, and modular electronics can enhance the feasibility of both automated and manual recycling processes

[100]. Integration of advanced technologies will continue to shape the future of e-waste valorization. Combining automated dismantling, sensor-based sorting, bioleaching, hydrometallurgical extraction, and thermal treatment methods can optimize material recovery and energy efficiency. Further development of digital platforms and IoT-based monitoring systems will improve traceability, regulatory compliance, and stakeholder coordination. Artificial intelligence and machine learning have the potential to optimize process parameters, predict material composition, and enhance operational efficiency, thereby reducing costs and environmental impacts [101].

Strengthening regulatory frameworks and international cooperation will be pivotal. Future policy development should focus on harmonized global standards, robust enforcement mechanisms, and support for EPR programs. International collaboration can prevent illegal e-waste trade, facilitate technology transfer, and enable developing countries to adopt safe and sustainable recycling practices. Policymakers must also integrate informal recycling sectors into formal systems, providing technical support, incentives, and regulatory recognition [3]. Capacity building and public education remain critical for future progress. Investing in training programs for workers, government officials, and recyclers, as well as expanding community education campaigns, can promote safe practices and increase participation in formal recycling schemes. Research institutions and NGOs will play a central role in raising awareness, developing new technologies, and providing evidence-based guidance to policymakers [102]. The challenges of e-waste management are multifaceted, involving technical, economic, regulatory, and social dimensions. Heterogeneous waste streams, informal recycling, regulatory gaps, economic constraints, technological limitations, and low public awareness continue to hinder effective management. However, future perspectives point toward integrated solutions that combine circular economy principles, advanced technologies, robust policies, stakeholder engagement, and public education [10]. By addressing these challenges, it is possible to transform e-waste from an environmental liability into a valuable resource, promote sustainable development, and contribute to global environmental stewardship. The convergence of innovation, regulation, and collaboration provides a pathway toward a resilient, efficient, and sustainable e-waste management ecosystem capable of meeting the demands of rapidly evolving technology and growing electronic consumption [19].

9 Conclusion

Electronic waste management remains a complex global issue due to its mixed composition of valuable materials and hazardous substances. This chapter discussed the nature of e-waste and emphasized the need for effective characterization to support safe and efficient recovery processes. Various valorization strategies – physical, chemical, biological, and energy-recovery techniques – offer opportunities to extract resources while reducing environmental harm. Economic and environmental assessments highlight that integrated approaches improve both sustainability and profitability, reinforcing the need for balanced, responsible management. Regulatory frameworks at international, regional, and national levels guide the proper collection, recycling, and disposal of e-waste. Emerging policies increasingly promote circular economy principles, innovation, and collaboration among stakeholders. Strong enforcement, compliance monitoring, and public awareness remain essential to addressing policy gaps and reducing unsafe informal recycling practices. Technological advancements such as automated dismantling, advanced sorting, bioleaching, hydrometallurgy, and digital tracking systems are reshaping the sector by enhancing efficiency and safety. Stakeholder engagement – including producers, consumers, recyclers, governments, NGOs, and educational institutions – is crucial for supporting formal recycling systems and ensuring responsible behavior. Future progress in e-waste management depends on strengthening regulations, advancing technology, integrating circular economy practices, and building capacity toward a more sustainable and resilient system.

References

[1] Priyashantha AKH, Pratheesh N, Pretheeba P, E-waste scenario in South-Asia: An emerging risk to environment and public health. Environ Anal Health Toxicol, 2022, 37, e2022022.

[2] Quinto S, Law N, Fletcher C, Le J, Antony Jose S, Menezes PL, Exploring the e-waste crisis: Strategies for sustainable recycling and circular economy integration. Recycling, 2025, 10, 72.

[3] Mohamed D, Fayad A, Mohamed AMO, Al Nahyan MT, The role of e-waste in sustainable mineral resource management. Waste, 2025, 3, 27.

[4] Andersen T, A comparative study of national variations of the European WEEE directive: Manufacturer's view. Environ Sci Pollut Res Int, 2022, 29, 19920–19939.

[5] Mathew JT, Inobeme A, Shaba EY, Musah M, Azeh Y, Abubakar H, Adam IB, Muhammad AI, Muhammad HA, Ismail H, Umar MT, Aliyu MS, Yisa SP, Ismaila AO, Etsuyankpa MB, Musa ST, Mamman A, Adsorptive removal of Cu^{2+}, Pb^{2+} and Cr^{6+} from pharmaceutical wastewater using graphene/rutile (TiO_2) nanocomposites. Sci World J, 2025, 20, 1263–1272.

[6] Lo Sardo C, Cacciatore G, Cappuccino G, Aiello D, Napoli A, Spent lithium battery recycling: Traditional and innovative approaches. Processes, 2025, 13, 950.

[7] Adetunji CO, Mathew JT, Singh KRB, Bodunrinde RE, Inobeme A, Olaniyan O, Ogundolie FA, Singh J, Nayak V, Singh RP, Molecular characterization of multidrug-resistant genes in wastewater treatment plants, In: Microbial Community Studies in Industrial Wastewater Treatment. Taylor & Francis eBooks, 2023, 127–141.

[8] Kálmán J, The role of regulatory sandboxes in FinTech innovation: A comparative case study of the UK, Singapore and Hungary. FinTech, 2025, 4, 26.

[9] Srivastava AN, Sikarwar VS, Bisen D, Fathi J, Maslani A, Lopez Nino BN, Barmavatu P, Kaviti AK, Pohořelý M, Buryi M, E-waste unplugged: Reviewing impacts, valorization strategies and regulatory frontiers for efficient e-waste management. Processes, 2025, 13, 2014.

[10] Ghulam ST, Abushammala H, Challenges and opportunities in the management of electronic waste and its impact on human health and environment. Sustainability, 2023, 15, 1837.

[11] Mathew JT, Shaba EY, Musah M, Azeh Y, Muhammad EB, TankoE, Tanko MS, Muhammad AI, Hussaini J, Mamman A, Aerogel nanocomposites: Properties and applications. Adv Appl Future Perspect Aerogels, IGI Global, 2025, 1, 25–62.
[12] Islam MK, Khatun MS, Mourshed M. An in-depth analysis and review of management strategies for E-waste in the south Asian region: A way forward towards waste to energy conversion and sustainability. Heliyon. 2024,10, e28707.
[13] Etsuyankpa MB, Augustine AU, Musa ST, Mathew JT, Ismail H, Salihu AM, Mamman A, An overview of wastewater characteristics, treatment and disposal. J Appl Sci Environ Manag, 2024, 28, 1553–1572.
[14] Liu M, Brandsma SH, Schreder E, From e-waste to living space: Flame retardants contaminating household items add to concern about plastic recycling. Chemosphere, 2024, 365, 143319.
[15] Etsuyankpa BM, Musa ST, Ambo AI, Sulaiman LA, Mathew JT, Removal of Cu^{2+}, Fe^{3+} and Pb^{2+} from abattoir wastewater using TiO_2/CdS nanocomposite: Isotherm and kinetics studies. Scholars Int J Chem Mater Sci, 2025, 8, 300–310.
[16] Nisha B, Shajil S, Dutta R, Jain T, Consumer awareness and perceptions about e-waste management in semi-urban area of northern Tamil Nadu. J Fam Community Med, 2022, 29, 132–137.
[17] Ali A, Chiang YW, Santos RM, X-ray diffraction techniques for mineral characterization: A review for engineers. Minerals, 2022, 12, 205.
[18] Musah M, Mathew JT, Azeh Y, Synthesis, characterization and application of ZnO/GO/Zeolite-A nanocomposite in the sorption of selected heavy metals from pharmaceutical effluent. Scholars Int J Chem Mater Sci, 2025, 8, 202–212.
[19] El Hosry L, Sok N, Richa R, Al Mashtoub L, Cayot P, Bou-Maroun E, Sample preparation and analytical techniques in the determination of trace elements in food. Foods, 2023, 12, 895.
[20] Anuar ST, Altarawnah RS, Mohd Ali AA, Lee BQ, Khalik WMAWM, Yusof KMKK, Ibrahim YS, Utilizing pyrolysis–GC/MS for monitoring microplastics in polychaete worms. Polymers, 2022, 14, 3054.
[21] Idris AY, Elele UU, Mathew JT, Preparation and characterization of MoO_3 nanoparticles for photocatalytic degradation of dyeing wastewater. Sci World J, 2024, 19, 1006–1011.
[22] Buchberger DA, Boczar M, Jasinski JB, Czerwiński A. Raman spectroscopy complemented with XRD and TEM for studying structural evolution in initial cycles of LiNi1/3Mn1/3Co1/3O2 cathode material. Discov Nano. 2025, 20, 189.
[23] Spanu D, Binda G, Marelli M, Rampazzi L, Recchia S, Monticelli D, Quantitative determination of surface distribution of supported metal nanoparticles by LA-ICP-MS. Chemosensors, 2021, 9, 77.
[24] Tonjes DJ, Wang Y, Firmansyah F, Manzur S, Johnston M, Walker G, Thyberg KL, Hewitt E, Similarities and differences in waste composition over time and space. PLoS One, 2025, 20, e0308367.
[25] Ferreira-Filipe DA, Hursthouse A, Duarte AC, Rocha-Santos T, Patrício Silva AL, E-waste plastics in the environment: Assessment, characterisation and bioprocessing. Appl Sci, 2025, 15, 2122.
[26] Chicardi E, Lopez-Paneque A, García-Orta VHG, Sepúlveda-Ferrer RE, Gallardo JM, Enrichment methods for metal recovery from WEEE. Metals, 2025, 15, 140.
[27] Oladele IO, Okoro CJ, Taiwo AS, Onuh LN, Agbeboh NI, Balogun OP, Olubambi PA, Lephuthing SS, Modern trends in recycling waste thermoplastics. J Compos Sci, 2023, 7, 198.
[28] Mathew JT, Hussaini J, Inobeme A, Adetunji CO, Tanko MS, Nanocarriers for delivering pesticides. Nanocarriers Plant Sci Agric, IGI Global, 2025, 273–296.
[29] Grdić D, Despotović I, Ristić N, Grdić Z, Ćurčić GT, Use of recycled CRT glass in concrete blocks and paving flags. Materials, 2022, 15, 1499.
[30] Hu D, Zeng X, Lin Y, Chen Y, Chen W, Jia Z, Lin J, High value-added reutilization of waste-printed circuit boards. Molecules, 2023, 28, 6199.
[31] Eze OO, Ogbuene EB, Ibraheem O, Küster E, Eze CT, Novel flame retardants in e-waste: Environmental burdens and health implications. Toxicology, 2025, 511, 154037.

[32] Frizziero L, Liverani A, Caligiana G, Donnici G, Chinaglia L, Design for disassembly and augmented reality: Gearbox case study. Machines, 2019, 7, 29.
[33] Mathew JT, Azeh Y, Adetunji CO, Monday M, Hussaini J, Tanko MS, Catalytic application of aerogels. Adv Appl Future Perspect Aerogels, IGI Global, 2025, 1, 147–178.
[34] Suzuki M, Tabata T, Role and limitations of in-store waste collection systems at supermarkets. Heliyon, 2024, 10, e29011.
[35] Seif R, Salem FZ, Allam NK, E-waste recycled materials as efficient catalysts for renewable energy technologies. Environ Dev Sustain, 2023, 1, 1–36.
[36] Gkika DA, Chalaris M, Kyzas GZ, Review of methods for obtaining rare earth elements from recycling and their impact on the environment and human health. Processes, 2024, 12(6), 1235.
[37] Bourlinos AB, Chalmpes N, Giannelis EP, Gournis DP, Moschovas D, Avgeropoulos A, Salmas CE, Karakassides MA, Total recycling of copper cable scrap and production of carbon using fast microwave technology. Eng, 2025, 6, 74.
[38] Takkar R, Birman K, Gao HO, Enhancing transparency in buyer-driven commodity chains for complex products: Extending a blockchain-based traceability framework towards the circular economy. Appl Sci, 2025, 15, 8226.
[39] Ramasubramanian B, Tan J, Chellappan V, Ramakrishna S, Recent advances in extended producer responsibility initiatives for plastic waste management in Germany and UK. Mater Circ Econ, 2023, 5, 6.
[40] Mathew JT, Hussaini J, Azeh Y, Musah M, Shaba EY, Tanko E, Muhammad AI, Etsuyankpa MB, Musa TS, Mamman A, Environmental and health impact assessment on aerogel application. Adv Appl Future Perspect Aerogels, IGI Global, 2025, 1, 179–202.
[41] Alhosani K, Leveraging public–private partnerships for a circular industry economy in Ajman, UAE. Challenges, 2025, 16, 31.
[42] Nisha B, Shajil S, Dutta R, Jain T, Consumer awareness and perceptions about e-waste management in semi-urban northern Tamil Nadu. J Fam Community Med, 2022, 29, 132–137.
[43] Mathew JT, Musah M, Azeh Y, Shaba EY, Muhammad EB, Hussaini J, Tanko MS, Mamman A, Exploring future perspectives in aerogel-based research. Adv Appl Future Perspect Aerogels, IGI Global, 2025, 1, 203–228.
[44] Ufitikirezi JDDM, Filip M, Ghorbani M, Zoubek T, Olšan P, Bumbálek R, Strob M, Bartoš P, Umurungi SN, Murindangabo YT, Heřmánek A, Tupý O, Havelka Z, Stehlík R, Černý P, Smutný L, Agricultural waste valorization for bioenergy conversion. Sustainability, 2024, 16, 3617.
[45] Morawa Eblagon K, Morais RG, Malaika A, Castro Bravo MA, Rey-Raap N, Pereira MFR, Kozłowski M, Production of 5-hydroxymethylfurfural from sucrose using doped hydrochars. Catalysts, 2025, 15, 656.
[46] Alizadeh P, Tabil LG, Mupondwa E, Li X, Cree D, Technoeconomic feasibility of bioenergy production from wood sawdust. Energies, 2023, 16, 1914.
[47] Mathew JT, Hussaini J, Inobeme A, Azeh Y, Musah M, Shaba YE, Etsuyankpa MB, Tanko E, Ndejiko JM, Adetunji CO, Muhammad AI, Musa ST, Mamman A, X-ray fluorescence analysis of vegetation tissues via chemometric tools, In: X-Ray Fluorescence Spectroscopy and Chemometrics. Springer, 2025, 209–231.
[48] Kaynak E, Piri IS, Das O, Revisiting the basics of life cycle assessment and lifecycle thinking. Sustainability, 2025, 17, 7444.
[49] Becker M, Lütz S, Rosenthal K, Environmental assessment of enzyme production and purification. Molecules, 2021, 26, 573.
[50] Yılbaşı Z, Biofuels, e-fuels and waste-derived fuels: Advances and challenges. Sustainability, 2025, 17, 6145.
[51] Lair A, Mansuy M, Romand C, Oberti O, Pradels C, Barina G, Denoun T, Venturini M, Trommsdorff C, Enhancing landfill efficiency to reduce greenhouse gases. Waste Manag Res, 2024, 42, 889–900.

[52] Blasi A, Verardi A, Lopresto CG, Siciliano S, Sangiorgio P, Lignocellulosic agricultural waste valorization: An overview. Recycling, 2023, 8, 61.

[53] Mathew JT, Inobeme A, Azeh Y, Musah M, Shaba EY, Tanko E, Etsuyankpa MB, Musa ST, Ndejiko JM, Hussaini J, Adetunji CO, Muhammad AI, Mamman A, X-ray fluorescence in analyzing agricultural contamination, In: X-Ray Fluorescence Spectroscopy and Chemometrics. Springer, 2025, 233–254.

[54] Alcocer-García H, Sánchez-Ramírez E, García-García E, Ramírez-Márquez C, Ponce-Ortega JM, Unlocking the potential of biomass resources. Resources, 2025, 14, 143.

[55] Wang Y, WangJ, Zhang X, Bhattacharyya D, Sabolsky EM, Environmental and economic impacts of porous activated carbon from biomass. Energies, 2022, 15, 351.

[56] Pal P, Singh AK, Srivastava RK, Rathore SS, Sahoo UK, Subudhi S, Sarangi PK, Prus P, Circular bioeconomy from food wastes. Foods, 2024, 13, 3007.

[57] Luo Y, Ye M, Zhou Y, Su R, Huang S, Wang H, Dai X, Environmental impact of municipal waste incineration using LCA. Toxics, 2024, 12, 786.

[58] Mathew JT, Inobeme A, Azeh Y, Musah M, Shaba EY, Tanko T, Etsuyankpa MB, Musa ST, Hussaini J, Adetunji CO, Muhammad AI, Mamman A, X-ray fluorescence spectroscopy for analyzing food products, In: X-Ray Fluorescence Spectroscopy and Chemometrics. Springer, 2025, 255–273.

[59] Nurzhan A, Ruan X, Chen D, Life cycle assessment in municipal waste management. Sustainability, 2025, 17, 302.

[60] Kitole FA, Ojo TO, Emenike CU, Khumalo NZ, Elhindi KM, Kassem HS, Impact of poor waste management on public health in Tanzania. Sustainability, 2024, 16, 10873.

[61] Gaviyau W, Godi J, Emerging risks in fintech-driven digital banking. Risks, 2025, 13, 186.

[62] Bagherifam N, Naghdi S, Ahmadian V, Fazlzadeh A, Baghalzadeh Shishehgarkhaneh M, Digital regulatory governance and RegTech. Int J Financ Stud, 2025, 13, 217.

[63] Van der Linden T, Shirazi T, Markets in crypto-assets regulation and adoption. Financ Innov, 2023, 9, 22.

[64] Mathew JT, Inobeme A, Azeh Y, Musah M, Shaba EY, Tanko E, Etsuyankpa MB, Muhammad AI, Musa ST, Ndejiko MJ, Hussaini J, Adetunji CO, Mamman A, Total reflection X-ray fluorescence: Developments and applications, In: X-Ray Fluorescence Spectroscopy and Chemometrics. Springer, 2025, 303–323.

[65] Jørgensen BN, Ma ZG, Impact of EU regulations on AI adoption in smart cities. Information, 2025, 16, 568.

[66] Di Febo E, Transition risk in climate change: A literature review. Risks, 2025, 13(4), 66.

[67] Hankendi C, Coskun AK, Sovacool BK, Why transparency matters for sustainable data centers and AI. iScience, 2025, 28, 113705.

[68] Adamyk B, Benson V, Adamyk O, Liashenko O, Risk management in DeFi. J Risk Financ Manag, 2025, 18, 38.

[69] Gasparri G. Risks and Opportunities of RegTech and SupTech Developments. Front Artif Intell. 2019, 2, 14.

[70] Sanz E, Levi Alfaroviz A, AI applied to compliance and fraud detection in crypto. J Risk Financ Manag, 2025, 18, 612.

[71] Pham T, Ethical and legal considerations in healthcare AI. R Soc Open Sci, 2025, 12, 241873.

[72] Mathew JT, Musah M, Azeh Y, Musa M, Removal of toxic metals using geopolymer/Fe_3O_4/ZnO nanocomposite. Confluence Univ J Sci Technol, 2024, 1, 50–58.

[73] Saenz J, Felsch T, Walter C, König T, Poenicke O, Bayrhammer E, Vorbröcker M, Berndt D, Elkmann N, Arlinghaus J, Automated disassembly of e-waste. Front Robot AI, 2024, 11, 1303279.

[74] Lubongo C, Bin Daej MAA, Alexandridis P, AI-driven technologies for plastic sorting and recycling. Recycling, 2024, 9, 59.

[75] Jaiswal A, Raj SI, Adetunji AI, Negadi L, Singh S, Tumba K, Bahadur I, Uddin I, Bioleaching as an eco-friendly nano-factory for sustainable inorganic waste management: Current advancements, challenges, and opportunities. ChemistryOpen, 2025, 14, e202500104.

[76] Rudnik E, Hydrometallurgical treatment of EAF by-products for metal recovery: Opportunities and challenges. Metals, 2025, 15, 914.

[77] Dharmaraj S, Ashokkumar V, Pandiyan R, Munawaroh HSM, Chew KW, Chen WH, Ngamcharussrivichai C, Pyrolysis as an effective technique for degradation of COVID-19 medical wastes. Chemosphere, 2021, 275, 130092.

[78] Mathew JT, Inobeme A, Azeh Y, Musah M, Abdulkadir A, Shaba EY, Etsuyankpa MB, Musa TS, Muhammad AI, Ismail H, Kanwa AM, Mamman A, Hussaini J, Carbon capture and storage via electrochemical and bioelectrochemical techniques: A review. Caliphate J Sci Technol, 2024, 2, 141–158.

[79] Akinbamini E, Vargas A, Traill A, Boza A, Cuenca L, Technologies enhancing supply chain collaboration in the food industry: A Nigerian survey. Logistics, 2025, 9, 8.

[80] Kassab A, Al Nabhani D, Mohanty P, Pannier C, Ayoub GY, Advancing plastic recycling through 3D printing and distributed recycling. Polymers, 2023, 15, 3881.

[81] Hossain R, Sahajwalla V, Recycling innovations for utilizing e-waste in green metal manufacturing. Philos Trans R Soc A, 2024, 382, 20230239.

[82] Nandan A, Suresh AC, Saole P, Jeevanasai SA, Chandrasekaran R, Meili L, Wan Azelee NI, Selvasembian R, Integrated approaches for electronic waste management. Sustainability, 2023, 15, 16946.

[83] Ameur S, Tabaa M, Hidila Z, Hamlich M, Karboub K, Bearee R, The future of robotic disassembly in the age of AI. Front Robot AI, 2025, 12, 1584657.

[84] Mathew JT, Inobeme A, Abdulkadir A, Etsuyankpa MB, Shaba EY, Musah M, Azeh Y, Adetunji CO, Ibrahim AM, Tanko E, Mamman A, Properties and characteristics of nanocoatings for medicinal applications. Sustain Approach Protect Nanocoatings, IGI Global, 2024, 1, 203–227.

[85] Rendón-Castrillón L, Ramírez-Carmona M, Ocampo-López C, Gómez-Arroyave L, Bioleaching techniques for sustainable metal recovery. Sustainability, 2023, 15, 10222.

[86] Kishiko LLV, Nheta W, Ntumba EM, Solvometallurgical processing for recovery of valuable metals from sulfide ores. Minerals, 2025, 15, 576.

[87] Dennison MS, Paramasivam SK, Wanazusi T, Sundarrajan KJ, Erheyovwe BP, Marshal Williams AM, Pyrolysis for waste-to-energy conversion in Africa. Clean Technol, 2025, 7, 20.

[88] Pracucci A, Giovanardi M, Sensor-based digital product passport for low-tech manufacturing. Sensors, 2025, 25, 5653.

[89] Mathew JT, Inobeme A, Adetuyi BO, Falana YO, Adetunji CO, Shahnawaz M, Application of microplastics in toiletry products, In: Microplastic Pollution. Springer, 2024, 73–84.

[90] Masefield SC, Msosa A, Chinguwo FK, Grugel J, Stakeholder engagement in health policy processes in Malawi. BMC Health Serv Res, 2021, 21, 984.

[91] Abdulkadir A, Musah M, Lakan II, MathewJT, MgO/ZnO nanocomposite for toxic metal removal from tannery wastewater. Scholars Int J Chem Mater Sci, 2025, 8, 250–262.

[92] Prebanić KR, Vukomanović M, Stakeholder engagement as a success factor for infrastructure projects. Buildings, 2023, 13, 1785.

[93] Si W, Jiang C, Meng L, Environmental awareness and pro-environmental behavior of residents. Int J Environ Res Public Health, 2022, 19, 13253.

[94] Pérez Jolles M, Willging CE, Stadnick NA, Crable EL, Lengnick-Hall R, Hawkins J, Aarons GA, Implementation research collaborations from a co-creation lens. Front Health Serv, 2022, 2, 942658.

[95] Murphy J, Qureshi O, Endale T, Esponda GM, Pathare S, Eaton J, De Silva M, Ryan G, Barriers and drivers to stakeholder engagement in global mental health. Int J Ment Health Syst, 2021, 15, 30.

[96] Inobeme A, Natarajan A, Safalmani P, Adetunji CO, Ariola D, Chetan C, Mathew JT, Inobeme J, Kaushik P, Kshitij RBS, Shyam SP, Heavy metals in tropospheric particulate matter. J Mol Liq, 2024, 412, 125822.

[97] Maccioni S, d'Angella F, De Carlo M, Sfogliarini B, Stakeholder engagement for sustainable development in fragile ecosystems. Sustainability, 2024, 16, 9879.

[98] Moyen Massa G, Archodoulaki VM, Electrical and electronic waste management problems in Africa. Environments, 2023, 10, 44.

[99] Inobeme A, Mathew JT, Devolli A, Adetunji CO, Sharma N, Maliki N, Hussaini J, Metal Components in Industrial Wastes and Recovery Methods. Elsevier, 2025, 1–15.

[100] Awan U, Sroufe R, Sustainability and circular business model design. Appl Sci, 2022, 12, 1521.

[101] Kamanga TW, Chitete MM, Kamanga BC, Damazio C, Yafeti Y, Sibande M, Sustainable solid waste management systems in Malawi. Environ Health Insights, 2024, 18, 11786302241255800.

[102] Anuardo RG, Espuny M, Costa ACF, Espuny ALG, Kazançoğlu Y, Kandsamy J, De Oliveira OJ, Transforming e-waste into opportunities for sustainable development goals. Sustainability, 2023, 15, 14150.

Anisha Arora*

Click, share, recycle: social media strategies for smarter e-waste management

Abstract: The increasing rate of digital technology has resulted in a significant increase in electronic waste (e-waste), which causes environmental and health challenges. Conventional management approaches are not enough, and it is necessary to develop innovative solutions that are community-based. In this chapter, the author explores the way social media enhances e-waste management by creating awareness, behavior change, and responsible disposal. Social media, including Facebook, Instagram, and YouTube, allow running targeted initiatives, engaging with influencers, gamifying, and communicating in real time to promote involvement in recycling. The topic of applying data analytics and AI to enhance the effectiveness of the campaign is mentioned in the chapter as well, along with global case studies. It ends with strategic frameworks and future opportunities, such as AR tools, blockchain traceability, and a digital reward system that facilitates circular economy practices and appropriate Sustainable Development Goals.

Keywords: E-waste management, social marketing, circular economy, sustainability, computerized crusade

1 Introduction

The high pace of development of digital technologies has radically transformed the manner in which people communicate, work, learn, and use services. Meanwhile, it has also caused an all-time high consumption of electronic products – smartphones, laptops, and smart appliances at home, wearable devices, and Internet of Things (IoT) systems [1]. With the shrinking lifecycle of these gadgets as a result of the rapid and accelerating technological change and upgrades at the behest of consumers, electronic waste (e-waste) has become one of the most common and dangerous sources of waste in the world. E-waste generation in the world is steadily high, as reported by the United Nations Global E-waste Monitor, but recycling is comparatively low [2]. The imbalance is raising serious environmental and communal health risks such as poisonous chemical contamination, unhealthy informal reprocessing, pollution of soil and water forms, and the loss of precious resources, namely gold, copper, and rare earth components.

The conventional e-waste management systems have been mostly dependent on regulatory systems, recycling facilities, producer responsibility programs, and formal

***Corresponding author: Anisha Arora,** University Centre for Research and Development, Chandigarh University, Mohali, Punjab, India, e-mail: anishaarora0199@gmail.com

 | https://doi.org/10.1515/9783112248072-012

collection centres. Although these strategies are still critical, they are limited in their effectiveness due to low public awareness, poor involvement, and resistance of consumers to change [3]. Some people are ignorant of appropriate disposal methods, do not know the effects of discarding unutilized gadgets in the house on the environment, or they do not find recycling e-waste a personal issue. In this regard, social media marketing integration offers a strong, yet affordable, and community-oriented opportunity to change the population perception and encourage the community to engage in sustainable disposition behaviors. Social media networks have become such an inseparable part of daily existence that they provide real-time communication, disseminate information at high speed, and interactivity, and it is possible to use them to overcome the problem of global e-waste by means of innovative solutions [4].

Having billions of users across the globe, social networks like Facebook, Instagram, YouTube, TikTok, and LinkedIn provide unprecedented possibilities for organizations, governments, and businesses to create a specific digital campaign, reinforce the work with the population, and encourage mass actions [5].

The delivery of content via social media is individualized and hence easier to target the diverse demographic clusters with customized messages concerning the importance of responsible disposal of e-waste and how individual contributions can be made toward wider sustainability targets. Complex issues related to recycling can be reduced in complicated forms with visual storytelling, short videos, infographics, live demonstrations, and interactive polls. Besides, collaborating with influencers, online activists, and content creators can raise the awareness of e-waste projects through the use of social capital and community confidence online [6].

Since participation is the key component of a new communication approach, the new technologies, e.g., gamification, various challenges that use a reward, buy-back announcements, and recycling contests, help to make the users more engaged. As an illustration, environmental responsibility brings a sense of fun and accomplishment through digital badges, point systems, and competitions on leaderboards. Meanwhile, the emergence of social analytics and sentiment analysis driven by AI allows organizations to measure the response of users, define the patterns of behavior, monitor the campaign success, and optimize strategies to achieve the best outcomes. Through these pieces of data, policymakers, recyclers, and supply chain managers develop more responsive, participatory, and inclusive e-waste management models.

Nonetheless, there are no challenges that the use of social media in e-waste management cannot overcome. Misinformation, digital gaps, media literacy disparities, algorithm biases, and changing levels of online engagement may undermine a long-term change in behaviors. Quick campaigns rarely lead to a long-lasting change without a constant message, joint ventures, and available recycling systems [7, 8]. Also, organizations have to negotiate privacy, ethical communication practices, and information oversaturation, fighting to command the users. Despite these obstacles, strategic planning (which involves stakeholder mapping, KPI-based monitoring, and cross-sector collaboration) significantly improves the process of integrating social media into official waste management systems.

New opportunities for the future of e-waste management are also presented by the new convergence of technology, sustainability, and e-communication. It is possible to create immersion in educational experience through augmented reality (AR) filters; provide transparency to track e-waste flows through blockchain; and promote the idea of buying refurbished devices to consumers to extend the life of the device and thus the principle of the circular economy. The models of social commerce, peer-to-peer interactions, and digital repair communities show how online networks can create responsible consumption and limit unneeded disposal [9–11].

The chapter places social media marketing not as a promotional device, but as a revolutionary process of facilitating more intelligent, scalable, and community-focused e-waste management. The ability to use the strength of social media would add to Sustainable Development Goal 12 (Responsible Consumption and Production) and SDG 13 (Climate Action) and provide an avenue to a cleaner, greener digital future [12, 13].

2 Literature review

The expanding pace of electronic waste has grown into a focal point of environmental concerns, and the amount of e-waste is set to grow to over 74 million metric tons by 2030 worldwide [14]. Researchers highlight that e-waste is particularly dangerous because of its toxic elements, lead, mercury, cadmium, and flame retardants that harm the soil, air, and waterways [15]. Whereas the conventional paradigm of management focuses on regulatory interventions and recycling infrastructure [16, 17], studies are placing more importance on consumer behavior as the key determinant of disposal behavior. A significant number of studies show that households usually keep unused electronics at home, or are not aware of recycling opportunities, or dispose of e-waste through informal means, which only serves to increase environmental and health risks [18]. In turn, researchers claim that behavior change models, communication interventions, and community engagement systems should be incorporated into e-waste management systems.

2.1 E-waste management and behavioral perspectives

Theory of planned behavior (TPB), norm activation model (NAM), and the value-belief -norm (VBN) frameworks are some of the behavioral theories that have been used to explain consumer choice on recycling. Research shows the strong predictive power of a willingness to dispose of e-waste responsibly by the attitudes and the perceived level of behavioral control, as well as social norms [19]. Also, recycling intentions are affected by moral obligation, environmental values, and risk perceptions. Studies reveal that consumers who already know about the dangers of environmental pollution are not motivated to dispose of their used products properly due to the unavailability of collection points, incentives, and an insufficient perceived value [20]. Therefore, researchers point out the necessity of motivational, awareness-making, and socially integrated interventions – spheres where social media is transformative [21].

2.2 The social media as an eco-communication tool

Social media has become a widely used form of mass communication because of the high levels of interaction, participatory affordance, and real-time dissemination features. Research shows that social media platforms such as Facebook, Instagram, YouTube, TikTok, and Twitter are powerful media that can be used to influence the population on their attitudes toward the environment [22]. Social media allows interactive storytelling, community building, and direct communication between institutions and individuals over and above the traditional media one-way communication model. According to the studies of environmental activism, online campaigns are able to stimulate viral activism, rally individual action, and reinforce knowledge exchange [23]. Digital platforms have been employed in areas of waste management to promote recycling, minimize plastic consumption, and promote sustainable consumption, which indicates their applicability to e-waste behavior change.

Some of the studies emphasize the impact of social media in encouraging recycling behavior by visual content, infographics, challenge campaigns, and influencer campaigns [24]. Social media is participatory, and hence users are encouraged to like, comment, share, and produce content, thus increasing the volume of sustainability messages. In addition, online platforms will enable environmental organizations and local authorities to engage directly with the citizens, explain the myths, and offer the most recent news on collection efforts, buy-back, and new recycling centers.

2.3 Social media marketing and e-waste management

The specific connection between social media marketing and e-waste management has not yet been thoroughly studied, yet the literature regarding the transformative potential of social media marketing is developing. The case studies of such countries as India, China, Kenya, and the United States demonstrate that social campaigns have a great effect on raising awareness about the risks of e-waste and involvement in the recycling programs [25]. Facebook pages and WhatsApp groups have been utilized by municipal corporations, NGOs, and individual recyclers to announce e-waste drives, receive feedback, and disseminate educational messages. Researchers also mention that personalized advertisements and AI-driven audience profiling can be used to increase the relevance and visibility of the message [26].

The use of influencer-based campaigns has been successful in attracting youth to active participation. The study demonstrates that the perceived credibility and social worth of the recycling process are promoted when the environmental content is promoted by familiar influencers or local online creators [27]. Equally, environmental responsibility should be made fun and competitive through gamified campaigns that include challenges, rewards, and interactive quizzes; these elements increase participation.

An interesting line of research is that which concentrates on how social media encourages a circular economy mentality. With the help of tutorials on repairing devices, marketplaces of refurbished products, and the ideas of upcycling, platforms help people increase product life and decrease disposal [28, 29]. As an example, YouTube and Instagram have flourishing communities of repair makers, DIY hobbyists, and sustainability educators who un-twist the secrets of fixing a broken device or knowing about the dangers of e-waste. Such Internet communities intensify the power of shared knowledge and provide fewer obstacles to responsible consumption.

Table 1: Role of social media strategies in e-waste management.

Social Media strategy	Platform example	Purpose	Expected outcome	Supporting studies
Awareness campaigns	Facebook and Instagram	Inform users about e-waste hazards	Increased awareness	Lopez J, Bastein T, Tukker A, 2021
Influencer marketing	YouTube and Instagram	Build trust and credibility	Positive attitude formation	Wang and Wong, 2020
Gamification	Instagram and Apps	Encourage participation through rewards	Higher engagement and intention	Holmes A, Shand J, 2020,
Educational videos	YouTube and TikTok	Simplify recycling procedures	Improved PBC	Bressanelli G, Perona M, Saccani N, 2020
Social listening and analytics	Twitter and Meta	Monitor public sentiment	Optimized campaigns	Kapoor K, Tamilmani K, Rana N P, Patil P, Dwivedi Y K, Nerur S, 2022,

Table 1 presents a structured overview of key social media strategies employed in the context of e-waste management and their corresponding platforms, objectives, and expected behavioral outcomes. The table highlights how different digital approaches, such as awareness campaigns, influencer marketing, gamification, educational content, and social listening, contribute to shaping consumer awareness, attitudes, perceived behavioral control (PBC), and recycling intentions. By mapping specific strategies to their functional outcomes and supporting literature, the table demonstrates the strategic role of social media not only as a communication tool but also as a behavioral intervention mechanism. This synthesis provides a practical framework for policymakers, environmental organizations, and marketers to design targeted, data-driven social media initiatives that support sustainable e-waste management practices.

2.4 Data analytics and AI in social media campaigns

Recent publications indicate the applicability of data analytics and machine learning to the optimization of environmental campaigns. Sentiment analysis, the measurement of engagement, and social listening tools enable companies to gauge how people respond and refine their message to the greatest effect [32]. Artificial intelligence-based tools can detect what is trending, forecast behavior, and tailor content to particular demographic groups. Predictive analytics in the context of e-waste management enables policymakers to know which areas have a lack of awareness and what audiences need specific interventions. Such a data-driven strategy guarantees efficiency and increases the scale of initiatives that social media leads [33–35].

2.5 Problems in the execution of social media-based e-waste policies

Although there is proximal evidence, researchers highlight several issues regarding applying social media in environmental governance. The sustainability messages may be warped by misinformation, algorithmic biases, and echo chambers, or made less efficient [35]. The participation of older adults, rural populations, and marginalized groups is also frustrated by digital divides, or disparities in access, internet literacy, and familiarity with the platforms. It has been found that brief social media campaigns are ineffective in creating a long-term behavioral change unless there is ongoing interaction, institutional reinforcement, and easy recycling facilities [18].

Furthermore, critics also note that social media campaigns usually emphasize awareness, but structural solutions. As an illustration, an improved knowledge about the issue of e-waste might not result in a concrete change of behavior of recycling without proper e-waste collection centers or logistic support. In such a way, literature emphasizes the importance of the combination of digital outreach with infrastructural and policy-based changes.

2.6 New trends and future

The literature on the recent advancements in the area discusses novel technologies, such as AR, virtual reality (VR), blockchain, and social commerce, to improve e-waste management. Educational applications built with AR help simplify the difficult recycling process, and blockchain will allow tracking the flows of e-waste between consumers and recyclers [15]. Social commerce websites foster second-hand gadgets and promote circular consumption. These trends are indicators of a time in which digital ecosystems have an increasingly key role in sustainability transitions [16].

3 Research methodology

The sample size of 160 respondents, including primary data on the awareness, attitudes, and behaviors of e-waste and use of social media, was gathered by a cross-sectional survey technique. A mixed purposive and stratified method was applied by the researcher so that there was a representation in terms of age, gender, and location (urban/rural). The sample of respondents comprised students, the working population, and household decision-makers who used electronic devices.

A structured questionnaire was used to collect the data of the study and addressed the demographics of respondents, patterns of use of social media, awareness of e-waste, the main constructs of the study, attitude, subjective norms, PBC, recycling intention, and actual behavior. The measurements of all variables were made with the use of a 5-point Likert scale supplemented with several open-ended questions to gather additional insights and pilot-tested on 20 participants to verify the tool. A mixed-method approach to data collection was used, which consisted of online surveys that would be conducted using Google Forms via social media and e-mail, and paper-based surveys that would be distributed in colleges and local communities. Informed consent was obtained, and anonymity was ensured by all the respondents.

The dataset was filtered through missing data, inconsistency, and outliers before analysis. These were then followed by descriptive statistics, correlation analysis, and regression to determine the relationship between social media exposure and e-waste recycling behavior. The study was conducted in a manner that ethical considerations were addressed, and this participation was voluntary, confidential, and devoid of any personal identifiable information.

4 Results and analysis

The chapter summarizes the findings and reports results of the primary data that were gathered on 160 individuals to determine their awareness, attitude, PBC, and behavioral intentions of the examined phenomenon. A structured questionnaire was used to collect the responses and was analyzed using descriptive statistics, reliability testing, and interpretation analysis. The purpose of this section is to point out the major trends and patterns that the dataset has brought, including insights, and justify how these findings lead to the overall goals of the study.

4.1 Demographic profile of respondents

Understanding the demographic characteristics of the respondents is essential because individual attributes such as age, gender, and education influence awareness, attitude formation, and behavioral intentions. The demographic distribution of the sample is presented in Table 2.

Table 2: Demographic distribution of respondents ($n = 160$).

Variable	Category	Frequency	Percentage (%)
Gender	Male	82	51.3%
	Female	78	48.7%
Age group	18–25	64	40.0%
	26–35	57	35.6%
	36–45	27	16.9%
	46+	12	7.5%
Education level	Undergraduate	58	36.3%
	Postgraduate	79	49.4%
	Other (diploma/doctorate)	23	14.3%

Demographic basis shows that there is almost an equal division of both men and women which will mean that there is hardly any gender bias in the interpretations. Most of the respondents are in the age group of 18–35, and this indicates that the sample will be made up of young and middle-aged people who are more likely to be more sensitive and technology-conscious when it comes to research on digital behavior and decision making. Moreover, the share of postgraduates is comparatively high, which means that there is a well-educated sample that can represent the informed decision-making patterns.

4.2 Descriptive analysis of key constructs

Descriptive statistics, including mean and standard deviation, were used to summarize the central tendencies of the major constructs: awareness, attitude, PBC, and behavioral intention. Table 2 presents the aggregated results.

The average scores demonstrate some significant results. General awareness ($M = 3.84$) is relatively high and this means that the respondents have a reasonable amount of knowledge with regard to the topic of investigation. One of the strongest constructs is the attitude ($M = 4.02$), which implies that the respondents have relatively positive perceptions and beliefs, in general. Behavioral intention ($M = 4.01$) is also high, which means that there are high chances of the respondents acting on the desired behavior or embracing the corresponding technology/practice.

Table 3: Mean and standard deviation of construct variables.

Construct	Mean	Standard deviation
Awareness	3.84	0.71
Attitude	4.02	0.68
Perceived behavioral control (PBC)	3.48	0.74
Behavioral intention	4.01	0.65

PBC has the lowest score (*M* = 3.48) which implies that despite the respondents having awareness and positive attitudes, there are some challenges or obstacles that may restrain them from executing the behavior comprehensively. The barriers may be in terms of access, resources, time, knowledge, or external constraints.

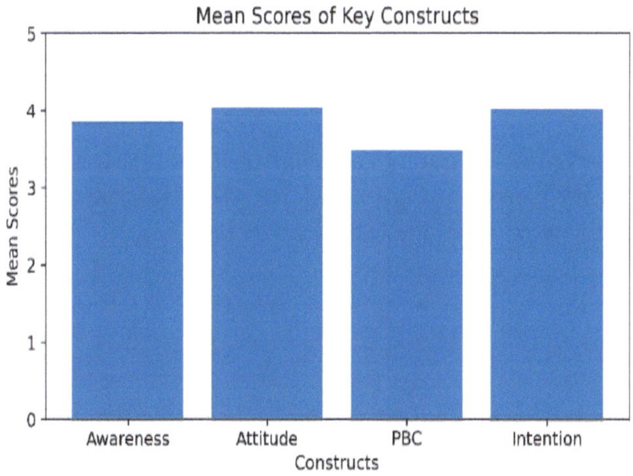

Figure 1: Mean scores of key constructs.

Figure 1. shows the bar chart shows the mean scores of four key constructs: Awareness, Attitude, Perceived Behavioral Control (PBC), and Intention. Attitude and Intention have the highest mean scores, both around 4.0, indicating stronger responses in these areas. Awareness has a slightly lower mean of about 3.8, while PBC is the lowest at roughly 3.4. Overall, the results suggest generally positive responses, with comparatively weaker perceived behavioral control.

4.3 Reliability Analysis

Before interpreting the constructs, it is crucial to ensure that the items used to measure them are reliable. Reliability was tested using Cronbach's alpha, and the results are displayed in Table 3.

A Cronbach's alpha value above 0.70 is considered acceptable for social science research. All constructs in this study exceeded the threshold, indicating strong internal consistency and supporting the reliability of the dataset. The highest reliability is observed for behavioral intention ($\alpha = 0.84$), demonstrating that respondents answered consistently across the items measuring this variable Table 4.

Table 4: Reliability statistics.

Construct	No. of items	Cronbach's alpha
Awareness	4	0.78
Attitude	5	0.82
PBC	4	0.75
Behavioral intention	4	0.84

4.4 Findings analysis and interpretation

In this section, the researchers explain the findings by referring to the research objectives and theoretical framework. The results are indicative of evident patterns, as well as give hints on respondent perceptions and behavioral inclinations.

4.4.1 Awareness levels

The overall score of the moderately high awareness scale ($M = 3.84$) is a pointer that the respondents possess a general knowledge about the issue. Since the process of shaping the attitude and intentions typically begins with awareness, the specified finding helps to reach the conclusion that the sample in question is highly informed. There is also a heightened level of awareness that will bring receptivity and involvement of people toward programs, technologies, or social programs.

Additionally, the awareness whose mean is close to 4 will imply that the respondents are not conscious but also understand the importance of the phenomenon under investigation. This is also very handy where the research is done to promote the adoption of technology or modify consumer behavior or participation in social courses.

4.4.2 Respondents' attitudes

One of the constructs that came out quite strongly was that of attitude ($M = 4.02$). Positive attitude means that the respondents tend to think in favor of the benefits, utility, or relevance of the behavior/ technology under study. The correlation is more often positive between positive attitudes and behavioral intention, and the values in this dataset confirm the existence of this correlation.

Positive attitude implies the presence of potential communication interventions, awareness, or marketing campaigns that can be effective within the population. When people have positive perceptions and attitudes toward the benefits, they tend to participate in the desired behavior.

4.4.3 Perceived Behavioral Control

The PBC shows a relatively low mean score ($M = 3.48$), indicating that many respondents feel some level of limitation or difficulty in performing the desired behavior. Since PBC plays a crucial role in determining whether intentions are successfully translated into actual behavior, a lower score suggests the presence of several constraints. Such reduced PBC may stem from a lack of necessary resources, structural or procedural barriers, difficulty navigating processes, low confidence in performing the behavior, or external environmental factors beyond individual control. These findings imply that policymakers and service providers need to focus on reducing practical obstacles, improving accessibility, simplifying processes, and offering clearer guidance and support systems to empower individuals and enable smoother behavioral execution.

4.4.4 Behavioural intention

The score of behavioral intention ($M = 4.01$) is also high and this means that the respondents are ready to use the behavior or technology desired. A powerful intention demonstrates the willingness to take action as long as people possess enough resources and have solvable challenges. Nevertheless, the high behavioral intention and the relatively low values on PBC indicate that even though respondents are willing to be involved in the behavior, they might require some support or enabling factors that will allow them to effectively be involved in the behavior. Improvement of PBC would thus increase the chances of such intentions being translated into actual behavior.

4.5 Integrated interpretation

Combined, the findings demonstrate a high degree of psychological readiness of the respondents, high awareness, positive attitudes, and a definite desire to take measures. However, real or perceived obstacles still pose a hindrance to behavioral performance. The interaction between the constructs indicates that awareness creates the cognitive base required to comprehend the problem, attitude influences positive perception and receptivity, PBC defines the perceived practicability of performing the behavior and behavioral intention is the most motivational force that is just short of action. This correspondence corresponds to behavioral models like TPB and the technology acceptance models, in which intention and perceived control have a combined impact on behavior.

4.6 Practical implications

The results have valuable implications for practitioners and policymakers. The awareness and positive attitudes are high which means that campaigns and interventions conducted by the organization aimed at the population are likely to be positively received. But there should be an attempt to overcome these obstacles that are associated with PBC by increasing access, affordability, ease of use, and simplifying the process. Since the behavioral intention is already high, organizations and policymakers need to take advantage of this willingness by establishing points that will translate the intention into actual participation or adoption. Making support systems more effective, increasing capacity-building efforts, and simplifying processes are critical steps toward increasing perceived control and promoting coherent behavioral responses. The findings indicate the existence of strong awareness, positive attitudes, high intention, and moderate perceived control trends. These trends ensure that there is a significant preparedness among the respondents, but some feasible obstacles might require consideration. Overall, the results provide helpful information that can be used in the strategy formulation, policy formulation, communication design, and intervention development.

5 Conclusion and future scope

The study aimed to analyze the formation of the engagement of people in the specified phenomenon by means of the awareness, attitudes, perception of behavioral control, and behavioral intention, using the data obtained from a group of 160 respondents. The results indicate that the overall population is well-informed, positive, and motivated to implement the desirable behavior or practice. The high awareness shows that the respondents are informed of the form of the issue and its meaning, and a high positive attitude shows that they have positive perceptions about the benefits and the importance of the issue. Behavior intention was also rated as high, which shows a willingness to act and a willingness to engage in the related activities or initiative.

The lower score on PBC, however, indicates that respondents might encounter practically, structurally, or resource-based obstacles which might restrict their capacity to translate intention to behavior. This highlights the necessity of supportive structures, capacity building, information availability, and facilitating structures to address the perceived limitations. People become empowered and believe that they can do something and their chances of executing their behavior are high.

The findings support the need to integrate awareness-building with practical assistance. The awareness and attitude are not sufficient without the interventions that could reinforce the perceived control and decrease the real-world obstacles. The dataset also demonstrates that the population is young and educated as well, which implies the possibility of a specific approach, taking advantage of digital communication channels, tailored messages, and interactive content.

5.1 Future scope

Although the present research offers substantial information about the behavioral factors that have an impact on the target population, there are still several directions that can be pursued in the course of further investigation and developmental progress:

The score for behavioral intention ($M = 4.01$) is also high, indicating that the respondents are ready to adopt the desired behavior or technology. A powerful intention demonstrates the willingness to take action as long as people possess enough resources and have solvable challenges. Nevertheless, the high behavioral intention and relatively low values on perceived behavioral control (PBC) indicate that, although respondents are willing to be involved in the behavior, they may require some support or enabling factors to effectively participate in it. Improvement of PBC would thus increase the chances of such intentions being translated into actual behavior.

The findings demonstrate a high degree of psychological readiness of the respondents, high awareness, positive attitudes, and a definite desire to take measures. However, real or perceived obstacles still pose a hindrance to behavioral performance. The interaction between the constructs indicates that awareness creates the cognitive base required to comprehend the problem, attitude influences positive perception and receptivity, PBC defines the perceived practicability of performing the behavior and behavioral intention is the most motivational force that is just short of action. This correspondence corresponds to behavioral models like the theory of planned behavior (TPB) and the technology acceptance models, in which intention and perceived control have a combined impact on behavior.

5.2 Practical implications

The results have valuable implications for practitioners and policymakers. The awareness and positive attitudes are high, which means that campaigns and interventions conducted by the organization aimed at the population are likely to be positively received. But there should be an attempt to overcome these obstacles that are associated with PBC by increasing access, affordability, ease of use, and simplifying the process. Since the behavioral intention is already high, organizations and policymakers need to take advantage of this willingness by establishing points that will translate the intention into actual participation or adoption. Making support systems more effective, increasing capacity-building efforts, and simplifying processes are critical steps toward increasing perceived control and promoting coherent behavioral responses. Additional research can be done in the future, which will involve a bigger sample and a more diverse sample that will cover various geographical areas, socioeconomic status, and academic education. This would assist in affirming the external validity of the results and revealing more demographic trends.

5.3 Longitudinal studies

A longitudinal design would be able to look at the change in awareness, attitudes, and behavior over time. This would enable the researcher to assess the effects of prolonged communication campaigns, policy adjustments, or social transformations on the engagement and adoption rates.

5.4 The use of behavioral outcomes

Although this research aimed at using psychological constructs and intention, subsequent studies can incorporate real behavioral outcomes. To have closer evidence regarding the intention-behavior gap and the influence of PBC on the development of real actions, real participation tracking would be stronger.

Subsequent research examines other variables, such as trust, perceived risk, social influence, or digital literacy, as intermediaries or even moderators. These other elements are usually very important influences on behavior but this study did not cover them. Scientists and professionals can develop and implement certain interventions, including awareness campaigns, training workshops, streamlined service processes, or incentives, to determine their impact on PBC and behavior. The evidence-based guidance might be provided by randomized controlled trials (RCTs).

Opportunities exist in the incorporation of data analytics, AI, and digital platforms to comprehend user behavior better and streamline engagement strategies. The impact of real-time insights and personal communication on awareness, attitudes, and participation can be studied in the future.

The research provides valuable data regarding the influence that the psychological constructs have on the engagement and intention of the population. The recognition of the strong awareness, positive attitudes, and high intention, and the understanding of the moderately perceived control are the pivotal points that will help the research to establish more informed strategies, policy development, and targeted interventions. As these will be explored and put into practice further, these insights can be used to create more effective, inclusive, and impactful programs in the future.

References

[1] Ajzen I, The theory of planned behavior. Organ Behav Hum Decis Process, 1991, 50, 179–211.

[2] Amin S, Hasan M, Hassan R, Social media marketing and consumer environmental behavior: A systematic review. J Clean Prod, 2021, 280, 124137.

[3] Badni K, Virtual reality design techniques for web-based historical reconstructions. Virtual Real, 2006, 9, 215–225.

[4] Baldé CP, Forti V, Gray V, Kuehr R, Stegmann P, The Global E-Waste Monitor 2020: Quantities, Flows and the Circular Economy Potential. United Nations University, Bonn, 2020.

[5] Batrinca B, Treleaven PC, Social media analytics: A survey of techniques, tools and platforms. AI Soc, 2015, 30, 89–116.

[6] Bhatnagar P, Kim J, Influencer-driven environmental communication and its impact on consumer recycling behavior. Sustainability, 2022, 14, 3821.

[7] Chung SS, Murakami-Smith A, Circular economy strategies for e-waste management: A global perspective. Resour Conserv Recycl, 2021, 170, 105548.

[8] Cowls J, Tsamados A, Taddeo M, Floridi L, A definition, benchmark and database of AI for social good initiatives. Nat Mach Intell, 2021, 3, 111–115.

[9] Deepa S, Pradeepkumar V, Saranraj R, Sivaperumal M, Nanthakumar S, IoT based garbage monitoring system. Int J Psychosoc Rehabil, 2020, 24, 1682–1689.

[10] Dugdhe S, Shelar P, Jire S, Apte A, Efficient waste collection system, In Proc Int Conf Internet of Things Appl (IOTA). IEEE, Pune, 2016, 143–147.

[11] Dwivedi YK, Ismagilova E, Hughes DL, Carlson J, Jacobson J, Jain V, et al, Setting the future of social media research: Perspectives and research propositions. Int J Inf Manag, 2021, 59, 102168.

[12] Ferreira ME, Caires C, Pitarma R, Health and sustainable behaviours: A pedagogical and didactic experiment in preschool teaching. Int J Knowl Learn, 2015, 10, 238–257.

[13] Forti V, Baldé CP, Kuehr R, E-Waste Statistics Guidelines. United Nations University, Bonn, 2018.

[14] Ghobakhloo M, Industry 4.0, digitization, and opportunities for sustainability. J Clean Prod, 2020, 252, 119869.

[15] Guiñez-Cabrera N, Mansilla-Obando K, Jeldes-Delgado F, Advertising transparency in social media influencers. Retos, 2020, 10, 265–281.

[16] Guttentag D, Virtual reality: Applications and implications for tourism. Tour Manag, 2010, 31, 637–651.

[17] Holmes A, Shand J, Gamification for environmental behavior change: A review. J Environ Manag, 2020, 258, 110.

[18] Kapoor K, Tamilmani K, Rana NP, Patil P, Dwivedi YK, Nerur S, Advances in social media research: Past, present, and future. Inf Syst Front, 2022, 24, 221–245.

[19] Kirchherr J, Reike D, Hekkert M, 2017, Conceptualizing the circular economy: An analysis of 114 definitions. Resources, Conservation and Recycling, 127, 221–232.

[20] Kumar A, Pandey N, AI-driven sentiment analysis for improving sustainability communication on social media. J Bus Res, 2023, 158, 113675.

[21] Li J, Zeng X, Chen M, E-waste recycling and circular economy: A global perspective. Renew Sustain Energy Rev, 2021, 137, 110554.

[22] Liu Z, Sustainable tourism development: A critique. J Sustain Tour, 2003, 11, 459–475.

[23] Lu QS, Seah ZY, Social media influencers and consumer online engagement management, In: Dwivedi YK, Eds, Digital Marketing and Consumer Engagement. IGI Global, Hershey, 2018, 83–104.

[24] Mangla SK, Luthra S, Jakhar S, Gandhi S, Muduli K, Kumar A, Using theory of planned behavior to explain consumer intention for e-waste recycling. Technol Forecast Soc Change, 2021, 163, 120458.

[25] Nielsen KR, Mathiassen L, Social media as a tool for public awareness and engagement. Gov Inf Q, 2021, 38, 101531.

[26] Nnorom IC, Osibanjo O, Overview of electronic waste management practices and legislations. Waste Manag, 2020, 109, 1–12.
[27] Pinto AS, Guerreiro J, Rita P, Augmented reality for a new reality: Using UTAUT-3 to assess adoption in tourism. J Inf Syst Eng Manag, 2022, 7, 1–13.
[28] FAO, Communication and Sustainable Development. Food and Agriculture Organization of the United Nations, Rome, 2007.
[29] Schivinski B, Dabrowski D, The impact of brand communication on brand equity through Facebook. J Mark Commun, 2015, 21, 2–24.
[30] Lopez J, Bastein T, Tukker A, Social media-based environmental campaigns and recycling behaviour. J Clean Prod, 2021, 280, 124137.
[31] Bressanelli G, Perona M, Saccani N, The role of digital platforms in circular economy strategies for electronic products. Resour Conserv Recycl, 2020, 162, 105–118.
[32] Syed Ali SA, Khan MS, Ahmed S, Khan NA, Packaging plastic waste from e-commerce sector: The Indian scenario. J Clean Prod, 2024, 447, 140729.
[33] Tomašev N, Cornebise J, Hutter F, Mohamed S, Picciariello A, Connelly B, et al, AI for social good: Unlocking the opportunity for positive impact. Nat Commun, 2020, 11, 2468.
[34] UNEP, A New Circular Vision for Electronics: Time for a Global Reboot. United Nations Environment Programme, Nairobi, 2019.
[35] Vanninen H, Mero J, Kantamaa E, 2023, Social media influencers as mediators of commercial messages. Journal of Internet Commerce, 22, S4–S27.
[36] Wang S, Wong F Y, 2020, Social media influencers' impact on purchase intention: The moderating effect of brand engagement. Journal of Marketing Advances and Practices, 3(2), 1–15.

Neelam Sidhu*, Khushboo, Sakshi Sharma, Vishali, Purnima Chaudhary, Abhay Dogra

Case studies from developing and developed nations on techno-economic analysis and life cycle assessment of e-waste removal and recycling

Abstract: A rapid increase in electronic waste (e-waste) has emerged as one of the most urgent environmental and public health challenges globally. Addressing this issue involves sustainable strategies that can trade off between technology, economics, and ecological considerations. This review summarizes global evidence for e-waste management, with a focus on the two research methodologies of Techno-Economic Analysis (TEA) and Life Cycle Assessment (LCA). Through combining a systematic and scientific approach (PRISMA-informed screening and bibliometric mapping), we summarize peer-reviewed articles, government frameworks and national case studies published until 2025 in comparing technological trajectories, economic feasibility, and environmental impacts in developed (EU, USA, Australia, Switzerland, Japan, Korea, Belgium) and developing (India, Nigeria, Brazil, Ghana, Egypt) contexts. Based on the systematic literature review and cross-case comparisons, the study outlines that developed countries generally rely on formalized infrastructures, innovative recovery technologies (mechanical, hydrometallurgical, and emerging bio-electrochemical processes), and well-enforced policies (EPR, MFA, MCDM approaches), enabling a high level of resource utilization and material recovery. In contrast, in emerging economies TEA and LCA studies revealed that informal practices achieve lower net material value and substantially greater emissions and toxic releases. There are significant health concerns, environmental deterioration, and a lack of policy enforcement. Cross-case synthesis reveals prospects for scalability to promote circularity and bridge formal and informal divisions. Intermediary organization models, digital traceability, micro-recycling, and targeted technology transfer are used. Major research needs were found by the findings, including insufficient socioeconomic integration in current models, uneven LCA system boundaries, and a lack of real-world TEA datasets. The review concludes with actionable policy direction for alignment of TEA and LCA evidence spectrum to form equitable and implementable e-waste strategies in order to improve global progress toward sustainable development goals.

***Corresponding author: Neelam Sidhu,** University Centre for Research and Development, Chandigarh University, Mohali 140413, Punjab, India, e-mail: neelam.e18886@cumail.in
Khushboo, Sakshi Sharma, Vishali, Purnima Chaudhary, Abhay Dogra, Department of Environment Studies, Panjab University, 160014 Chandigarh, India

 | https://doi.org/10.1515/9783112248072-013

Keywords: E-waste, life cycle assessment, techno-economic analysis, circular economy, extended producer responsibility, formal-informal recycling

1 Introduction

The exponential growth of electronic and electrical devices over the last two decades has generated one of the most salient environmental and resource management problems of the twenty-first century. Globally, e-waste generation has been increasing at an alarming rate of nearly 3–5% annually, reaching 53.6 million tons in 2019 and projected to surpass 74.7 million tons by 2030 [1, 2]. The improper disposal of e-waste is risky to health and the environment through the presence of hazardous materials, while it also has potential for secondary resources recovery and economic growth through circular solutions [3, 4].

The two most comprehensive and evidence-based frameworks for evaluating e-waste management solutions are life cycle assessment (LCA) and technological economic analysis (TEA). LCA calculates how e-trash over its whole life cycle, tracking emissions, toxicity, and resource consumption from manufacturing and production to disposal [5]. It provides a comprehensive understanding of how different management approaches impact environmental degradation, energy needs, and the potential for global warming. In order to determine overall viability, TEA simultaneously assesses the technical and financial viability of recycling agreements, taking into account infrastructure investment, operational effectiveness, and external socioeconomic costs [2]. These frameworks enable a thorough comprehension of the economic and environmental performance throughout the e-waste management value chain.

Sophisticated collecting techniques, integrated logistics, and the application of automated sorting, shredding, and chemical separation technologies characterize e-waste infrastructure in developed nations [6]. In order to increase material recovery rates and encourage design for recyclability, they have strict Extended Producer Responsibility (EPR) legislation, which mandates that producers manage products after their useful lives [7]. Technological advancements like robotic disassembly, sorting automation, hydrometallurgical recovery, and urban mining technologies are made possible by large capital investments. Sustained circular economy methods with effective infrastructure minimize the extraction of virgin materials while reducing the environmental impact and net greenhouse gas (GHG) emissions. Additionally, closed-loop supply chains encourage continuous material flow, enabling sustainable production and consumption habits.

However, chronic e-waste problems in the developing countries are caused by disorganized recycling, lack of e-waste regulations, and geographically dispersed logistics networks [8]. Insufficient technological proficiency, weak regulatory oversight, and collection infrastructure lead to extensive activity in the unorganized sector, where open burning and hand disassembly are prevalent. TEA results highlight that such systems have low capital intensity and major health effects since pollutants are usually released untreated into local ecosystems. LCA outcomes in such environments show higher emissions of heavy metals, dioxins, and particulate matter, leading to extreme air, water, and soil pollution [3]. This degrades environmental quality, creates excessive public health problems, and reduces the economic value of material recovery. In order to reduce these effects, enhanced policy implementation, transfer of technology, and circular economy alignment are important toward development of sustainable and equitable global e-waste management. While there are many case-based studies, there has not yet been a comprehensive systematic synthesis that combines the findings from TEA and LCA studies and compares differences in national contexts [6, 4].

This review aims to address this gap by synthesizing research evidence from around the globe to conduct a comparative analysis of the techno-economic-environmental relationship within e-waste management systems in developed and developing countries. This review has the following objectives: (1) To synthesize research that has applied TEA and LCA approaches to assess e-waste recycling, recovery, and policy interventions. (2) To compare the economic and environmental outcomes of e-waste recycling across developed and developing countries to identify differences in the technological systems, economic viability, and environmental performance. (3) To identify challenges and options for policy convergence, circular economy, and sustainable innovation. (4) To identify areas for future research and policy which will improve e-waste management globally in alignment with the Sustainable Development Goals (SDGs), specifically, SDG 12 (Responsible Consumption and Production) and SDG 11 (Sustainable Cities and Communities).

This review's combined analytical approach integrates data from TEA and LCA to provide a comprehensive view of e-waste management performance worldwide. This review's case studies include the United States, Australia, India, Nigeria, Brazil, Ghana, Egypt, Korea, Switzerland, Belgium, and Japan. Each of these countries has unique implications for the development of policies, technology, and social inclusion values in e-waste systems. These nations were selected particularly because they offer extensive variations in e-waste management contexts, ranging from huge informal recycling sectors with emerging policies in developing countries to high e-waste generation and formal recycling systems in developed nations. This global perspective is necessary in developing reliable LCA and TEA. Bibliometric mapping supported this review to illustrate the evolution of research from 2000 to 2025, while systematic review principles were guided by the PRISMA protocol to provide a synthetic and structured analysis of the evidence.

2 Methodology

This study will employ a systematic literature synthesis methodology that will employ bibliometric mapping and thematic content analysis during the time frame of 2000, in accordance with the Preferred Reporting Items for Systematic Reviews and Meta-Analyses (PRISMA) standard 2025. To cover the international e-waste management research, a multi-stage strategy is created, with a focus on the most recent TEA and LCA frameworks. A triangulation of quantitative and qualitative methods will allow for a structured review of technological, economic, and environmental insights [9, 2]. A flowchart of the methodology used and the data source is shown in Figure 1.

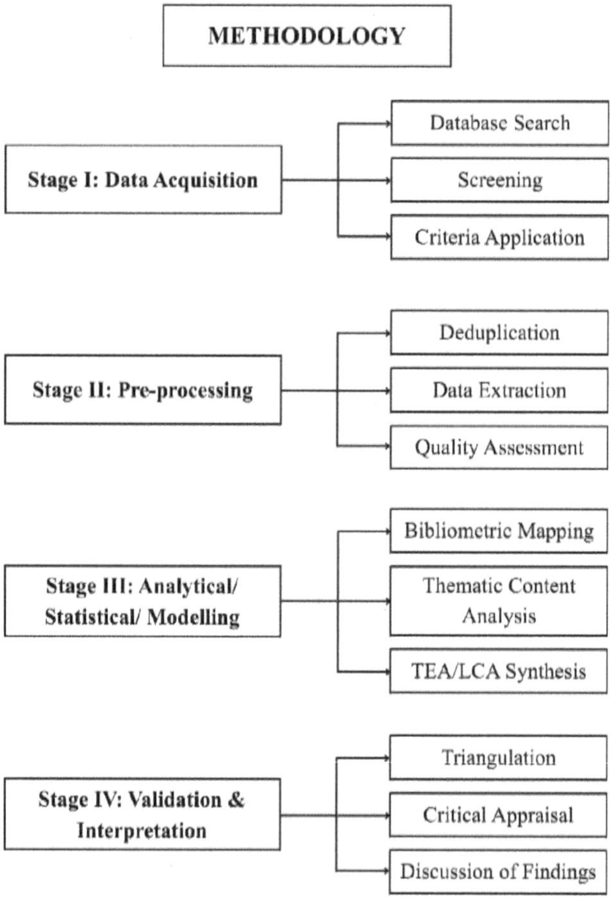

Figure 1: Flowchart of the methodology used and the data source.

2.1 Literature search and data sources

The review was first initiated with extensive research of the literature, carried out in multiple databases of academic literature, including Scopus, the Web of Science (WoS), ScienceDirect, SpringerLink, and Wiley Online Library. The review was conducted from January to September 2025, on articles published from 2000 to 2025, which we saw was a time of rapid e-waste generation, developments in discourse and policy on e-waste, and the incorporation of the circular economy frame of reference globally. The original search resulted in 612 publications. Following the removal of duplicates (n = 104) and screening of titles and abstracts for relevance (n = 508), there were 186 studies that were relevant to the inclusion criteria for review of the full text. Of this, 84 publications defined or utilized in their study TEA, LCA, or a hybrid TEA–LCA methodology and 102 provided contextual or case studies on factors related to implications on policy, technological, and socioeconomic factors related to e-waste management.

2.2 PRISMA screening process

The screening process was comprised of four distinct stages outlined by the PRISMA framework: Identification, Screening, Eligibility, and Inclusion. Identification: 612 records in all were obtained from scholarly databases. Screening: The titles and abstracts of 508 publications were used to determine their relevance. Eligibility: 186 complete texts were evaluated for topic coverage and methodological rigor. Inclusion: The comprehensive synthesis and comparative analysis comprised 84 main papers. While a graphical PRISMA flow diagram is not presented here, the process followed the Systematic Reporting process of the PRISMA framework to ensure transparency in the selection and inclusion of the literature.

Thematic classifications were based on three axes: Technological readiness and the efficiency of the system; economic feasibility and policy incentives; and environmental performance and circularity outcomes.

2.3 Analytical framework

In the analytical phase, integration of bibliometric perspectives with qualitative cross-case synthesis was done. Each study was situated across a comparative framework that distinguished between developed and developing nations, and the comparative perspective was the basis for identifying patterns within system performance, policy delivery, and research modes. A comparative perspective will facilitate understanding of the ways TEA–LCA results are impacted by governance maturity, technological advances, and the market. To identify and understand patterns qualitatively, coding of data was completed using NVivo 14 software, and quantitatively using Microsoft Excel to compare findings.

3 Results and discussion

3.1 Comparative analysis of advanced economies

In terms of more stringent regulations, cutting-edge recovery technology, and circular economy approaches, advanced economies are thought to be at the forefront of e-waste management systems. Their methods show a complex interplay between technology, regulatory frameworks, and analytical evaluation instruments, including LCA and TEA. In order to operationalize sustainability through data-driven decision-making and environmental-economic balance, the EU, USA, Australia, Japan, Korea, Switzerland, and Belgium are examined in this part.

3.1.1 European Union: circular economy leadership and multi-model decision analysis

With the help of the Waste Electrical and Electronic Equipment (WEEE) Directive, the RoHS Directive, and other regulations, the European Union (EU) remains a global leader in e-waste governance. Regulations known as the EPR formalize producer duty throughout the life cycle of a product. As a result, formal collection rates have risen to about 42.5%, the highest in the world [1].

The study by Baars [10] looks at the risks associated with the rapid use of lithium-ion batteries (LIBs) for electric vehicles (EVs), as the need for essential raw materials like cobalt is both very taxing on the environment to extract and is increasing. The excessive amount of primary cobalt production and the vulnerability of European supply chains to supply disruptions have been recognized as the main problems. The authors use a material flow analysis (MFA) approach to examine the full life cycle of cobalt found in EV batteries inside the European Union (EU) in order to address these problems. The research simulates various scenarios to estimate the circular economy strategies' (CES) ability to mitigate primary cobalt demand, specifically substitution, technology reduction, reuse, and recycling.

The key findings indicated that technological substitution and reduction approaches produce the highest reductions in the EU's primary cobalt dependency in relation to EV batteries but those methods may increase the dependency on nickel and lithium. This is more about shifting environmental burdens than alleviating them. Recycling and reuse approaches supported by policy only dent primary cobalt demand, although they are essential for long-term supply chain material security and circularity. The research concluded that only an integrated approach to addressing cobalt supply risk and transitioning toward a more sustainable EV supply chain in Europe could include advanced technology, effective capacity and infrastructure for recycling the materials, and strict and comprehensive policy.

Cottes [11] focuses on decision-making and sustainability issues in optimizing WEEE treatment plants for an efficient design, with a special interest in Greece in the context of the European circular economy strategy. The issue targeted is the increasing amount of electronic waste that either represents an environmental risk or an economic opportunity, depending on how well resources are reclaimed. The authors establish a multi-criteria decision-making (MCDM) model, a quantitative approach that combines technical, economic, environmental, and social criteria to evaluate the viability of building a WEEE treatment plant. In contrast to conventional cost-benefit analysis, their approach considers various conflicting goals and stakeholders' priorities, thus offering a more holistic decision-support tool for policymakers and investors.

The study highlighted that combining techno-economic assessment with environmental and social indicators provides a useful basis for optimizing investment in e-waste treatment. It reveals that considerations of techno-economic feasibility in Greece, as a case study, are particularly sensitive to material market prices, policy incentives, and facilities' scale. Notably, in contexts where recovery efficiency is acceptable and net present value is positive, these models are most advantageous. The findings also point to the role that combining economic, environmental, and social criteria, i.e., policy-related modeling, can have in providing viable waste management solutions that are more resilient and favorable than those relying on economic information only, in relation to circular economy goals in the EU context. Collectively, the two studies represent European research leadership in circular economy practice, highlighting how cutting-edge modeling and scenario analysis may inform sustainable transitions across high-impact sectors, from electric cars to e-waste.

3.1.2 United States: techno-economic optimization and critical material recovery

The Global E-waste Monitor 2020 provides the overall setting, stating that e-waste generation across the world had risen to 53.6 million metric tons in 2019, 7.3 kg per capita, and was set to surge well over 74 Mt by 2030. The rising resource contained in e-waste is worth \$57 billion annually, with merely 17.4% is still accounted for as properly recycled, with the rest unaccounted for or recycled informally, usually within developing nations with ineffective environmental protection measures in place. Formal recycling allows the net capture of \$10 billion worth of secondary raw materials and prevents 15 Mt of CO_2 emissions mainly by substituting iron, aluminum, and copper, but informal flows present critical environmentally poisonous dangers because of sustained hazardous materials such as mercury, brominated flame retardants, and CFCs.

They highlight notable disparities in e-waste regulatory frameworks and recycling rates, with Europe having a statutory rate of 42.5% and Asia, the Americas, Oceania, and Africa trailing behind at 11.7%, 9.4%, 8.8%, and 0.9% [1].

In order to close the growing circularity gap, supportive legislation, efficient infrastructure, and targeted public awareness are crucial. Additionally, it describes the complex structure of e-waste, which includes everything from large appliances to ICT hardware, and highlights the importance of material centered design, ethical procurement, and harmonized policy implementation (e.g., through Basel Convention regulations and EPR) in order to facilitate product lifetime extension, reduce illegal transboundary movements, and facilitate global progress toward SDGs such as SDG 11 (sustainable cities) and SDG 12 (sustainable consumption and production) (The Global E-waste Monitor 2020).

Based on global interests the recovery of critical materials (CMs) from e-waste in U.S. supply chains is the focus of USEA (2023), which emphasizes the strategic importance of CMs in the defence, clean energy, and technology sectors. According to the study, recovery of rare earths and speciality metals is still very low because of logistical, financial, and technological limitations, even though metals like nickel and aluminum benefit from established recycling systems. It also points out that while new biometallurgical and electrochemical techniques show promise but lack commercial scalability, traditional recovery techniques (pyrometallurgy and hydrometallurgy) are energy-intensive and ineffective for low volume, high value elements. The lack of affordable, scalable technology and infrastructure that supports the circular economy to improve domestic CM recovery and lessen reliance on imports is the primary research need (USEA, 2023).

Concentrating on the implementation of operations and environmental analysis within the context of the US, Olafasakin [12] critically assesses standalone single stream material recovery facilities (MRFs) using TEA and LCA. Their findings show that the MRFs, which sort 120,000 tons annually at a baseline capital expense of $23 million USD, create high net present value (NPV) and environmental impact if the composition of the waste has valuable recyclables such as aluminum, HDPE, and PET. Based on the US national average MSW makeup as their benchmark, NPV can vary from $3.57 million to $60 million, 100-year global warming potential (GWP) between 5.98 and 8.53 kg CO_2 eq per ton waste significantly lower compared to landfilling options. Sensitivity and uncertainty analysis reveal the substantial impact on total profitability and sustainability of local waste composition, recovered material market price fluctuations, facility capacity, and tipping fees. The research determines that MRF design and performance need to be optimized locally and be more resistant to economic cycles and changes in waste composition to maximize profitability as well as environmental effectiveness. They also establish the need for mechanized sorting and sophisticated separation technologies to enhance material recovery efficiency at a lower cost and environmental footprint, advising future buildout of infrastructure to take into account both financial and environmental life cycle metrics. Finally, their results support the necessity of combined MRF operations, advanced separation/sorting technologies, and responsive business models in facilitating the transition of the circular economy in US municipal solid waste management [12].

Overall, the three studies together demonstrate that although the potential for the recycling of significant and valuable materials from e-waste is vast and of strategic importance for economic, environmental, and security purposes, actual levels are weakened by fragmentation of policy, immature infrastructure, constrained technological scale-up, supply chain disconnections, and weak economic and behavior incentives. Addressing these challenges will need harmonized global-regional policy, R&D and infrastructural transformative investment, extensive application of circular economy concepts (such as EPR, eco-design of products, and repair/refurbishment), and increased public and industry participation to enhance recovery, close the material loop, and provide both protection of health and the environment amid fast growing e-waste development (The Global E-waste Monitor 2020; USEA, 2023) [12].

3.1.3 Australia: behavioral integration and reverse logistics optimization

In Achieving Circular Economy for the E-Waste Sector in Australia, a dissertation done at Macquarie University in 2022, Islam [13] examines the operational and policy-level barriers preventing Australia's transition from a linear to a circular economy in e-waste management. The study identifies fragmented policy implementation, inadequate recycling infrastructure, and limited stakeholder collaboration as major impediments, emphasizing the need for integrated circular strategies and stronger regulatory enforcement to achieve sustainable e-waste governance [13]. The research points out the quick increase of e-waste production, triple that of municipal solid waste, opposed to the narrow and disjointed system of existing management strategies, which primarily address a narrow group of products under the National Television and Computer Recycling Scheme (NTCRS). Despite increased public interest and policy awareness, there are still significant gaps in accurately estimating e-waste flows, expanding product coverage, and capturing consumer behavioral dynamics in collection and recycling structures, all of which prevent a true closed-loop economy from being realized [13]. Australian scholars have taken the lead in multidisciplinary techniques that integrate behavioral modeling, mathematical optimization, and LCA to improve circular economy outcomes while being aware of these limitations. In order to maximize cost-effectiveness and minimize environmental degradation, Islam [13] created a reverse logistics optimization model for waste solar photovoltaic (PV) systems that combined transportation routes, collection nodes, and recovery facilities. The work of Islam [14], which used empirical consumer data and a life cycle simulation approach to show elements of policy misalignment and consumer involvement. The study's findings showed that collection and routing, whether done locally or remotely, might result in an overall recycling cost reduction of up to 25–30% while recovering higher recycling efficiencies.

Additionally, using SimaPro and the ReCiPe 2016 approach, Islam and Iyer Raniga [15] conducted an LCA of recycling waste printed circuit boards (PCBs) recycling systems. They discovered that regional integrated recycling programs were able to reduce GWP by 53% and fossil 98% less resources are needed when recycling waste PCBs locally as opposed to processing them abroad. An illustrative sample of local circular economy infrastructure is shown in the paper.

These studies further support the idea that for national e-waste management to be effective for both environmental and economic results, social behaviors, logistical planning, and life cycle optimization must coincide.

3.1.4 Japan and Korea: technological innovation and EPR implementation

Both South Korea and Japan are highly developed Asian economies with sophisticated recycling technologies and robust EPR programs. According to Itoh [16], Japan's "Urban Mining" method includes Rare Metal Recovery Complexes (RMRC), which seek to recover rare critical metals including neodymium, cobalt, and tungsten, with a recovery rate exceeding 90% by employing a variety of mechanical, physical, and chemical techniques like hydrothermal recovery, molten salt extrusion, and chloride evaporation. The Japanese example shows how mechanistically coupled hydrometallurgical and pyrochemical processes can be replicated in a controlled setting, with a strong emphasis on attaining circularity, particularly in high-value metal loops.

The EPR framework governs South Korea's e-waste management system, which requires producers to gather and recycle predetermined amounts of e-waste annually [17]. There are two ways to collect: (1) curbside collection and (2) retail take-back programs (local systems). In South Korea, the recycling stream uses eddy-current and magnetic separation after human disassembly. Large household appliances are the focus of South Korean recycling plants, which have above 85% material recycling efficiency for metals like copper and steel. Small electronics, however, have been slow to catch up due to the absence of collection incentives. In the future, opportunities exist for deploying IoT-based monitoring for the collection and tracking of toxic emissions during processing. Japan and Korea's examples demonstrate how rigorous regulatory enforcement plus technology can create consistent, systemic circular economy outcomes. The systems in both nations prioritize integration of efficient processes as well as data transparency and accountability for producers as part of national strategies.

3.1.5 Switzerland and Belgium: scientific precision and circular economy optimization

Switzerland and Belgium have led in the scientific standardization of the recycling of E-waste through state-supported R&D programs on LCA-based system assessment and electrochemical recovery optimization. Switzerland's Federal Laboratories for Materials Science and Technology (EMPA) focus on Eco-Material Characterization and high tech recycling processes that would allow recovery of value metals of palladium, platinum, and rare earths [18] The LCA modeling into national policy in Switzerland also means improvements of the system could be more advanced and a mechanism for technology transfer to the global south that is responsible and sustainable [19] EMPA have undertaken projects with programs in developing countries (Egypt and India) thus reinforcing Switzerland's microbial work of capacity building and technology proliferation.

Belgium also has mechanical electrochemical hybrid recovery systems to recover components of printed circuit boards and copper at high efficiencies [20]. The model for Belgium has been successful at optimizing economic environmental integration in achieving greater material recovery, and doing so with inherently lesser potential toxicity. Collectively, these applications have shown that high-tech circular systems can return a profit while being environmentally friendly, provided the system is developed based on sound TEA–LCA data.

3.2 Comparative perspectives from developed countries

Many similar trends are found across the developed world:

Institutionalized EPR: EPR schemes are mandated by law, with clear material and cost accountability cycles in place, creating a predictable and stable stream of funding for recycling infrastructure (EU, Japan, and Korea).

Technological Systems: Mechanized sorting, hydrometallurgical refining, and urban mining systems create recovery rates greater than other programs, while maintaining a lower environmental burden (Switzerland and Belgium).

Analytical Framework: The use of TEA–LCA hybrid models helps determine smart and intentional evidence-based policymaking and investment funding prioritization (EU, USA, and Australia).

Economic Feasibility: Positive NPV and IRRs for larger-scale facilities show circular systems can be economically feasible even in volatile commodity price environments. Residential Needs Gap: While progress has been made across many programs, challenges remain for small devices and ensuring consumer participation and meeting burden transfer and cost structure standards (Australia and Korea).

Therefore, these nations collectively demonstrate that it is indeed possible to combine technological innovation, economic feasibility, and environmental protection, provided there are regulatory frameworks and transparent evaluation systems in place.

3.3 Comparative review of developing nations

3.3.1 E-Waste management in India: scientific trends, policy barriers, and pathways to circular economy

India is among the world's fastest-growing e-waste-producing countries due to increasing consumption of electronic goods, fast urbanization, and short product life spans. India generates about 3.8 million tons of e-waste per year, with over 90% of it being processed in the informal sector under primitive processes, including manual dismantling, open burning, and acid leaching, giving rise to substantial environmental pollution and occupational health risks, such as higher blood lead levels and respiratory diseases for workers and surrounding residents [21, 22].

Indian studies have used varied methodological approaches. Phogat [9] performed a scientific and bibliometric analysis using VOSviewer and CiteSpace about a dataset of over one thousand publications (2000–2024) to map evolving developments in sustainable metal recovery. The researchers noted a shift from energy-intensive pyrometallurgical techniques to hydrometallurgical and bioleaching techniques, with an 85–95% metal recovery and much lower pollution emissions. Sinha [21] performed a systematic review in accordance with PRISMA, which used a PICO framework to categorize the main structural barriers such as weak enforcement, economic disincentives, and the cultural hoarding of discarded devices. Awasthi [22] supported these findings with a stakeholder mapping and econometric model that revealed informal practices included at least an annual ₹ 10,000 crore cost for health care expenditure. These studies converge to reveal three important dimensions: (1) expressed formal recycling capacity in India accounts for only about 5% of total generation, (2) informal recyclers reclaim 20–30% of value associated with materials but release thousands of tons of toxins each year, and (3) an opportunity exists to create 100,000 green jobs through enforcement of EPR and cluster-based formalization of informal recyclers. Collectively, Indian studies suggest that increased technological innovation paired with policy inclusion and financial mechanisms can help transition e-waste away from being an environmental liability and transform it into a socioeconomic asset.

3.3.2 E-Waste management in Nigeria: informal sector dynamics, policy gaps, and pathways toward sustainable formalization

Nigeria's e-waste scenario is distinguished by pervasive informal practices, illegal imports, and weak enforcement of current regulations. Domestic e-waste is estimated to amount to approximately 216,000 tons annually, and 60,000 tons of used electrical and electronic equipment (UEEE) is imported, of which 75% is broken and discarded. Informal recycling networks in Lagos and Abuja include a workforce of over 100,000 workers who recover valuable metals in unsafe environments, releasing heavy metals and dioxins into the soil and air [23, 24].

Ogungbuyi [23] integrated material flow modeling and stakeholder engagement within the Basel Convention's e-Waste Africa Project to assess inflows and formulate national control measures. Their analysis showed that recyclers only recover between 15% and 20% of recyclable metals, and open burning contributes about 40 tons of dioxins per year. Barnabas [24] conducted a comparison analysis of Nigeria's legal and policy frameworks including the 2011 National Environmental Regulations relating to the established templates from the EU and Ghana to determine the lack of systemic enforcement and to limit the perverse financial incentives not to comply with the existing regulations. Adeola and Othman [25] conducted an LCA study comparing the impact of recycling versus landfilling waste, conforming to ISO 14040 standards, which indicated that emissions could be reduced with sustainable recycling by 65% from that of landfilling; however, a high impact illustrated the potential for acidification effects from incineration.

Collectively, these studies demonstrate that processing e-waste is dominated by Nigeria's informal sector, with low material recovery and high ecological costs. However, policy simulations reveal that even hybrid EPR-type regulations using blockchain for tracking imports could formalize up to 50% of current informal activity, creating 50,000 jobs and USD 200 million in annual revenues. Formalizing regulatory oversight with these mechanisms within a cohesive national framework is essential for achieving sustainable resource recovery and SDG 12.

3.3.3 E-waste governance and circular transition in Brazil: assessing policy, technology, and informal sector integration

E-waste processes in Brazil reflect those in many emerging economies, including rapid growth in electronic consumption, inconsistent application of environmental legislation and plans, and reliance on informal waste recyclers. National generation was estimated to be 2.2 million tons in 2019, and only about 3% was collected and processed through formal channels [26]. While the National Solid Waste Policy (PNRS) established the framework for EPR since 2010, the accompanying laws are not yet uniformly applied throughout the states. Kiddee [3] offered a narrative meta-synthesis of the situation of e-waste in the world, and placed Brazil as part of the broader context of developing countries that have low technical capability and weak policy enforcement Leme (2014) [27] performed TEA and lifecycle assessment (LCA) on energy recovery through municipal solid waste (MSW) in Betim and found that biogas-to-energy routes have positive net present values (R$ 5–15 million) and an estimated 40,000 tons

CO_2e reduction in GHG emissions annually, while incineration was not economical without subsidies. Expanding on this understanding, Dias [26] completed a value chain mapping of 45 recyclers in major cities and also documented that 89% of firms only carry out pretreatment activities (sorting, manual dismantling) and export printed circuit boards to Europe or Asia for refining.

Though the Brazilian case highlights structural barriers to change, such as inadequate reverse-logistics networks and inconsistent adherence to EPR, it also reveals pathways to success through WtE(waste-to-energy) models and formally organizing informal collectors. If Brazil could integrate state-level programs under one federal program, it would be possible to recover up to 1.5 million tons of recyclable material each year and eliminate, at a minimum, R$ 2 billion in health-related costs each year.

3.3.4 Urban waste challenges and circular economy transitions in Ghana: insights from emerging waste-to-energy studies

Ghana, a rapidly developing nation grappling with urbanization rates of more than 4% annually, faces an increasingly difficult waste management dilemma that manifests in serious environmental degradation, threats to public health, and lost economic opportunities, where the generation of the amount of municipal solid waste (MSW) has increased to 12,710 tons per day, most of which is organic (53.91%). However, collection efficiency is stuck at 40–50%, leaving the remainder to unregulated landfilling and noxious open dumping, which releases methane emissions equal to 2.5 million tons of CO_2-equivalent annually, contaminates aquifers with leachates containing heavy metals and pathogens, and spreads vector-borne diseases that affect over a million people in places like Accra and Kumasi [28, 29].

Anokye [28] performed a narrative review amalgamating the global literature on waste activation and mineralization with literature specific to Ghana, developing pyrolysis and mineralization as opportunities to soil organic municipal solid waste (MSW) to biochar and sequestering heavy metals. Additionally, Osei-Appiah and Dioha [29] employed techno-economic modeling (HOMER and RETScreen) to consider four waste-to-energy (WtE) technologies in Accra, finding anaerobic digestion and landfill gas recovery provided the best returns, with an internal rate of return of 12–18% with partial subsidies. Findings suggest that Ghana's planned transition to circular economy solutions will rely on scaling capacity and incentives for private participation and enforcing the Hazardous and Electronic Wastes Control and Management Act (Act 917, 2016). Consequently, the country will continue to pay disproportionately high health and ecological costs and will miss economic opportunities in recovering renewable energy and materials.

3.3.5 Hybrid governance and public-private partnerships in Egypt's emerging e-waste circular economy

Egypt is an illustrative transitional case of an emerging economy that is moving from informal to semi-formal e-waste repair, refurbishment, recycling, and disposal. Through the Sustainable Recycling Industries (SRI) program, the formal processing capacity has increased from 1,900 tons in 2019 to 6,800 tons in 2023 [30]. Furthermore, 27 licensed processing facilities have been opened and vocational training has taken place. This has engaged 43% women and has shown increasing institutional support for sustainable recycling. According to Lotfy [30], the Intermediary Organization Model connects informal collectors and formal recyclers together, with guaranteed purchases, micro-insurance in case of quality or theft, and price matching. This method opens the door for all parties to work together rather than have informal collection illegally divert e-waste from the formal sector. It has also enhanced social inclusion in this value chain and enables the formal facilities to get stable feedstock, especially for the hydrometallurgical recovery of copper, gold, and silver. Although this is progressing well, Egypt and other emerging economies still face challenges. They do not have enough feedstock to feed the new facilities and there is not enough enforcement of environmental laws, and some communities still support the dismantlers who are not registered to handle e-waste. By providing examples from Egypt, it is clear that hybrid governance, donor-funded efforts, public-private partnerships, and including workers in sustainable developments can create operational pathways supporting the transition from informal, subsistence levels of recycling toward formal circular economy businesses.

3.4 Cross-country comparison

There are several similarities and differences of interest between these five developing countries. India has the highest volume of generation and a growing maturity of regulation, although fierce informal activity persists. Nigeria and Ghana are subject to grave environmental and health impacts because of weak regulatory controls over uncontrolled imports and basic recycling processes. Brazil has shown that recycling can be economically viable on a large scale but has yielded an inconsistent policy response, while Egypt has shown some incremental advancement through a combination of international engagement and models to formalize intermediaries. Analytically speaking, there are notable differences between the TEA and LCA in terms of policy execution and technology maturity. Egypt and India employ recovery methods based on bio-electrochemistry or hydrometallurgy, which show 95% or higher metal yields at reduced emissions, but Ghana and Nigeria's laborious disassembly methods result in high externalized costs. Economically, it has been noted that while their African counterparts typically depend on donor support or trial projects, India and Brazil are able to produce quantifiable returns on investments in circular economy projects. In terms of society, formalizing informal workers continues to be both a barrier and an opportunity, as their participation and contribution will be crucial to achieving a fair and just circular transformation.

The comparative data emphasizes the need for developing nations to work toward concurrent advancements in social inclusion, policy, and technology. Incorporating nonformal players, strengthening EPR programs, and encouraging Sustainable Development Goal 12: Responsible Consumption and Production outline the combined objectives of economic recovery and ecological preservation, both of which depend on environmentally sound technologies.

4 Cross-case synthesis and policy implications

There are notable differences between industrialized and developing nations in terms of socioeconomic participation, regulatory enforcement, and technical sophistication in the global e-waste scenario. Developed countries, including the European Union, the United States, and Australia, have a more institutionalized governance, standardized collection mechanisms, and EPR (extended producer responsibility) systems. In contrast, and as demonstrated by India, Nigeria, Brazil, Ghana, and Egypt, developing countries still rely heavily on fragmented institutional models and the informal economy.

4.1 Comparative structural and technological dynamics

A comparative overview of e-waste management practices in developing and developed countries is presented in Table 1. In developing countries, closed-loop production and eco-design practices have advanced through binding directives such as the EU's Waste Electrical and Electronic Equipment (WEEE) Directive and Restrictions of Hazardous Substances (RoHS) Directive. These directives mandate quantitative collection requirements, eco-labeling, and producer-funded recycling schemes. Some studies have reported metal recovery efficiencies to be above 85% in areas with an integrated hydro- and pyrometallurgical system with developments in automation and digital tracking [31, 3].

In contrast, developing nations show significant deficits in technology, even as hydrometallurgical and biotechnologically developed methods have been piloted in India and Egypt, the impediment to scale has been infrastructure costs, a lack of technical know how and continued inconsistencies in the supply of recyclable feedstock, while informal dismantling, open burning and acid leaching remain widespread in Nigeria and Ghana, where lead, mercury and brominated flame retardants were released into the environment uncontrollably. Brazil presents as an intermediate case, having both regulatory and industrial capacity but disparities across the region in operationalizing this regulatory structure.

Table 1: Comparative analysis of e-waste management in developing and developed countries.

Country/region	Practices	Economic insight	Environmental impact
Developing nations			
India	Mostly informal recycling.	High costs; limited formal capacity.	Health and supply-chain risks.
Nigeria	Burning and acid leaching.	Poor infrastructure viability.	Extreme pollution and toxicity.
Brazil	Mixed formal–informal system.	Policy gaps affect recycling.	Rising emissions from informal sector.
Ghana	Informal processing of imports.	Circular gold recovery potential.	High risk from unsafe processes.
Egypt	Informal, moving to regulation.	EPR can support financing.	Sustainability under growth pressure.
Developed nations			
EU (Belgium)	Organized EPR-based collection.	Cost-efficient systems.	Low emissions.
USA (California)	Formal recycling facilities.	Revenues offset costs.	Environmentally viable.
Australia	National regulated programs.	Lower landfill costs.	Policy-led low impact.
Japan	Mechanized, product-specific.	High recycling rates.	Full lifecycle control.
South Korea	State-guided formal systems.	High recovery efficiency.	Energy-efficient recycling.
Switzerland	Advanced regulated systems.	Precision recycling.	Minimal environmental impact.

4.2 Economic viability and circularity potential

TEAs from both categories of countries reveal stark differences in economic feasibility. Developed countries display huge capital investments but they are also able to obtain good financial returns from the secondary raw material markets and from integration into the circular economy. In the developed countries specifically, the pay-as-you-throw arrangements of EPR-based market structures reduced the financial burden on municipal waste management and have supported stable, long-term jobs in the recycling economy.

On the contrary, TEAs were carried out in developing countries, Leme [27] in Brazil, and Adeola and Othman [25] in Nigeria. indicated positive net present value only under subsidized initiatives or hybrid models of financing. In India, Ghana, and Nigeria, informal operations continue to dominate because of their low investment and fast cash turnover, despite the known health and environmental externalities. The lack of affordable credit, formal procurement arrangements, and reliable supply chains for feedstock places limits on achieving long-term scalability for climate-friendly technological solutions. Nonetheless, opportunities are emerging, India's open access online portal for EPR, Brazil's reverse logistics networks, and Egypt's intermediary organization illustrate ways that digital and institutional innovations can improve financial cost recovery and enhance operational transparency. If models are developed or adapted, they could improve operational performance and provide substantive gains to formally increase recovery from the present global recovery rate of 17% to above 40% by 2030.

4.3 Government and institutional progress

Agreements regarding policies are a key factor of success. In developed economies, lifecycle governance consists of consideration at design, consumption, and postconsumer life cycles of the products. Regulatory mechanisms in developed economies provide measurable targets, third-party audits, and public-private partnerships. For example, the most recent update to the WEEE Directive in the EU (2018) includes product takeback responsibilities for producers; meanwhile, some US states have instituted EPR programs which create standardized collection networks and public messaging campaigns about e-waste collection.

In comparison, developing economies exhibit some but not complete policy convergence. India has demonstrated progress in the E-Waste Management Rules (2022) through the amalgamation of EPR, digital tracking, and incentive structures to keep count of recyclers and inform consumers. Egypt has also made infrastructure changes by benefitting from SRI support and representation from civil society, through capacity building and social inclusion. However, in locations such as Nigeria, Ghana, and some parts of Brazil, overlapping institutional jurisdictions, enforcement, and financial incentives still represent barriers to infrastructure development.

When developed economies have maximized e-waste collection through formal collection systems, in developing economies informal recyclers represent both an opportunity and a barrier to success. Policy approaches will need to formalize informal practices by providing credits toward microinsurance and capacity building if they wish to cost-optimally through inclusion in EPR initiatives to fairly transition toward more equitable circular economies.

4.4 Environmental and social dimensions

The environmental impacts of e-waste are also distinctly different. In many developed countries, heavy metal pollution, GHG emissions, and reliance on landfilling have declined significantly through standardized recycling technologies. LCA show reductions in GWP from integrated recovery of 60–80% as compared to landfilling. Conversely, the informal processing of recycling e-waste in developing countries has led to localized environmental injustice. Studies of e-waste in Nigeria and Ghana demonstrate that exposure to e-waste is associated with contamination of soils; bioaccumulation of toxic metals; increased respiratory illness; and increased neurological disorders. Environmental burdens thus become entwined with public health issues and social inequalities which typically weigh disproportionately on informal workers and low-income communities. Socially, some countries, including India and Egypt, have begun to embed opportunities for gender and occupational safety into e-waste policy and programming in response to SDG 8 (Decent Work) and SDG 12 (Responsible Consumption). However, the majority of poor nations and unofficial economies still lack regulated or implemented regulations for labor rights, healthcare access, or protective gear for unofficial recyclers.

4.5 Strategic policy implications

According to the synthesis, developing nations' e-waste management issues need three-pronged approaches:

Technological Upgrading: Investment in biotechnological and hydrometallurgical recovery techniques, supported by regional innovation hubs and information transfer from industrialized nations.

Institutional Integration: Establishing producer responsibility funds, creating digital traceability systems, and combining disparate institutional duties for environmental management are all examples of institutional integration.

Social Inclusion and Just Transition: To reduce socioeconomic disruption, the informal sector is gradually formalized through cooperatives, incentives, and skill development. Furthermore, national frameworks can be coordinated to promote global policy coherence with SDG 12, the Circular Economy Action Plan (EU), and the Basel Convention. Green bonds, blended finance, and carbon credit trading are examples of funding methods that could encourage future investment in the construction of clean recycling infrastructure.

4.6 Toward a global convergence

Co-innovation and information sharing are necessary to close the gap between industrialized and developing nations. Experiences like the creation of EPR systems in the EU, material flow audits in Australia, and a decentralized collection strategy in the United States can be in favor of an adaptive model in the Global South. However, community-based recycling programs and unofficial repair economies in nations like Ghana, Brazil, and India may also offer socially ingrained instances of circularity that open up forums for international conversations on sustainability. In the end, synthesis indicates that more comprehensive governance that incorporates social, economic, and environmental aspects is needed; technology advancement alone is insufficient. Progress toward a globally equitable circular economy requires mutual cooperation, building upon capabilities, and contextually positive adaptations of a legitimized framework.

5 Challenges encountered and resulting outcomes

Advanced economies are challenged by scaling up emerging technologies such as biometallurgy and burden-shifting (e.g., the EU's nickel dependence). High capital costs and awareness deficiencies among consumers continue to exist, especially in the USA and Australia. Developing countries struggle with well-established informal sectors (India: 90%, Nigeria: dominant), illicit imports (Nigeria: 75% UEEE), and insufficient capital for infrastructure (Ghana: $200–500 M for WtE). Enforcement of regulation is poor in all developing settings, with Brazil's PNRS and India's 2016 Rules failing, and Egypt's formal facilities undermining feedstock access.

Developed economies are set to progress using R&D in green technologies (e.g., Japan's RMRC, Switzerland's EMPA), harmonized EPR, and digital monitoring for compliance. Emerging countries need technology transfers, such as India's microrecycling and Egypt's intermediary models, with international cooperation to prevent illicit trade (Nigeria). Investment in scalable WtE (Brazil, Ghana) and integration of formal and informal systems (India, Egypt) may be the driving factor for increased recycling rates, aiming for 50% formalization by 2030. Cross-border learning from the EU's policy maturity and Australia's behavior principles may close gaps, in line with SDG 12.

6 Conclusion

This review identifies the multidimensional complexity of e-waste management and a growing technology and institutional divide between advanced and developing economies. Countries, such as the member states of the European Union, the United States, and Australia, have developed closed-loop systems anchored on strong regulation, producer responsibility, and high-efficiency recovery technologies. The secret to their success is the synchronization of innovation and regulations, which has resulted in significant reductions in carbon emissions, recycling rates of over 85%, and progress toward circular economy objectives. In contrast, developing nations like India, Nigeria, Brazil, Ghana, Egypt, and other nations are still limited in their ability to invest in sound technology, recycle informally, and enforce environmental regulations. While national systems like Egypt's SRI-based reforms, Brazil's PNRS, and India's E-Waste Management Rules (2022) show significant progress, implementation in emerging economies is still hampered by a lack of financing and fragmented authority. Additionally, the majority of your electronic waste is still collected by unofficial methods, which have some commercial value but also serious negative externalities in terms of the environment and human health.

In both cases, the synthesis highlights that the transition to sustainable e-waste management necessitates an integrated governance approach to economic viability, environmental stewardship, and social inclusion rather than relying solely on technology innovation. Achieving an equitable circular economy transition requires developing policies that bridge the formal and informal divide, encourage green entrepreneurship, and incorporate transparency into traceability. The Global South must encourage global convergence, promote improved cross-regional collaborations, funding for capacity building, and testing of tried-and-true methods like LCA and EPR that are appropriate for regional settings. As an alternative, developed countries will have to facilitate technological transfer, funding, and simple secondary raw material trading.

By coordinating EPR, creating green technologies like hydrometallurgy and WtE, and encouraging stakeholder collaboration, the global community can reduce health and environmental costs, lessen reliance on virgin resources, and establish resilient, inclusive circular economies in line with Sustainable Development Goals 11 and 12 to offer a sustainable future for the electronics life cycle worldwide.

References

[1] Forti V, Baldé CP, Kuehr R, Bel G, The global e-waste monitor 2020: Quantities, flows and the circular economy potential. UNU/ITU/ISWA Rep, 2020, 120, 1–120.

[2] Dada SR, Okoye E, Hammed V, Omotosho SA, Environmental and economic impacts of sustainable e-waste management: A life cycle assessment approach. Int J Multidiscip Res Growth Eval, 2025, 6(1), 1851–1854.

[3] Kiddee P, Naidu R, Wong MH, Electronic waste management approaches: An overview. Waste Manage, 2013, 33(5), 1237–1250.

[4] Dutta D, Rautela R, Meena P, Cheela VRS, Dagwar PP, Kumar S, Life cycle assessment and techno-economic of e-waste recycling. Manage Electron Waste, 2024, 1, 173–190.

[5] Bisinella V, Götze R, Albizzati PF, Damgaard A, Astrup TF, Environmental impact quantification of e-waste management by life cycle assessment. Resour Conserv Recy, 2024, 180, 106216.
[6] Constantinescu M, Advanced e-waste management systems in developed countries: Technologies and policies. J Clean Prod, 2022, 344, 131056.
[7] Ou X, Tong X, Chen G, Extended Producer Responsibility (EPR) in e-waste management: Lessons from policy implementations. Environ Sci Policy, 2021, 123, 108–116.
[8] Sinha LN, Pundhir A, Jaiswal A, Kansal OP, E-waste management and its challenges in India: A systematic review. Int J Med Public Health 2025; 15(2): 1264–1271.
[9] Phogat P, Kumar S, Wan M, A scientometrics study of advancing sustainable metal recovery from e-waste: processes, challenges, and future directions. RSC Sustain. 2025; 3(6): 2434–2454.
[10] Baars J, Domenech T, Bleischwitz R, Melin HE, Heidrich O, Circular economy strategies for electric vehicle batteries reduce reliance on raw materials. Nat Sustain, 2021, 4(1), 71–79.
[11] Cottes M, Mainardis M, Simeoni P, Assessing the techno-economic feasibility of waste electric and electronic equipment treatment plant: A multi-decisional modeling approach. Sustainability, 2023, 15(23), 16248.
[12] Olafasakin O, Ma J, Bradshaw SL, Aguirre-Villegas HA, Benson C, Huber GW, Zavala VM, Mba-Wright M, Techno-Economic and life cycle assessment of standalone Single-Stream material recovery facilities in the United states. Waste Manage, 2023, 166, 368–376.
[13] Islam MT, Achieving circular economy for the e-waste sector in Australia. PhD diss, Macquarie University 2021.
[14] Islam MT, Iyer-Raniga U, Life cycle assessment of e-waste management system in Australia: Case of waste printed circuit board (PCB). J Clean Prod. 2023; 418: 138082.
[15] Islam MT, Huda N, Teh SL, Recycling perspectives of circular business models: A review. Recycling, 2022, 7(5), 79.
[16] Itoh H, Urban mining rare metal recovery complexes in Japan: Integrated chemical and mechanical processes. J Clean Prod, 2014, 72, 64–73.
[17] Yoon H, Jang YC, The practice and challenges of electronic waste recycling in Korea with emphasis on extended producer responsibility (EPR). Proc 2006 IEEE Int Symp Electron Environ. 2006; 1: 326–30.
[18] Ott D, Advanced material characterization and recycling technologies at EMPA, Switzerland. J Sustain Metall, 2013, 1(2), 89–95.
[19] Lee J, Life cycle assessment and environmental impact evaluation in Swiss e-waste recycling practices. Environ Sci Technol, 2024, 58(6), 3309–3319.
[20] Peeters JR, Vanegas P, Dewulf W, Duflou JR, Mechanical and electrochemical recycling of electronic waste in Belgium: Environmental and economic assessment. Resour Conserv Recy, 2014, 88, 43–53.
[21] Sinha LN, Pundhir A, Jaiswal A, Kansal OP, E-waste management and its challenges in India: A systematic review. Int J Med Public Health 2025; 15(2): 1264–1271.
[22] Awasthi AK, Zeng X, Li J, Environmental pollution of electronic waste recycling in India: A critical review. Environ Pollut, 2016, 211, 259–270.
[23] Ogungbuyi O, Nnorom IC, Osibanjo O, Schluep M, E-Waste Country Assessment Nigeria. BCCC-Africa/Empa, Ibadan, Nigeria/St Gallen, Switzerland, 2012.
[24] Barnabas SG, Towards an efficient e-waste management regime in Nigeria. J Envt Pol'y, 2025, 5, 117.
[25] Adeola FO, Othman MHD, Comparative life cycle assessment of sustainable recycling versus landfilling of municipal solid waste in Nigeria: Environmental impacts and acidification potential. J Waste Manag Environ Sustain, 2025, 7(1), 45–62.
[26] Dias P, Machado A, Huda N, Bernardes AM, Waste electric and electronic equipment (WEEE) management: A study on the Brazilian recycling routes. J Clean Prod, 2018, 174, 7–16.
[27] Leme MMV, Rocha MH, Lora EES, Venturini OJ, Lopes BM, Ferreira CH, Techno-economic analysis and environmental impact assessment of energy recovery from Municipal Solid Waste (MSW) in Brazil. Resour Conserv Recy, 2014, 87, 8–20.

[28] Anokye K, Darko AO, Agyemang P, Amuah EEY, Sodoke S, Agya BA, Douti NB, Kazapoe RW, Bentil J, Exploring waste activation and mineralization for environmental and economic sustainability in Ghana. Clean Waste Syst. 2025;11:100315.
[29] Osei-Appiah NA, Dioha MO, Techno-economic assessment of waste-to-energy technologies in Ghana. J Sustain Energy. 2019;10(2):112–21.
[30] Lotfy M, Bridging the gap between the formal and informal e-waste sectors in Egypt. Academic Report/Thesis 2024.
[31] Song Q, Liu Y, Zhang L, Xu Z, Selective electrochemical extraction of copper from multi-metal e-waste leaching solution and its enhanced recovery mechanism. J Hazard Mater, 2021, 412, 125191.

Rahul Sharma*, Nihal, J.K. Goswamy

Printable electronics: a pathway to green and sustainable e-waste management

Abstract: Advances in traditional electronics have resulted in an unprecedented rise in electronic waste, creating serious environmental and health concerns. However, printable electronics have shown promise as an alternative material for traditional electronics, given their low material requirements, additive processing, and ability to be processed on biodegradable materials. This chapter investigates printable electronics as an effective solution for managing electronic waste. It describes the printing process, ink, biodegradable materials, and their applications in developing environmentally friendly electronics. This chapter will further identify how printable electronics help in increasing the lifetime of electronics, making them recyclable and environmentally friendly, and making it an effective solution for managing electronic waste.

Keywords: Printable electronics, green electronics, sustainable materials, e-waste management, conductive inks, biodegradable substrates

1 Introduction

The exponential rise in the usage of electronic devices has brought a huge revolution in modern society and made it possible for people to have rapid communication, automation, advancements in healthcare, smart infrastructure, and adoption of digital technology [1, 2]. Ranging from consumer electronics like smartphones, laptops, and wearable tech to industrial and Internet-of-Things (IoT) platforms, electronics have become an integral part of modern lifestyle. However, the result of this exponential rise in technological developments has also brought an unintended and ever-increasing complication – the huge generation of electronic waste or e-waste, on a massive scale. Reports from around the world suggest that e-waste has become the fastest-growing waste stream globally, with tens of millions of tons of e-waste estimated every year [3, 4].

Conventional electronics are mostly produced by semiconductor technology involving complex manufacturing procedures that are multistep, energy-intensive, and difficult to execute, as reported in reference [5]. These manufacturing procedures include the need for high-vacuum conditions, higher processing temperatures, and the

***Corresponding author: Rahul Sharma,** UIET, Panjab University, Chandigarh 160014,
e-mail: rahulkumarsharma005@gmail.com
Nihal, UCRD, Chandigarh University, Gharuan, Mohali, Punjab 140413
J.K. Goswamy, UIET, Panjab University, Chandigarh 160 014

 | https://doi.org/10.1515/9783112248072-014

use of poisonous materials. Moreover, conventional electronics are mostly produced using nonbiodegradable materials like silicon wafers, glass, and plastics, which make the recycling process more difficult, as reported in reference [3]. Miniaturization and the combination of materials in one single product further make the recycling of materials more challenging and economically not viable, as stated in reference [6]. This leads to the disposal of a major portion of e-wastes in landfills and, in many developing countries, in informal sectors, resulting in soil, air, and water pollution.

Owing to such constraints, there has been an increasing need for new and alternative solutions for electronic manufacturing that highlight sustainability and environmentally friendly manufacturing concepts. This has led to the emergence of "printable electronics," which has proved to be a new and revolutionary manufacturing trend and has subsequently overcome various limitations linked to conventional electronic manufacturing techniques [7]. Unlike conventional electronic manufacturing techniques that rely upon "subtractive manufacturing" techniques, in "printable electronics," "additive manufacturing" techniques are employed in which materials are selectively placed only at desired locations. Printable electronics involve a number of printing methods such as inkjet, screen, gravure, flexographic, and aerosol jet printing, and each has specific applications, depending on the design architecture that is required [8]. The use of the aforementioned printing methods makes it possible for printable electronics to be printed on unconventional, flexible, and thin supports such as paper, textiles, biodegradable polymers, and recycled plastic, which would not be able to tolerate the conditions used in semiconductor manufacturing with printable electronics and semiconductor process conditions that require high temperatures and high-energy consumption [7, 9].

One of the major benefits of printable electronics lies in their ability to aid in the development of recyclable and biodegradable materials. Carbon-based conductive inks, conductive polymers, metal oxide-based inks, and two-dimensional nanomaterial-based inks can thus replace conventional conductive materials made from heavy metals [9]. Likewise, using a substrate made from cellulose paper, polylactic acid (PLA), polyvinyl alcohol (PVA), and other bio-based materials can also help in creating disposable and short-life electronic components and devices [9]. The use of such materials will help in reducing the overall ill-effect of persistent electronic wastes in the environment and thus follows the key aspect of green chemistry [10].

Printable electronics allow for the creation of electronics that have a designed lifespan and application-specific functionality. This is a requirement for many new usage models, for example, smart packaging, environment monitoring, medical diagnostics, and radio-frequency identification (RFID) tags, which must be functional only for a specified short time span [2, 8]. Such electronics, if produced with traditional electronics, will lead to unnecessary material usage and subsequently a large amount of waste for a considerably longer time span. This should allow for a paradigm shift from designing for durability to application and then be capable of degrading and recycling in a safe and efficient manner.

Figure 1 shows the comparison of printed energy storage and printable energy harvesting technologies for flexible electronics. In circular economy terms, printable electronics are also very appealing because they favor dismantling by design, recyclability, and material recovery [6]. The reduced material complexity and complexity of components in printed electronics make it easier to dismantle and recover valuable materials at the end of life. Moreover, roll-to-roll printing techniques are capable of manufacturing at a larger area with minimal waste and lesser energy consumption per item produced, which further adds to resource efficiency in printable electronics [8]. By incorporating sustainability aspects right from manufacturing to design, printable electronics favor the circular economy model of life cycles that differs from the materialistic "produce, use, and dispose" pattern of regular electronic devices. Although printable electronics offer many benefits, they also have limitations in terms of electrical characteristics, mechanical robustness, environmental stability, and mass standardization [7, 9]. However, advances in functional ink materials, printing resolution, encapsulation methods, and hybrid integration with traditional electronics are gradually enhancing device performances. It should be noted that in terms of performance for application areas aiming to counteract e-waste, printable devices are currently sufficient.

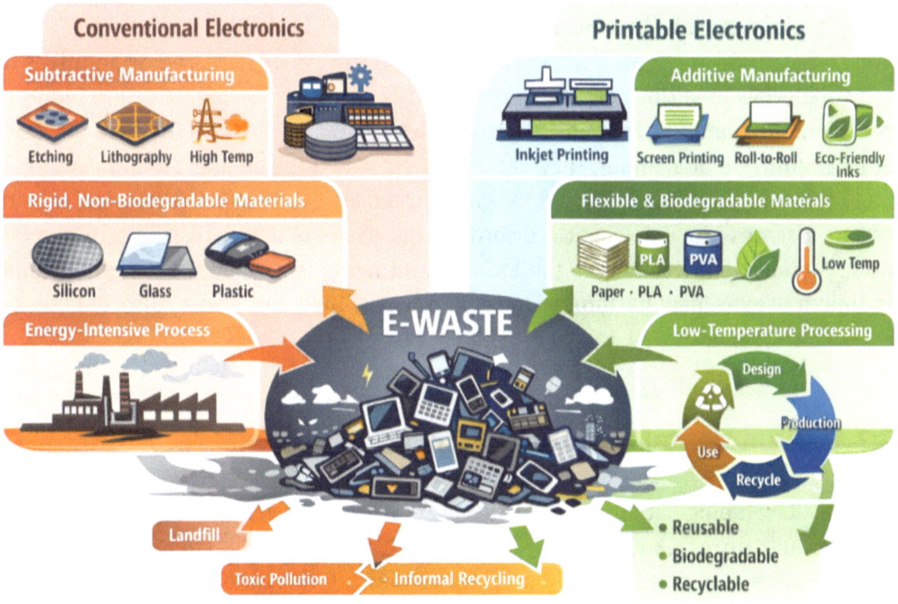

Figure 1: Conceptual illustration of e-waste generation in conventional electronics and the sustainable role of printable electronics.

2 Overview of e-waste: challenges and environmental impact

The fast pace of development in electronic technologies has made a huge difference in living standards, and there has been considerable innovation in communication, health care, transport, and industrial automation. Nevertheless, alongside this rapid pace of technological developments, there has been an unprecedented growth in electronic waste, or "e-waste." E-waste is defined as electrical and electronic equipment that has "reached the end of its life cycle." E-waste has been identified as one of the quickest growing types of waste worldwide, and it has become a source of grave concern for the environment and human health. Managing e-waste is an important challenge, and sustainable strategies must be adopted that minimize the consumption of materials and minimize impacts on the environment. Among these strategies, a huge potential exists for printable electronics in sustainable management of e-waste.

2.1 Hazardous nature of e-waste

E-waste is composed of intricate mixtures of materials like plastics, metals, glass, and many toxic materials. Poisonous materials including lead (Pb), mercury (Hg), cadmium (Cd), and brominated flame retardants (BFRs), which are used as additives in some materials in various devices, circuit boards, displays, and batteries, are found in considerable quantities in many electronic devices [11, 12]. In case they are disposed of in an unhygienic manner, they can find their way into soil and groundwater, pollute surface waters, and release toxic gases in the atmosphere. Lead can lead to dangerous diseases like neurological disorders and development defects, mostly in children. Mercury is poisonous to the kidneys and nervous system. On the other hand, cadmium is extremely harmful to lungs and bones. BFRs can lead to disorders in the endocrine system, as well as cancers [13].

2.2 Environmental and health impacts of informal recycling

In many developing nations, the practice of recycling e-waste is carried out in a disorganized and unregulated manner in terms of open burning, chemical extraction through acids, and disassembling. These activities are carried out in a bid to harvest precious metals such as copper, gold, and silver, which in turn result in the emission of toxic wastes into the environment [14]. Open burning of wires and plastic parts represents a source of the emission of dioxins and furans.

Workers engaged in the process of informal recycling are often handling toxic substances with minimal protection, thereby posing severe health threats to them in the form of respiratory ailments, dermatological disorders, and chronic illnesses. Furthermore, the local populations are also getting affected as a result of the toxic emission of air, water, and the food chain [14, 15]. All these crises shed serious light upon the need to have a safe and healthy approach towards managing the issue of e-waste.

2.3 Short product life cycles and technological obsolescence

One of the major causes of the global e-waste issue is the short lifespan of electronic products. This can be attributed to advances in technology, where software updates and consumer preferences lead to planned and perceived obsolescence. This causes products, for instance, smartphones, laptops, and wearable electronics, to become obsolete within a few years, despite their ability to function properly in many cases [16]. Such an upgraded turnover rate contributes to an increase in resource utilization, increased manufacturing demands, as well as an endless entry of disposed devices in the waste material stream. Consumer electronics, which are made to withstand longer but are not tailored to being recycled or dismantled, contribute to this problem by being difficult to recycle.

2.4 Limitations of conventional e-waste management

The traditional approaches for managing e-wastes basically involve collection, recycling, and disposal. Nevertheless, the recycling of the traditional type of electronics involves sophisticated steps in the process due to their use of a combination of various materials and components at a micron level. The separation of metals, polymers, and toxic components proves costly and technically not viable [17]. Furthermore, the current recycling process extracts a very low percentage of the useful materials found in e-wastes. Moreover, the amount of plastics as well as composites that are disposed of in landfills as well as the amount of plastics that are burned in trash-to-energy stations is extremely high. This shows that the end-of-life approach for e-wastes faces many hurdles and hence a focus on sustainability in the design and manufacturing of electronics is needed.

2.5 Sustainable design principles for e-waste reduction

For sustainable e-waste management, there should be a comprehensive approach to material reduction, eco-design, recyclability, and proper end of life disposal. Material reduction refers to the effort to use fewer materials in electronic components to lower the amount of raw materials in the products using techniques such as lightweight design and additive manufacturing. Eco-design targeting e-wastes focuses on using eco-friendly materials in the manufacturing of electronic components. The recyclability can be increased by reducing material types, excluding poisonous additives, and in device design, adopting a modular approach. Environment-friendly disposal methods ensure that nonrecyclable parts are handled properly, reducing the effects on nature and human life [17, 18].

2.6 Role of printable electronics in sustainable e-waste management

Printable electronics have great potential in overcoming most of the sustainability issues that arise from the generation of e-wastes. Unlike conventional electronics, printable electronics have low-temperature manufacturing processes that consume less energy and generate less waste material. On the other hand, in conventional electronics, excess material is required to be removed from the substrate. Moreover, issues of e-wastes are compounded due to the use of rigid substrates in conventional electronics. This aspect of printable electronics makes it possible to design electronics that have a controlled lifespan in the range of a few days to a few weeks. Moreover, the generated electronics can be safely degraded. Printable electronics make use of substrates that are recyclable and biodegradable. For example, they can be made from paper, cellulose materials, and biodegradable plastics. Printable electronics are suitable for the development of sensors that are useful for a short period, smart packaging systems, RFID systems, and diagnostic instruments that are disposable after use. With the implementation of eco-design in the fabrication process, printable electronics embrace the principles of the circular economy, with a view to recycling and reusing. This, in turn, indicates that printable electronics signify a move away from the traditional model of resource-wasting electronics and toward eco-friendly options.

3 Fundamentals of printable electronics

Printable electronics are an emerging technology that presents an alternative solution for conventional silicon microelectronics. This technology involves the production of electronic components and circuits on substrates using printing methods like inkjet, screen, gravure, flexographic, and aerosol jet printing. These electronics are unique in their ability, since unlike conventional semiconductor manufacturing, in which complex subtractive processes are employed, printable electronics utilize an additive manufacturing process. This manufacturing technique involves adding materials only in defined areas, resulting in minimized material waste, lower power consumption, and ease of manufacturing. Such properties have made printable electronics favorable for the creation of large-area, lightweight, and flexible electronics, especially when conventional electronics are not applicable.

3.1 Additive manufacturing approach in printable electronics

One major difference between printable electronics and traditional microelectronics lies in manufacturing paradigm. Traditional semiconductor manufacturing is mostly subtractive. It includes photolithography, etching, and chemical processing, which produce considerable material waste. On the other hand, printable electronics involve an additive deposition technique in which conductive, semiconductive, and dielectric inks are deposited directly on to the substrates without any material removal [20]. The additive process, therefore, not only leads to reduced waste but also ensures that the processing and related energy requirements are simplified. The low-temperature processing, therefore, makes it possible to work with temperature-sensitive materials like polymers, paper, and textiles. Thereby, printable electronics prove to be very much in line with green processing methods and scalable for large-area applications.

3.2 Printing techniques for electronic fabrication

There are various printing processes used in printable electronics, each with its own advantages based on the requirements of the application concerned. Inkjet printing is a noncontact and digital process that allows accurate deposition of picoliter-sized droplets of ink. This process is especially suited for rapid prototyping as well as custom designs since it is maskless and has low material waste as compared to other printing technologies [21]. Screen printing is an economical method of depositing thick functional layers. This process is very much suited to the mass production of conductive tracks, electrodes, and sensors.

Gravure and flexography printing are high-speed methods that can be used in roll-to-roll production. These techniques can be used for mass production of printed electronics because of their high print speeds and quality of prints that can be achieved on a mass scale [22]. The aerosol jet printing technique is a high-resolution printing technique that focuses droplets of ink in the aerosol form on a substrate. It can print on a nonplanar and three-dimensional surface, and this increases the design capabilities of printable electronics [23].

3.3 Functional materials for printable electronics

The performance of printable electronics is strongly determined by the characteristics of functional inks applied during the printing process. Most commonly, conductive inks incorporate silver, copper, carbon-based materials, conductive polymers, and emerging two-dimensional materials. Dielectric inks serve for insulation and capacitive elements, whereas semiconducting inks allow creating active components like, for example, transistors and sensors. Recent research has focused on the development of eco-friendly and low-toxicity inks to further enhance the sustainability of the printable electronic. Carbon-based materials and conductive polymers have been getting much attention as alternatives to precious metal inks due to their lower environmental impact and cost. Compatibility of such inks with low-temperature processability will be crucial in the coming times for enabling flexible substrates and biodegradable ones.

3.4 Flexible and large-area electronics

One of the major strengths of printable electronics is that they are amenable to flexible and light substrates. Because of their incompatibility with silicon wafers, printable electronics are different from traditional electronics in that they can be produced on polymers, paper, fabrics, and elastomers. This allows them to be used in making flexible, extensible, and moldable electronics that are amenable to integration into everyday items and surfaces [25]. Large-area processing is another major characteristic of printable electronics. Roll-to-roll processing makes it easy to process large areas of electronics in continuous rolls of meters long, making them useful in smart packaging, public displays, and flexible electronics. Such capabilities are simply not available in traditional electronics.

3.5 Applications of printable electronics

Printable electronics have been applied in a broad range of industries owing to their versatility and low costs. Regarding sensing technology, printable sensors have found use in environmental sensing, health care diagnostics, and wearable technology. Due to their low weight and flexibility, they are easily integrated into clothing, packaging, and biomedicine. Regarding display technology, printable electronics have found use in foldable displays, electronic paper, and signage. Although they do not have the performance levels of silicon technology displays in cognition, their performance is adequate for low-energy and low-cost disposable technology.

Printable electronics are also an essential area in terms of devices that interact with energy, in which printable batteries, supracapacitors, and energy harvester devices like solar cells fall. Such devices are amenable to low-cost processing and can also work on flexible substrates [22]. Then there is smart packaging, along with RFID. Printable RFID and sensors allow for real-time tracking, anti-counterfeiting, along with reducing material as well as an amount of waste in electronic waste when considered in relation to traditional RFID.

3.6 Advantages over conventional electronics

Printable electronics can provide a number of advantages over conventional semiconductor-based electronics, including low material usage, low energy usage, flexibility, and use of flexible and unconventional substrates. Furthermore, the simplicity in the manufacturing process brings down the cost and makes it possible through a decentralized approach, especially in emerging nations [20, 25]. From a green perspective, printable electronics can help with ecologically friendly designs in a number of ways, including use of recyclable and biodegradable materials, having a short product life based on product requirements, and manageable disposal and recycling at the end-of-life of products.

4 Printing techniques for sustainable electronics

Printable electronics insert the use of diverse printing technologies that make it possible to perform an additive process of low-temperature fabrication of electronic components. Among these printing technologies are those that are most used and include: inkjet printing, screen printing, and gravure/flexographic prints. These approaches have different advantages that include: the degree of resolution or speed of processing or material use or applicability depending on the material needed. This topic is discussed based on its application in the manufacture of green electronics.

4.1 Inkjet printing

Inkjet printing is one of the most applied printing technologies in the field of printable electronics applications, especially in research and development and low volume printing. Inkjet printing is a noncontact digital printing method that allows the precise dispensing of functional ink in the form of picoliter-sized droplets. Inkjet printing is based on the principle of ink droplets being ejected through the nozzle with a typical dimension of a few micrometers. Inkjet printing allows the ink droplets to be directly printed without the use of masks and printing plates [26]. One of the biggest advantages of inkjet printing is its additive printing process in which ink is deposited only where needed.

Inkjet printing is especially advantageous in the context of reducing the usage of expensive functional materials with the potential for waste generation with other technologies. Inkjet printing offers excellent sustainability in manufacturing processes since it is capable of reducing both material usage and waste generation [27].

Figure 2 shows the "precise deposition of functional inks (inkjet printed electronics)." Inkjet printing is especially suited to rapid prototyping and customized electronics, since design changes are easily effected digitally without the need to change hardware. It thus speeds up the development process with less consumption of resources. Because processing temperatures are low, a broad variety of substrates can be used, including flexible polymers, paper, textiles, and biodegradable materials. This makes inkjet printing highly applicable in areas such as printed sensors, thin-film transistors, RFID antennas, and flexible circuits. Inkjet printing has further advantages from an environmental point of view: no vacuum systems and no high-temperature processing are involved. However, there are a few problems, such as nozzle clogging, a limited ink viscosity range, and relatively low printing speed, which inhibit its large-scale manufacturing applications. Despite those factors, inkjet printing remains one of the most important technologies in the development of sustainable electronics, especially at the research stage and in small-batch fabrication [28].

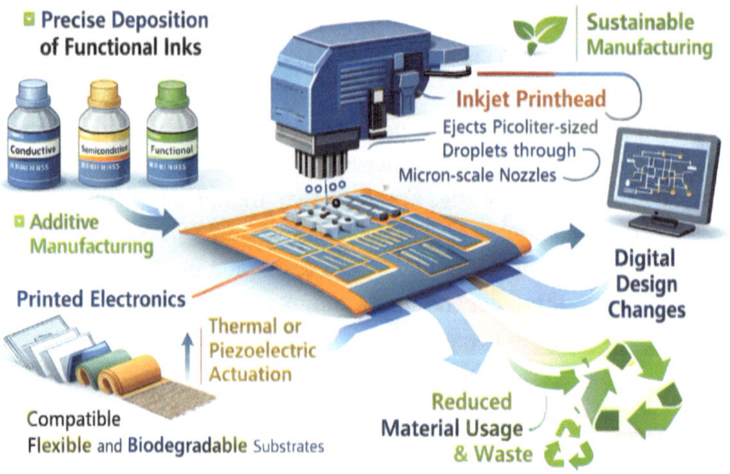

Figure 2: Precise deposition of functional inks (inkjet-printed electronics schematic).

4.2 Screen printing

Screen printing is an established and relatively cheap printing method, which has been widely accepted for the manufacturing of printed electronics. This technique requires ink being pressed through a screen mesh with a pattern corresponding to the printed design and a squeegee, facilitating the creation of thick ink layers. It is worth mentioning here that screen printing is very effective for creating thick electrodes and conducting tracks, as is mostly required in sensors, batteries, and interconnects for other electronics as well [29]. The other significant benefit of screen printing is its simplicity and scalability. In addition, screen printing is quite flexible and amenable to a broad range of ink viscosities. These make screen printing a better process for large-area and large-volume printing. In addition, screen printing has lower environmental costs since less energy is required compared to more complex processes used in semiconductor processing. Screen printing is important in the production and manufacture of green electronics. In this case, screen printing offers the ability to process and print using environmentally friendly inks. Moreover, screen printing is suited for processing with carbon-based inks, conductive polymers, and metal-oxide inks. These materials are environmentally friendly since they are not precious or toxic [30]. Another important benefit with screen printing is that it is well-suited for roll-to-roll processing. This increases productivity and decreases energy per unit [30].

Figure 3 shows the schematic representation of screen-printing technology for printed electronics, illustrating conductive ink deposition on flexible substrates and roll-to-roll processing. Despite the many advantages that screen printing presents, the screen printing process has a lower resolution in comparison with either inkjet or aerosol jet printing, thereby restricting its use in specific applications that could benefit from smaller features. However, for many applications, such as printed batteries, supercapacitors, heaters, and sensors, screen printing offers the best compromise between performance and sustainability. The use of screen printing in the production of printable electronics is a clear indicator of its importance in environmentally responsible manufacturing.

Figure 3: Schematic illustration of screen printing-based printed electronics showing patterned mesh screen, conductive ink deposition, substrates, and roll-to-roll manufacturing.

4.3 Gravure and flexographic printing

Gravure printing and flexographic printing are high-speed printing processes with a high production rate, well-suited for mass production of printable electronics. These processes have widely been used in the industry for printing purposes and have since then been adopted for the production of electronics because they are suitable for roll-to-roll processing. These production processes are suitable for continuous processing of flexible materials on a large scale [31]. In gravure printing, printing is done using engraved cellular patterns on a cylinder that rotates to transfer a pattern to a substrate. These printing processes produce high-quality printing at a fast printing speed. In flexographic printing, a flexible plate with an elevated surface is used to print patterns on a substrate. These processes are fast production processes and save a significant amount of energy per unit produced [32].

Figure 4 shows the roll-to-roll gravure and flexographic printing processes for high-throughput printed electronics, showing ink transfer mechanisms, recyclable substrates, and applications such as RFID tags and flexible displays. The continuous printing process in gravure and flexographic printing provides great benefits for environmental sustainability. The continuous processing approach leads to less material waste, faster processing, and, overall, a lower carbon footprint for producing electronic devices. The process can be effectively employed for applications like smart packaging, printed RFID, displays, and disposable sensors, where high volumes and lower costs are very important. Gravure and flexographic printing can help combine eco-design concepts because they are compatible with recyclable material and non-toxic ink. Their ability to develop light and flexible electronic components helps reduce material consumption and disposal difficulties. The high start-up investment for gravure printing can be a drawback, but overall, it can be a very successful approach for eco-friendly electronic device production, especially with respect to overall energy savings and scaling.

4.4 Comparative perspective and sustainability implications

With each of the printing processes considered – inkjet, screen, and gravure/flexographic – there are distinct advantages that make them better-suited to distinct phases of the RD of printable electronics. Inkjet is suited to precision printing and low waste in order to serve as a tool in the R&D of printed electronics. Screen printing is better suited to robust mass printing and is suited to the development of printed electronics in the area of "green electronics." Gravure and flexographic printing are suited to mass printing and are used in the mass production of printed electronics. Regarding sustainability, each of these processes is suited to material conservation, use of renewable resources, and the principles of eco-design that are critical to making the use of electronics more environmentally sustainable. Printable electronics can

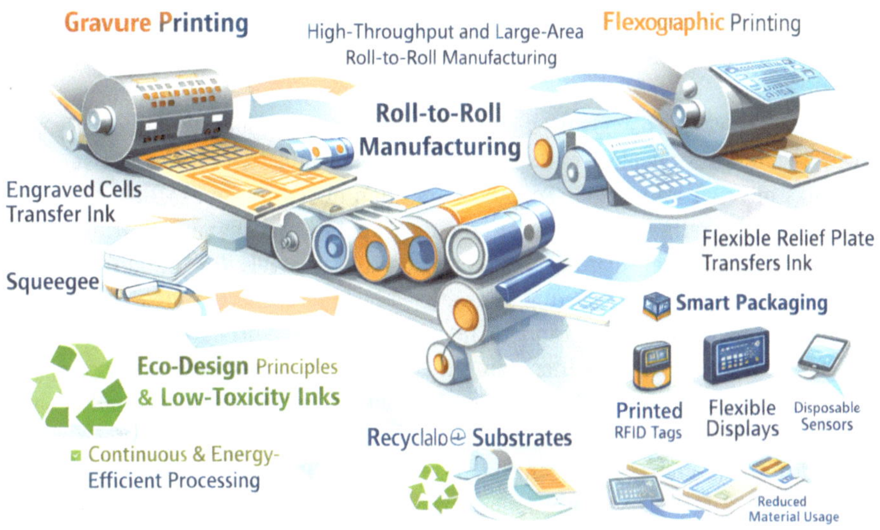

Figure 4: Schematic illustration of roll-to-roll manufacturing using gravure and flexographic printing for large-area printed electronics.

be considered environmentally sustainable as they can eliminate traditional subtractive electronic manufacturing processes by the use of print technologies.

5 Eco-friendly functional materials

5.1 Conductive inks

Conductive inks are basic materials in printable electronics that help produce electric interconnects, electrodes, or conductive patterns in electronic devices. Conventional materials for conductive inks include silver inks that are superior because of their excellent electric conductivity property. Inexpensive and environmentally nontoxic alternatives to silver-based inks are desirable because of high silver processing and mining environmental concerns. Silver nanoparticles also pose toxic risks to human and environmental safety when they leak into environmental interfaces. Henceforth, there has been intense scientific inquiry to explore cost-effective alternatives to conductive inks that are environmentally friendly and nontoxic to human life.

Figure 5 shows the comparative schematic of silver-based inks and eco-friendly conductive inks (carbon-based and polymer inks) for printed electronics, showing trade-offs in conductivity, sustainability, and applications. Among these alternatives, carbon-based inks have proven very promising. Carbon black, graphite, graphene, and carbon nanotubes (CNTs) are relatively conductive; they are highly stable and much less toxic compared to precious-metal-based inks [33]. Carbon-based inks are very appealing materials in the context of green electronics owing to the natural abundance, relatively low cost, and environmentally nontoxic nature of carbon. Carbon-based inks are extensively used in printed sensors, electrodes used in energy storage devices, and flexible electronics where very high conductivity is not an issue. Graphene inks have also attracted significant attention owing to the excellent electrical, mechanical, and thermal properties offered by graphene. Graphene inks printable via inkjet printing or screen printing have been used extensively for the fabrication of flexible and transparent conducting films at relatively low temperatures. Another significant advantage offered by graphene inks is the ability to prepare the ink in water or eco-friendly solvents, thus increasing the sustainability index of the ink significantly [34]. CNT inks have also been found to offer excellent electrical percolation paths at relatively low filler concentrations. These materials have been used extensively in wearable sensors. Another significant set of sustainable CPs is given by conductive polymers, with poly(3,4-ethylenedioxythiophene):polystyrene sulfonate (PEDOT:PSS) being a significant example. Conductive polymers have moderate electrical conductivity, good elasticity, and can be easily processed from solutions at low temperatures. Specifically, PEDOT:PSS can be dispersed in water and easily printed, representing one of the reasons for choosing this option in green printable electronics applications [35]. In contrast to metal inks, conductive polymers do not require high-temperature sintering for their preparation, hence their production.

The shift from silver-based inks to carbon-based materials and conductive polymers not only cuts the costs of the materials used in the production of printed electronics but also makes the production of printed electronics environment-friendly. Although the conductivity of the new materials lacks the conductivity of silver ink, they are quite adequate for the production of sensors, RFIDs, antennas, and energy-related devices. Further research and developments are bridging the gaps in the performance of the sustainable conductive inks to make them an alternative to the printable electronics.

5.2 Dielectric and semiconducting materials

Besides conductive inks, dielectric materials, as well as semiconducting materials, play a significant part in the operation of printable electronics. Dielectric materials find importance in the area of insulation, capacitance, and gate dielectrics in the transistor. Semiconducting materials, however, offer the ability for active device functioning, such as switching and sensing. In relation to sustainability, the choice of dielectric materials and semiconducting materials plays a significant role, especially since conventional inorganic semiconducting materials are typically high in energy requirements, considering the high temperatures needed during fabrication. Organic semiconducting materials have lately been identified as one of the most significant

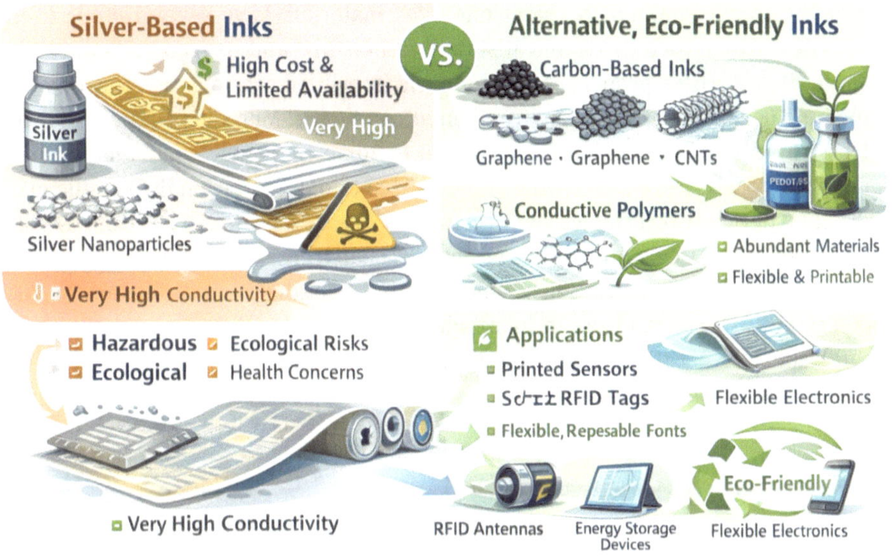

Figure 5: Comparison of silver-based conductive inks and alternative eco-friendly inks.

materials aligned to the sustainability required in printable electronics. These materials, which involve small molecules or conjugated polymers, have the advantage of being fabricated at reduced temperatures through various printing processes, such as inkjet or gravure printing. These semiconducting materials present notable flexibility, lighter materials, and degradability through biodegradable substrates, making them pertinent in the area of green electronics [36].

Organic semiconductors have found broad applications in printing thin-film transistors, sensors, and flexible electronics. Ever-improving molecular engineering, in spite of relatively low carrier mobility, has reached levels sufficient for various low-power and large-scale electronics. More importantly, organic semiconductors do not require toxic heavy elements, thus promoting a positive impact on the environment during both the manufacturing process and the end-of-life stage. Another class of metal oxide-based semiconducting inks has gained increased significance in the area of printable electronics. Zinc oxide (ZnO), indium oxide (In_2O_3), or tin-doped tin oxide (SnO_2) metal oxides have relatively desirable electrical properties, along with robust environmental stability. Innovative developments in the field of solution chemistry have helped process these materials at relatively lower temperatures, typically below 300 °C, which, in turn, reduces energy expenditure remarkably, considering typical high-temperature semiconductor device manufacturing processes [37]. This class of metal-oxide printable electronics inks is flexible, thus compatible with polymeric materials.

Figure 6 shows the comparative schematic of organic semiconductor inks and low-temperature metal-oxide inks (ZnO, In_2O_3, and SnO_2) for printed electronics, emphasizing processing temperature, stability, and dielectric compatibility. In regard to the environment, metal-oxide semiconductor materials have several other useful properties, which include chemical stability, moisture and oxygen resistance, and high durability. These properties are desirable in printed sensing and electronics designed for use or exposure to environmental conditions. Moreover, there is growing interest in minimizing the need for limited or toxic elements used as materials in semiconductor oxides, which involve the use of abundant, eco-friendly metallic materials [38]. There are also changes in dielectric inks used in printable electronics. There are developments involving the use of polymeric dielectrics, bio-based dielectrics, or water-processable dielectric inks, which are eco-friendly alternatives to the high-temperature process required in inorganic dielectrics. These dielectrics have appropriate voltage levels, flexibility, and abilities related to energy efficiency and environmental sustainability.

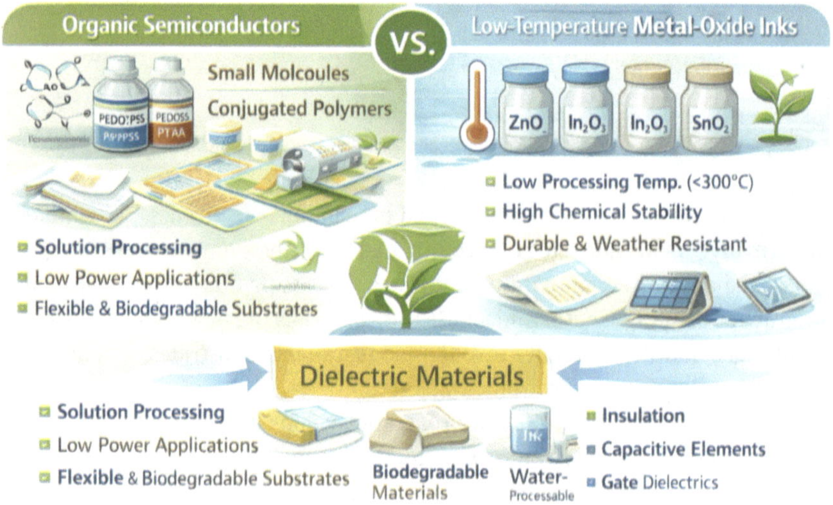

Figure 6: Comparison of organic semiconductors and low-temperature metal-oxide inks for printed electronics.

5.3 Sustainability implications of functional material selection

Material choices for conductive, dielectric, and semiconducting materials directly affect the sustainability of printable electronics. The replacement of silver ink with carbon materials and conductive polymers and the use of organic and low temperature metal oxide semiconductors help reduce energy consumption in the field of printable electronics. These material technologies act as a key enabler of eco-design in the field of electronics. For instance, the technology helps in improving the recyclability and disposal rates of electronics. In addition, the materials are suitable for biodegradable substrates. With the sustainability of electronics manufacturing becoming a major challenge in the industry, the emergence of "green" functional materials is set to play a critical role in the mass adoption of printable electronics.

6 Sustainable substrates for printable electronics

Substrates are very essential for defining the mechanical specifications, environmental durability, and recyclable or biodegradable nature of the printed electronics. Traditionally in the electronics industry, the most used substrates are rigid silicon wafers, glass, and plastic materials of the petroleum variety. These materials are among the most significant sources of e-waste as they are not biodegradable. In recent years, the need to use environmentally compatible electronics has led to an increased use of environmentally sound substrates. These are biodegradable polymers such as PLA and PVA, paper-based materials, biodegradable plastics such as PLA and PVA, and even recycled plastics rather than the traditional materials used in the printed electronics industry.

6.1 Paper-based substrates

Paper is another of the most widely explored sustainable materials for printable electronics due to its abundance, low costs, and ability to be recycled and biodegrade. Moreover, the cellulose fibers in the paper create a porous substrate that can be suitable for a variety of printing methods such as inkjet printing, screen printing, or even gravure printing. Figure 7 shows the schematic representation of paper substrates used in sustainable printed electronics, showing compatibility with multiple printing techniques and benefits such as biodegradability, recyclability, and reduced electronic waste. Moreover, paper electronics are very popular in applications that demand smaller operational lifetimes. Examples of such electronics include smart packaging electronics, RFID electronics, and even disposable electronic sensors [39]. From an eco-friendly standpoint, one of the most important factors that paper electronics possess is the ability to be recycled or disposed of in the proper manner due to the existing paper recycling process. Moreover, paper manufacturing processes are more eco-friendly compared to plastic electronics. This is due to the sustainable forestry practices that papers are harvested from. On the other hand, the potential technology limitations include the porous nature of the substrate.

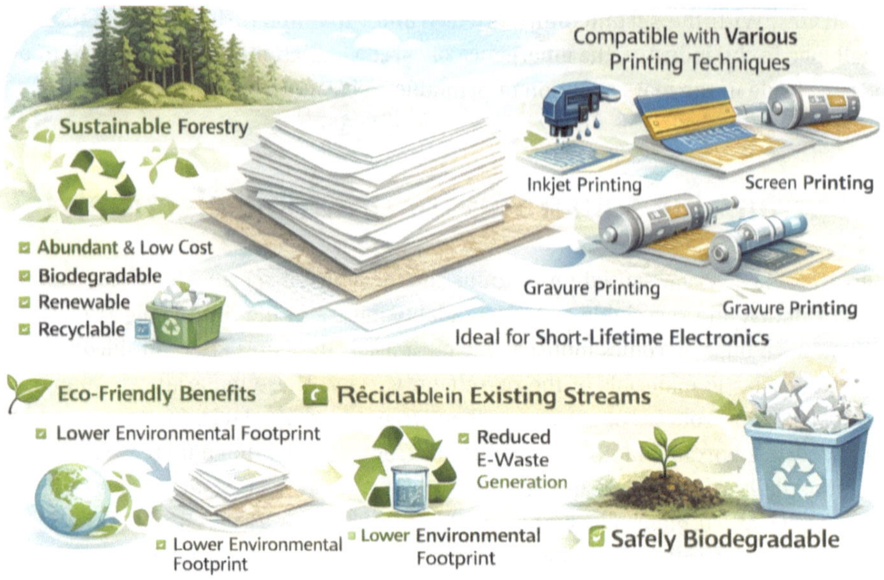

Figure 7: Schematic illustration of paper-based substrates for sustainable printed electronics.

6.2 Cellulose and cellulose-derived substrates

In addition to traditional paper, other cellulose-based materials like nanocellulose films and regenerated cellulose are emerging as promising substitutes in high-performance green electronics. Nanocellulose films have superior mechanical properties, optical clarity, and flexibility, which make them very useful in advanced printable electronics technology [40]. The comparative schematic of nanocellulose films and regenerated cellulose substrates for printed electronics emphasizes renewable resources, mechanical stability, barrier properties, and low-temperature printing compatibility. Indeed, these cellulose-based materials leverage the biocompatibility of cellulose, with added advantages of smoothness and stability in terms of size, as compared to traditional paper. Biocompatibility and renewability of cellulose make it easily disposable and sustainable in the green electronics context. Further, cellulose materials can be developed with strategically balanced barrier properties, such as water and oxygen, which play very crucial roles in deciding the sustainability of devices.

6.3 Biodegradable polymer substrates: PLA and PVA

Biodegradable polymers, like PLA and PVA, have emerged as key substrates for sustainable printable electronics. PLA is derived from renewable resources such as corn starch or sugarcane and is compostable under industrial conditions. It has good mechanical strength and thermal stability, which qualify it for use in flexible electronics applications. Water-soluble and biodegradable, PVA is well known for excellent film-forming properties and compatibility with aqueous inks. PVA substrates hold a particular attraction for transient and disposable electronics, where controlled dissolution or degradation is desired after use. Comparative schematic of biodegradable polymer substrates (PLA and PVA) for printed electronics shows material properties, processing compatibility, and applications such as transient electronics and environmental monitoring. PLA and PVA can be processed at fairly low temperatures, thus enabling energy-efficient fabrication and compatibility with the additive printing technique. Thus, the utilization of biodegradable polymers offers an opportunity to design electronic devices with predetermined lifetime, reducing their long-term environmental accumulation. This is very helpful for devices made for medical diagnostics, environmental monitoring, and single-use sensors, among other applications, where retrieval and recycling of devices is not practical.

6.4 Recycled plastic substrates

Although biodegradable materials have several advantages from an environment perspective, recycled plastics act as a transitional alternative for plastic sustainability. Recycled polyethylene terephthalate, polyethylene, and polypropylene films are moving toward commercialization due to their usage as printable electronics substrates [43]. The recycled plastic substrates and biodegradable polymers for printed electronics, highlighting closed-loop life cycle, low-temperature processing, and reduction of plastic waste. Recycled plastics act as a method to recycle plastic waste, thus reducing its usage and disposal in landfills and oceans. Recycled plastic materials have properties similar to those of normal plastics, which include flexibility, strength, and water resistance. Recycled plastic materials can be produced on a commercial scale using roll-to-roll printing and large area processing for applications ranging from plastic flexible electronics to energy devices. Although recycled plastics are not biodegradable, reusing plastic materials helps form a closed-loop material system, which is a guiding principle for sustainable manufacturing.

6.5 Role of sustainable substrates in reducing e-waste

The use of sustainable materials in printable electronics is a direct response to the challenges posed by e-waste. Firstly, the project helps in the conservation of materials since the use of thin and lightweight electronics is promoted. Secondly, the materials are easily recycled or degradable. As a result, there is a reduction in the formation of e-waste [44]. Disposable and recyclable electronics produced using paper or biodegradable polymers are best applicable in environments where the useful life span is shorter. Rather than forming inconsequential e-waste materials, the electronics can easily be recycled. This is a major improvement from the existing electronics that are often recycled inappropriately before disposal.

6.6 Challenges and future perspectives

Despite their numerous benefits, sustainable substrates also pose challenges in terms of their strength and compatibility with high-performance electronic components. They may be sensitive to moisture and lack high temperature resistance and compatibility with inks to a certain extent. Researchers continue to work on possible solutions in terms of modifying their surfaces, combining different materials, and using protective materials in an effort to alleviate these challenges and ensure ecological advantages in the end [45]. In the future, it is predicted that attempts will be made to develop a completely biodegradable system for electronics in a bid to ensure that the substrates, inks, and encapsulation materials do not pose any danger to the environment.

7 Printable electronics and e-waste reduction

The increased use of consumer electronics at a rapid rate has led to the generation of considerable amounts of electronic waste, thereby making it one of the fastest-growing wastes in the world. The manufacturing of consumer electronics in their classical form requires heavy materials and extensive amounts of energy, as well as being unrecyclable, thereby making their disposal at the end of their useful life more complex. Printable electronics were introduced as a promising technology to overcome the difficulties in the disposal of consumer electronics as they offer prospects in additive manufacturing, biodegradable technology, low-energy processing, short-life electronics, and circular economy.

7.1 Reduction of material usage through additive manufacturing

One of the most valuable roles of printable electronics in managing e-waste is in the method of additive manufacturing. Unlike traditional manufacturing of semiconductors, which involves extensive material subtraction through etching, printable electronics add material only in desired places. Such addition of material results in a remarkable reduction in raw material usage and, in effect, minimizes material wastage [46]. Additive manufacturing methods also help in achieving lighter and more compact designs in their manufacturing. Schematic comparison of traditional subtractive manufacturing and additive printed electronics explored material efficiency, reduced waste, and fabrication of sensors, antennas, and interconnects. Printable electronics, in their role with sensors, antennas, and interconnects, consume only a small fraction of what traditional electronics do. Such methods and designs not only minimize the exploitation of resources but also ensure that less waste is generated at the end-of-life disposal of the products. Additive manufacturing, therefore, acts as a major factor in controlling manufacturing wastes as well as post-use e-wastes.

7.2 Biodegradable and recyclable device architectures

Printable electronics have opened possibilities for the creation of biodegradable and recyclable electronics architecture. Indeed, with the introduction of eco-friendly substrates in printable electronics technology in the form of paper, cellulose, biodegradable polymers, and recycled plastics and conductive inks, the challenge of electronic waste is hoped to be solved since the issue of disposal after use will be eliminated [47]. Biodegradable electronics are useful in applications of a shorter life span where the aim is not recovery. For example, sensors in environmental or health applications can be made biodegradable since after the completion of the task of monitoring or testing health conditions necessary in health care or environmental conditions in environmental monitoring applications, the sensors will easily biodegrade without causing harm or creating pollution in the form of landfills. On the other hand, electronic devices with the capacity for recycling will have materials easily separated for reuse in manufacturing.

7.3 Low-energy fabrication processes

Energy usage in the manufacturing sector is a significant component in the environmental performance of electronic devices. Typical microelectronics technology requires the use of high temperatures in the manufacturing processes. Also, the manufacturing of microelectronic devices requires the use of vacuum systems and clean rooms. As a result of these processes and operations, a high usage of energy is observed. On the contrary, printable electronics require low temperatures in their processes. As a result, the usage of energy is low [48]. Inkjet printing, screen printing, and gravure printing are done in low temperatures of 150–300 °C depending on the materials. Processing in low temperatures allows the printing technology to support the use of flexible and biodegradable materials. Both processes reduce the carbon emission in the manufacturing sector.

7.4 Short-life and application-specific electronics

Printable electronics make it easy to produce short-life application-specific devices. Indeed, one of the most critical factors in lessening the amount of waste in our environment is through application-specific devices that serve for some short period [49]. In many new applications, like smart packaging, RFID tags, disposable sensors, or health care diagnostics worn on clothing, it is necessary to produce devices that will serve for some short period. Traditional devices today serve much longer than necessary. This creates unnecessary material waste in addition to waste accumulated over many years. Indeed, printable electronics can save material and can also be made biodegradable. Containment or recycling is possible after they serve the purpose for which they are made. Such purposeful devices are an entirely different paradigm in design and usage of devices.

7.5 Design-for-disassembly and circular economy

A major tenet of sustainable management of e-wastes is the application of the circular economy. This is a strategy that focuses on resource effectiveness, reuse, recycling, and the minimized generation of waste. Printable electronics aggrandize this strategy by promoting easy disassembling and simple device designs [50]. The printed devices are made up of fewer layers, in contrast to regular electronics. Moreover, printed devices do not have complex parts, thus it becomes easy to separate different materials. In addition, it is easy to recycle this material. Also, there are roll-to-roll printing and modular designs, and these facilitate the application of the closed-loop system, whereby materials for new production can be reused. Through the application of the strategy for a circular economy, there is an end to the take-make-dispose approach by the electronics industry.

7.6 Environmental and societal implications

The application of printable electronics in managing e-waste has various implications. Nontoxic materials associated with printable electronics can prevent environmental pollution and exposure to toxins. Moreover, printable electronics can alleviate some environmental problems caused by unethical recycling activities in developing countries [51]. On social grounds, printable electronics can provide inexpensive access to technology in managing health, environmental monitoring, and smart infrastructure. Combining sustainability with affordability, printable electronics promote responsible technology development that focuses on environmental conservation.

7.7 Challenges and future directions

Although printable electronics have numerous benefits, they still have some challenges with respect to performance, reliability, and large-scale commercialization. Printed electronics reported relatively lower performance in terms of electric current conduction compared to silicon chips. Nevertheless, innovation in functional materials, printed resolution, and hybrid integration is continuously enhancing the reliability and performance of printed electronics [52]. Future studies will include the development of fully biodegradable electronics, innovative methods of recycling, and sustainability metrics. When these challenges are overcome, printable electronics will be one of the fundamental building blocks of sustainable electronics and electronics recycling.

8 Applications supporting green electronics

Printable electronics have made it possible for a new class of flexible, lightweight, and relatively inexpensive electronics, which have found significant uses in sustainable applications. As printable electronics leverage additive manufacturing processes, eco-friendly materials, and device designs, they make it possible to create new architectures designed in a way that conserves materials, energy, and environmental impact during disposal. This has significantly contributed to the fast adoption of printable electronics in various domains, which range from flexible and disposable sensors, smart packaging, wearable/physio-devices, printed batteries, IoT devices, among others. This topic covers significant aspects related to these domains.

8.1 Flexible and disposable sensors

Flexible and disposable sensors are currently one of the major application areas of printable electronics. Sensors can be widely utilized in environment-related applications, health diagnostics, food analysis, and in process control in industrial setups. Printable electronics help in preparing sensors on flexible surfaces like paper, plastics, and biodegradables using low-temperature methods of printing [53]. Figure 8 shows the schematic illustration of application areas for flexible and biodegradable printed electronic sensors, including environmental monitoring, healthcare diagnostics, and food safety, emphasizing lightweight design and controlled device lifetime. In today's world, traditional silicon-based sensors would be an over-kill in applications in which sensors would not need a long lifetime, like in single-use biosensors and environment-related monitoring applications wherein the sensors would be discarded and replaced with new ones in a short span of time. Along with a higher cost and resource consumption, traditional silicon-based sensors also emit a huge amount of waste in the environment in their life cycle, whereas printable electronics help in minimizing their footprint in the environment through controlled usage and disposal.

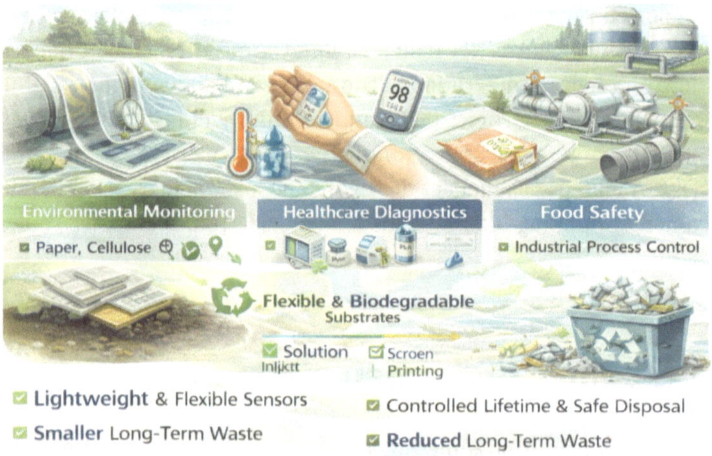

Figure 8: Various applications of flexible and biodegradable printed electronics.

8.2 Smart Packaging and RFID Tags

"Smart" packaging and RFID technology have greatly benefited from the advancements in printable electronics. Roll-to-roll printing of RFID tags, antennas, and sensors is now economical and can be applied on a massive scale for the management of the overall supply chain, inventory control, and anti-counterfeiting processes [54]. Printed electronics help in the development of smart packaging that can provide information on temperature, humidity levels, gas composition, or product freshness. Such an advantage increases the shelf life of food products and increases efficiency in the whole logistical system. Figure 14 shows the schematic illustration of smart packaging and printable RFID tag applications enabled by printed electronics, highlighting sensing, inventory tracking, anti-counterfeiting, and supply-chain optimization on flexible substrates. Notably, RFID tags and smart labels can now be printed on paper or recycled materials. As a result, the overall negative impact posed by an estimated "nine billion tags produced each year" is mitigated. The overall impact of smart packaging is "neutral" as regards the "growing e-waste problem" since the materials are biodegradable or recycled.

8.3 Wearable and biomedical devices

Another fast-growing application domain for printable electronics is wearable and biomedical devices. By integrating flexible printed circuits, sensors, and electrodes into textiles, patches, and wearable accessories, physiological parameters such as heart rate, body temperature, motion, and biochemical markers can be continuously monitored [55]. The advantages brought by printable electronics to biomedical applications are very important and relate to their mechanical flexibility, lightweight, and compatibility with soft and biocompatible substrates. Figure 15 shows the schematic illustration of flexible printed electronics for wearable healthcare monitoring, smart packaging, and printable RFID applications, emphasizing biocompatibility, lightweight design, and low-cost sustainable materials. By making the design for short-time or even one-time use of a printed biomedical device, the risk of cross-contamination can be avoided and the sterilization requirements can be significantly relaxed. Moreover, the use of low-cost and potentially biodegradable materials supports the demand for sustainable healthcare solutions, especially in resource-limited settings where device reuse and/or limited disposal options are challenging.

8.4 Printed energy storage and harvesting devices

Energy storage and harvesting are cardinal components of autonomous and sustainable electronic systems. Printing enables the fabrication of printed batteries, supercapacitors, and energy-harvesting devices like printed solar cells and piezoelectric generators [56]. Energy storage and harvesting devices can be fabricated on various flexible substrates by low-temperature additive manufacturing techniques and thus easily integrated into sensors, wearables, and IoT systems. Printed energy devices have several advantages associated with the lower consumption of materials, low-temperature processing, and scalability. Figure 16 shows the schematic comparison of printed energy storage devices (flexible batteries and supercapacitors) and printable energy harvesting systems (printed solar cells and piezoelectric generators) for self-powered flexible electronics. Although energy density is smaller compared to the conventional systems of energy storage, printed batteries and supercapacitors already have enough performance to operate low-power applications. From a sustainability point of view, appropriate design of printed energy devices using environmentally benign materials and compatible substrates allows for recyclable or biodegradable system architecture.

8.5 Internet of things (IoT) components

IoT is based on a huge infrastructure that involves billions of sensors, communication modules, and data-processing units. Printable electronics will be extremely important for providing low-cost and scalable IoT components in applications like smart cities, environmental monitoring, and industrial automation. Printed IoT components are sensors, antennas, interconnects, and simple logic elements that can be fabricated in huge quantities at very low cost. Because of their compatibility with flexible and lightweight substrates, their deployment can be easily considered in places where conventional electronics would be impractical. Moreover, printable electronics can enable application-specific IoT devices with limited lifetime, which reduces the pile-up of obsolete electronics and is in line with sustainable deployment strategies.

8.6 Sustainability and e-waste reduction implications

Given these numerous application domains, printable electronics offer significant contributions toward sustainability and e-waste reduction. In spite of that, printable electronics minimize manufacturing waste and post-consumer e-waste by allowing material-efficient fabrication, low-energy processing, and short-life device architectures [58]. Recyclable and biodegradable substrates simplify end-of-life management by minimizing environmental persistence. Importantly, printable electronics support decentralized manufacturing and localized production, a notion that is believed to reduce transportation-related emissions while simultaneously enhancing supply chain resiliency. These attributes make printable electronics an important technological enabler for sustainable development and circular economy models.

8.7 Challenges and future outlook

Despite their increasing popularity, printable electronics still face issues in terms of their limitations, device longevity, and standardization. Printable electronics tend to have relatively inferior electrical properties than their silicon-based counterparts, which might hinder their application for high-end devices. However, ongoing improvements in material properties, resolution, and combining different technologies will increase device capabilities continuously. Looking forward, R&D will be focused on developing a completely integrated printed system, innovative energy solutions, and a sustainability index. Once these issues are overcome, printable electronics will be fundamental for sustainable electronics and IoT infrastructure.

9 Challenges and limitations

The printable electronics technology is seen as a promising new alternative technology that can benefit from its additive manufacturing method, low-energy process, and ability to use environmentally friendly materials, compared with the risks and drawbacks of traditional silicon-based electronics. Nevertheless, in spite of its huge potential, there are a number of critical challenges that the printable electronics technology currently faces, limiting its widespread use and applications. Challenges in relation to the use or performance, longevity, standardization, availability of a manufacturing infrastructure for mass recycling, and the longevity versus biodegradability of the devices remain significant research and technological considerations for the widespread use and integration of this technology in this technological environment.

9.1 Performance limitations

One of the major difficulties in printable electronics is that these materials have relatively lower electricity and functional performance compared to traditional microelectronics. This is mainly because printed materials have higher resistivity compared to metals, and printable semiconductors have lower carrier mobility compared to their silicon counterparts [60]. These factors are mainly restrictive for printable electronics in terms of using them for applications that demand a higher speed and power performance, as well as computations. This performance deficit mostly lies in the choice of materials and processing methods. Solution-processable inks, although printable at relatively low temperatures, consist of materials like binder additives, wetting agents, and solvents, inhibiting their conductivity. Likewise, semiconductors in printable electronics, although very elastic and environmentally favorable, are largely not as effective as semiconductor silicon materials in terms of performance.

9.2 Mechanical and environmental durability

Durability is another major limitation with respect to the performance and longevity of printable devices. Most printed devices, particularly the early models, will be made of flexible or biodegradable materials, which will be more prone to mechanical stress, water, air, and temperature compared with other materials, such as silicon, which is relatively more durable and less sensitive to environmental and mechanical factors [61]. Although bending, stretching, and other environmental strains and exposure might pose serious challenges to the functionality of most printed devices, their short durability and environmental stability, particularly in outdoor and environmental sensor systems, and other IoT devices, pose major challenges in regards to their performance. Biodegradable and organic materials in printable devices might experience degradation in humidity and UV exposure, thus shortening the durability of printable devices.

9.3 Lack of standardization

The lack of standardized material, processing, and test methods is a major hindrance to the commercial scale-up of printable electronics. In contrast to the established semiconductor industry with standardized processing and the development of reliable standards to ensure the quality and compatibility of processed devices on a global scale [62], the relatively recently emerged printable electronics community deals with materials and technologies that are still very diverse [62]. This hinders compatibility and the approval of regulatory agencies for medical and safety-related applications of the new technologies on a global scale. It seems that the development of global standards for printable electronics in all aspects of environmental studies would facilitate mass production of the technologies and device commercialization on global markets.

9.4 Challenges in large-scale manufacturing

While the printing methods like gravure and flexographic printing are capable of high throughput and can be done on a roll-to-roll basis, large-scale printable electronics face difficulties in high-throughput manufacturing due to reasons like overall printing uniformity, control of defects, and handling and stabilization of the inks, among others [63]. Even minute variations in the viscosity of inks, the properties of the substrate surface, and other variables related to the printing process could cause deviations in performance in large-area electronics. In addition, combining printed components with other electronic components would involve hybrid manufacturing methods, resulting in higher complexity and cost, losing the sustainability benefits of printable electronics.

9.5 Recycling infrastructure and end-of-life management

Though there are views that printable electronics are environmentally friendly, there is still a major challenge in that there are no appropriate recycling systems for printed devices in place. The current recycling systems for e-waste are still suited for regular electronic gadgets rather than printed ones and are also not appropriate for biodegradable materials [64]. These printed electronic materials could be multilayers made up of polymers, metals, and functional inks that make it difficult to recycle and harvest materials from them. For example, in some situations, they could be made up of biodegradable materials that degrade, but there could be semiconducting or conductive materials that are still in the environment.

9.6 Balancing device lifetime and biodegradability

One of the main challenges in printable electronics, which has been under investigation, is the ability to strike a balance between device lifetime and biodegradability. There are various demands in several applications, which require these devices to operate in a particular manner, offering a well-defined lifetime before the device degrades or becomes disposable. Materials that degrade readily may lack stability, while more stable materials could hinder environmental sustainability [65]. This particular concern assumes utmost importance in several other usages, which involve biomedical implants, environmental sensing, or smart packaging. Scientists have been exploring various novel ideas related to transient electronics, programmable degradation, or bio-encapsulant layers, which will enable device lifetime control.

9.7 Economic and market barriers

Apart from the technical difficulties that printable electronics must overcome, there are economic and market-related constraints as well. Although printing technologies minimize material waste and energy usage, when it comes to high-quality functional inks and equipment, it can be relatively costly, especially in preliminary development stages [66]. Market acceptance is also constrained by traditional industry methods and existing semiconductor technologies. Addressing these constraints would necessitate innovative responses to technical requirements, as well as an enabling policy environment. It would also require that printable electronics demonstrate clear environmental and economic advantages in order to promote acceptance in consumer markets.

9.8 Future research directions

In addressing all these issues, a research combination in fields such as material science and engineering, environmental science, and policy formulation will be needed. The broad research fields in the coming future could incorporate research in high-performance sustainability materials, encapsulation techniques, and test and recycle techniques tailored for printing electronics. Advances in "life cycle assessment and sustainability analysis" will assist in taking the right decisions and maximizing the effectiveness of technology.

10 Future perspectives

With the spate of global challenges from e-waste (or electronic waste), it is now an imperative to ensure that technologies in the realm of electronics are both environment-friendly and sustainable. Printable electronics, in governance of e-waste through being manufactured in an additive manner and being compatible with bio-friendly materials, are poised to make an impact. Ongoing developments in bio-inks, fully biodegradable electronics, and recycling loops are also poised to make printable electronics play an even vital part in managing e-waste in an environment-friendly manner. Green chemistry and life cycle assessment will also play an integral part in maximizing the positive environmental influence.

10.1 Bio-based and environmentally friendly inks

One of the most promising areas of the future of printable electronics is the emergence of bio-based and nontoxic functional inks. Traditional metallic inks are typically based on rare or toxic substances that are not easily recyclable and pose toxicity concerns to the environment. On the other hand, bio-based inks based on renewable materials such as carbon-based substances, conductive polymers, and bio-based binders are relatively nontoxic and more environmentally sustainable [67]. Bio-based inks can be printed using low-temperature processing techniques and are also biodegradable substrate-friendly, making them amenable to the rapid manufacture of environmentally sustainable electronics. Further advances in the area of bio-based inks are also expected to improve their electrical performance characteristics without sacrificing environmental sustainability.

10.2 Fully biodegradable electronic systems

Aside from the materials themselves, the idea of fully biodegradable electronics is a paradigm shift in the design of sustainable electronics. Here, all the parts of the circuit – substrates, conductive materials, semiconductors, and encapsulants – will be designed to degrade in their environment after a certain lifespan of use [68]. This technology will be most suitable for short-term use electronics such as disposable sensors, medical diagnostic equipment, and smart packaging. Biodegradable electronics can halt the generation of e-wastes in the future and there are no complex processes required for the recycling of electronics.

10.3 Closed-loop recycling and circular economy

The development and deployment of closed-loop recycling processes is another key technology with immense potential to further improve the sustainability of printable electronics. Closed-loop recycling operations focus on material retrieval and reuse from the end-of-life stage of devices, thereby reducing material waste and conserving the use of virgin materials [69]. Printable electronics, with their simplified material structures and additive manufacturing, have been found to be well-suited to these closed-loop recycling methods. Strategies of design disassembly and using recyclable materials in manufacturing, therefore, hold immense potential in terms of supporting the reuse-recycle-reduce approach and enabling the circular economy model in the electronics industry.

10.4 Integration of green chemistry and life cycle assessment

However, for printable electronics and e-waste management to succeed in the future, their incorporation with the principles and idea of green chemistry will be very crucial. Green chemistry deals with the idea of using safer materials, biodegradable resources, and environmentally more-efficient processes, all of which are closely associated with the idea of printable electronics [70]. Life cycle assessment will help in creating an environment of measurement for identifying the effect of e-wastes on the environment, right from the processing of materials in electronic products to their disposal activities at the end of their lives. This will help in further improving printable electronics in terms of their environmental impact as well [71].

11 Conclusion

Printable electronics offer a revolutionary strategy toward achieving green and sustainable electronics manufacturing by drastically focusing on rethinking how consumer electronics should be conceptualized, processed, and recycled. Unlike how semiconductor technology is conventionally performed and executed, which requires complex, energy-intensive, and material-wasteful methods, printable electronics employ additive manufacturing methods whereby functional materials are printed only in desired and necessary positions. Consequently, this leads to a drastic reduction in raw material usage and material wastage, thus directly attacking one of the major causes of increased generation of electronic waste (e-waste). The major benefit of printable electronics is their capacity to function at low temperature and in ambient conditions, thus leading to decreased energy demands during their manufacturing. Printing methods such as inkjet, screen, and gravure printing methods completely eliminate the need for high temperature processing and cleanroom facilities, which are required in conventionally used semiconductor technology. Consequently, this leads to the drastic reduction of the carbon footprint of consumer electronics manufacturing. Printable electronics offer further benefits in their capacity to be easily printed on flexible, low-weight, and unconventional materials, thus leading to innovative device design and further addressing material efficiency.

Printable electronics can thus be used for the creation of recyclable and biodegradable electronics by using biodegradable materials. Paper, cellulose, and other biodegradable polymers, as well as recyclable plastics, can be coupled with printable conductive and semiconducting materials to make devices for a specific life span. Such application-specific and disposable electronics will be ideal for sensors, smart packaging, RFID, and wearable electronics, since they do not need a longer life span. This will also ensure that there is no hazardous electronic waste created. Despite everything, there is a need for continued innovation for the large-scale implementation of printable electronics. Sustainable materials, device development, durability, and recycling processes are some critical factors for which innovations will be necessary for the large-scale introduction of printable electronics and the resultant reduction in the electronic waste crisis.

References

[1] Robinson BH, E-waste: An assessment of global production and environmental impacts. Sci Total Environ, 2009, 408, 183–191.

[2] Grant K, Goldizen FC, Sly PD, et al, Health consequences of exposure to e-waste. Lancet Glob Health, 2013, 1, 350–361.

[3] Widmer R, Oswald-Krapf H, Sinha-Khetriwal D, et al, Global perspectives on e-waste. Environ Impact Assess Rev, 2005, 25, 436–458.

[4] Sepúlveda A, Schluep M, Renaud FG, et al, Environmental fate of hazardous substances released from e-waste. Environ Impact Assess Rev, 2010, 30, 28–41.

[5] Baldé CP, Forti V, Gray V, et al, The Global E-waste Monitor. United Nations University, 2017.

[6] Cucchiella F, D'Adamo I, Koh SCL, Rosa P, Recycling of WEEEs: An economic assessment. Waste Manag, 2015, 45, 263–272.

[7] Magdassi S, The Chemistry of Inkjet Inks. World Scientific, 2010.

[8] Cummins G, Desmulliez MPY, Inkjet printing of conductive materials. Circuit World, 2012, 38, 193–213.

[9] Kamyshny A, Magdassi S, Conductive nanomaterials for printed electronics. Small, 2014, 10, 3515–3535.

[10] Nathan A, Ahnood A, Cole MT, et al, Flexible electronics: The next ubiquitous platform, Proc IEEE, 2012, Vol. 100, 1486–1517.

[11] Kiddee P, Naidu R, Wong MH, Electronic waste management approaches. Waste Manag, 2013, 33, 1237–1250.

[12] Tansel B, From electronic consumer products to e-waste. Waste Manag, 2017, 68, 333–346.

[13] Li J, Lu Y, Yin J, et al, Health risks of informal e-waste recycling. Environ Sci Technol, 2015, 49, 731–738.

[14] Ongondo FO, Williams ID, Cherrett TJ, How are WEEE doing?. Waste Manag, 2011, 31, 714–730.
[15] Babbitt CW, Kahhat R, Williams E, Evolution of product lifecycles. Resour Conserv Recycl, 2009, 53, 478–491.
[16] Ardente F, Mathieux F, Environmental assessment of recycling electronics. J Clean Prod, 2014, 64, 46–58.
[17] Irimia-Vladu M, Green electronics. Chem Soc Rev, 2014, 43, 588–610.
[18] Someya T, Bao Z, Malliaras GG, The rise of plastic bioelectronics. Nature, 2016, 540, 379–385.
[19] Forrest SR, The path to ubiquitous and low-cost electronics. Nature, 2004, 428, 911–918.
[20] Derby B, Inkjet printing of functional materials. Annu Rev Mater Res, 2010, 40, 395–414.
[21] Perelaer J, Smith PJ, Van den Bosch E, et al, Printed electronics by inkjet printing. Adv Funct Mater, 2008, 18, 2101–2107.
[22] Krebs FC, Fabrication and processing of polymer solar cells. Sol Energy Mater Sol Cells, 2009, 93, 394–412.
[23] Secor EB, Principles of aerosol jet printing. Flex Print Electron, 2018, 3, 035002.
[24] Kamyshny A, Steinke J, Magdassi S, Metal-based inks for printed electronics. Open Appl Phys J, 2011, 4, 19–36.
[25] Rogers JA, Someya T, Huang Y, Materials and mechanics for stretchable electronics. Science, 2010, 327, 1603–1607.
[26] Khan S, Lorenzelli L, Dahiya RS, Technologies for printing sensors, IEEE Sens J, 2015, Vol. 15, 3164–3185.
[27] Arapov K, Rubingh E, Abbel R, et al, Inkjet-printed RFID antennas. IEEE Trans Antennas Propag, 2014, Vol. 62, 1970–1978.
[28] Trung TQ, Lee NE, Flexible and stretchable sensors. Adv Mater, 2016, 28, 4338–4372.
[29] Torrisi F, Hasan T, Wu W, et al, Inkjet-printed graphene electronics. ACS Nano, 2012, 6, 2992–3006.
[30] Park S, Vosgueritchian M, Bao Z, PEDOT:PSS for printed electronics. Nanoscale, 2013, 5, 977–988.
[31] Sirringhaus H, Organic field-effect transistors. Adv Mater, 2014, 26, 1319–1335.
[32] Fortunato E, Barquinha P, Martins R, Oxide semiconductor thin-film transistors. Adv Mater, 2012, 24, 2945–2986.
[33] Nogi M, Iwamoto S, Nakagaito AN, Yano H, Transparent nanocellulose paper electronics. Adv Mater, 2009, 21, 1595–1598.
[34] Martins R, Nathan A, Barros R, et al, Paper electronics. Adv Mater, 2011, 23, 4491–4496.
[35] Irimia-Vladu M, Sariciftci NS, Bauer S, Exotic materials for bioelectronics. Adv Mater, 2011, 23, 2506–2523.
[36] Bressanelli G, Adrodegari F, Perona M, Saccani N, Circular economy strategies for electronics. J Clean Prod, 2018, 178, 224–245.
[37] Korhonen J, Honkasalo A, Seppälä J, Circular economy: Concept and limitations. Ecol Econ, 2018, 143, 37–46.
[38] Liao C, Zhang M, Yao MY, et al, Flexible organic electronics. Adv Mater, 2015, 27, 7493–7527.
[39] Sekitani T, Zschieschang U, Klauk H, Someya T, Flexible organic transistors. Nat Mater, 2010, 9, 1015–1022.
[40] Sondergaard R, Hösel M, Angmo D, et al, Roll-to-roll fabrication of polymer solar cells. Mater Today, 2012, 15, 36–49.
[41] Mensing JP, et al, Defects in printed electronics. Microelectron Eng, 2016, 156, 1–7.
[42] Hwang SW, Tao H, Kim DH, Rogers JA, Transient electronics. Science, 2012, 337, 1640–1644.
[43] Bettinger CJ, Bao Z, Biodegradable electronics. Adv Mater, 2010, 22, 651–655.
[44] Dahiya R, Mittendorfer P, Valle M, et al, Large-area electronics for the internet of things. Proc IEEE, 2019, Vol. 107, 2015–2032.
[45] Kim DH, Lu N, Ma R, et al, Epidermal electronics. Science, 2011, 333, 838–843.
[46] Anastas PT, Warner JC, Green Chemistry: Theory and Practice. Oxford University Press, 1998.

[47] Finnveden G, et al, Life-cycle assessment of electronic products. J Environ Manage, 2009, 91, 1–21.
[48] Irimia-Vladu M, Biodegradable electronics for a sustainable future. Nat Commun, 2012, 3, 1–9.
[49] Cheng H, et al, Fully transient electronics. Adv Mater, 2015, 27, 1573–1580.
[50] Krebs FC, Jørgensen M, Recycling of printed electronics. Energy Environ Sci, 2013, 6, 775–785.
[51] Riede M, Spoltore D, Leo K, Organic solar cells. Adv Energy Mater, 2021, 11, 2002653.
[52] Tao H, et al, Materials for transient electronics. Adv Mater, 2014, 26, 393–406.
[53] Trung TQ, Lee NE, Printed wearable electronics. Adv Mater, 2016, 28, 4338–4372.
[54] Salmeron JF, et al, Printed RFID technologies. Sensors, 2014, 14, 20059–20085.
[55] Stoppa M, Chiolerio A, Wearable electronics. Sensors, 2014, 14, 11957–11992.
[56] Pikul JH, et al, Printed supercapacitors. Nat Commun, 2013, 4, 1732.
[57] Khan Y, et al, Flexible hybrid electronics. Adv Mater, 2018, 30, 1707029.
[58] Pei Q, et al, Green printed electronics. Adv Funct Mater, 2019, 29, 1903862.
[59] Sondergaard R, et al, Life cycle of printed electronics. Energy Technol, 2015, 3, 22–35.
[60] Sirringhaus H, Reliability of printed transistors. Adv Mater, 2018, 30, 1707236.
[61] Kim S, et al, Encapsulation strategies for printed electronics. Adv Electron Mater, 2017, 3, 1600440.
[62] ISO/IEC 62899. Printed electronics standards. International Organization for Standardization, 2016.
[63] Allen MP, et al, Manufacturing challenges in printed electronics. Flex Electron, 2019, 1, 012003.
[64] Schluep M, et al, Recycling of emerging electronic waste. Waste Manag, 2018, 76, 1–9.
[65] Koo J, et al, Transient bioelectronics. Adv Funct Mater, 2020, 30, 1908008.
[66] IDTechEx. Printed Electronics Market Report. IDTechEx, 2022.
[67] Irimia-Vladu M, Green materials for electronics. Chem Soc Rev, 2014, 43, 588–610.
[68] Hwang SW, Rogers JA, Transient electronics. Science, 2012, 337, 1640–1644.
[69] Bressanelli G, et al, Circular electronics. J Clean Prod, 2020, 269, 122118.
[70] Anastas PT, Zimmerman JB, Design through the principles of green engineering. Environ Sci Technol, 2003, 37, 94A–101A.
[71] Hauschild MZ, Rosenbaum RK, Olsen SI, Life Cycle Assessment. Springer, 2018.

Prakash Ray*

Sustainable e- waste management in India in the emerging global context: policy, practice, and ethical pathways

Abstract: The path to Viksit Bharat (Developed India) by the year 2047 is inevitably tied to finding a solution to the most serious environmental and economic problem, known as electronic waste (e-waste). India is one of the largest manufacturers of e-waste in the world due to its rapid digitalization and economic growth. This paper theoretically and practically deconstructs the sustainable e-waste management approach in India, analyzed through the dual lenses of the circular economy (a concept formulated by Walter Stahel and the Ellen MacArthur Foundation) and extended producer responsibility (EPR) (theorized by Thomas Lindhqvist). It contends that the legislative evolution in India, namely the E-Waste (Management) Rules, 2022, constitutes a complicated policy transition to a formalized, producer-centric framework. Still, the informal sector's continued existence, together with the environmental problems and health dangers it poses to the population, shows a severe gap in practice that could endanger the long-term strategy.

Given that India aspires to be a developed nation by 2047, the country's policy structure must be multifaceted, encompassing integration, innovation, and regulatory enforcement. Programs that run continuously are helping to root this. The unified EPR platform ensures compliance for producers, tracks waste movements, and brings about digital accountability and control. A dedicated market for effective collection and channeling between producers and recyclers is created as a result of the promotion of producer responsibility organizations (PROs). Access to formal banking, safety equipment, and skill-building programs like Skill India are all ways to help formalize informal waste pickers, which is in line with inclusive development principles, and is vital for a fair transition.

The success of the country in the long-term would lie in encouraging local green technology in recycling technology, decreasing dependence on importation of the processes, and keeping the valuable materials in the country's economy. This demands that there should be an incentive in the form of policies to support R&D and investment into sophisticated recycling infrastructure. Thus, the realization of a sustainable e-waste management by 2047 is dependent on a synergistic model that stringently imposes EPR, technologically empowers the formal sector, socio-economically integrates the informal work force, and inculcates a culture of responsible consumption and recycling into the

*Corresponding author: Prakash Ray, Syamsundar College (Under the University of Burdwan), Bardhaman Road via Pahalanpur, Shyamsundar, Purba Bardhaman 713 424, West Bengal, India, e-mail: prakashray2014@gmail.com

 | https://doi.org/10.1515/9783112248072-015

citizens. This combined strategy is not only a waste management plan but a base requirement toward the construction of a circular, fair and fully developed India.

Keywords: E-waste management, circular economy, Viksit Bharat, extended producer responsibility, policy framework, recycling technology

1 Introduction

The twenty-first century has been witness to unprecedented technological revolution, driven by digitalization, consumerism, and the proliferation of electronic and electrical equipment (EEE). While this revolution has elevated living standards and enhanced productivity, it has also generated a parallel and rapidly expanding stream of waste known as **electronic** waste (e-waste), comprising discarded computers, mobile phones, televisions, household appliances, lighting, medical devices, and other electronic products. The Global E-Waste Monitor 2024 estimates that the world produced over 62 million tons of e-waste in 2023, marking a hazardous 82% rise since 2010 [1]. Only about 22% of this waste was formally collected and recycled, leaving the remainder to be processed under unsafe, informal conditions or disposed of in landfills, causing significant ecological and public health hazards.

India, the world's most populous nation and a fast-growing digital economy, lies at the center of the e-waste challenge. The Central Pollution Control Board (CPCB) reported that India generated 1.6 million tons of e-waste in 2022–23, ranking third globally after China and the United States [2]. Major contributors include ICT devices (computers, mobiles, printers) and consumer electronics (TVs, refrigerators, and washing machines). Urbanization, shorter product lifecycles, and rapid technological shifts have intensified this problem. Sustainable e-waste management is vital to achieving "Viksit Bharat @ 2047," India's vision of becoming a developed and environmentally secure nation by its centenary of Independence. The path forward involves integrating Extended Producer Responsibility (EPR), green technologies, and inclusive formalization of the informal sector into a unified circular economy strategy.

Although the E-Waste (Management) Rules 2022 have strengthened accountability through digital EPR systems, nearly 85% of e-waste is still handled informally by around 300,000 workers, often without safety measures. Bridging this gap demands policy innovation, technological investment, and ethical governance.

This chapter aims to:

- Contextualize India's e-waste management in the global sustainability framework
- Assess alignment with the UN Sustainable Development Goals (SDGs)
- Trace India's policy evolution in EPR and circular economy
- Present state- and category-wise e-waste data
- Showcase success stories and best practices
- Propose an inclusive roadmap for sustainable management by 2047

Ultimately, the chapter argues that e-waste management is not merely an environmental concern but a cornerstone of India's economic progress, social inclusion, and environmental justice.

2 Global scenario of e-waste generation and management

2.1 Dynamics of global e-waste production

The digital world has begun a period of hyper-obsolescence, more than it has ever been; products become obsolete. Electronic and electrical equipments' end-of-life management has become integral to the *UN Agenda 2030 for Sustainable Development.* E-waste intersects directly with SDGs 3, 6, 8, 9, 11, 12, and 13, covering health, clean environment, responsible production, and climate action. Sustainable e-waste systems therefore address not only pollution but also decent work, innovation, and inclusive growth. The Global E-Waste Monitor 2024 (ITU, UNU, and ISWA) reports that there was 62 million tons (MT) of e-waste produced in the world in 2023 or 7.8 kg per capita [4]. It is estimated that by 2030, e-waste around the globe will exceed 75 MT, exceeding the combined growth of human population and the level of recycling. China (12 MT) is the largest producer, followed by United States (7.6 MT), India (1.6 MT), Japan (1.2 MT), and Brazil (1.1 MT). The European Union has an average production of approximately 10 MT, but is still the only region where recycling and collection are above 40%. In comparison, Africa and South Asia formally recycle less than 5% of their e-waste, with the remainder of the material informally addressed or on illegal export [5]. The increase in the quantity of e-waste in the world is a result of several intertwined factors:
- Quick consumerism and reducing product sizes, resulting in a reduced replacement cycle.
- Growth of smart and IoT products, which increases the amount of electronic products in a household.
- The world is moving towards a renewable energy source, which produces new types of waste (e.g., solar panels and EV batteries).
- Lack of awareness of consumers on repairability and safe disposal.

This rising trend shows that the e-waste issue is not only about worn out devices, but production design, consumption culture, and linear economy processes, which are still prevalent in most of the world.

2.2 Statistical characterization and GIS categorization of e-waste around the globe

E-waste materials and components are more than a thousand in kind (Table 1). They are majorly divided into six streams (ITU 2024) [5].

Table 1: Characterization of e-waste around the globe.

Category	Examples	Global share (%) 2023	Key recyclable/hazardous materials
Large equipment	Washing machines, refrigerators, and air conditioners	30 (%)	Steel, aluminum, plastics, and refrigerants
Small equipment	Vacuum cleaners, toasters, and cameras	20 (%)	Copper, aluminum, and mixed plastics
Small IT and telecom	Mobile phones, computers, and routers	17 (%)	Gold, palladium, copper, and lithium
Screens and monitors	TVs, LCDs, and laptops	14 (%)	Lead glass, indium, and plastics
Lamps	CFL, LED, and fluorescent tubes	6 (%)	Mercury, glass, and phosphor
Temperature-exchange equipment	Cooling and heating devices	13 (%)	Refrigerants, compressors, and steel

Source: ITU [5].

2.3 Environmental and human health implications

The environmental impact of e-waste that is not controlled is extensive. Electronic products contain lead, cadmium, mercury, arsenic, and brominated flame retardants (Table 2), which contaminate the soil, water, and air when dumped or burnt [6]. Informal recycling, a practice that mostly involves the use of non-protective gear, renders the workers, even children, to serious health dangers like neurological disorders, respiratory diseases, and bio-disruption. The World Health Organization (WHO, 2024) reports that about 18 million child laborers and teenagers take part in the informal recycling of e-waste around the world. Long-term exposure has been associated with decreased IQ, pulmonary inflammation, or damage of DNA [7]. The communities around large informal recycling centers (e.g., Guiyu in China, Agbogbloshie in Ghana, and Seelampur in India) have levels of heavy metals several times above the WHO safety limits. In addition to health risks at the local level, e-waste is the source of climate change. Malpractice in the maintenance of cooling and IT equipment emits gases of fluorine, which has a global warming potential (GWP) that is thousands of times more powerful than CO_2. It is estimated that in 2023, the mishandling of the world e-waste emitted approximately 98 MT of CO_2-equivalent – almost the yearly emissions of Belgium.

2.4 Development of the global recycling gap and resource opportunity

Ironically, e-waste is an environmental risk, yet, it is also a uranium mine. In 2023, the total e-waste value of raw materials neared over US $90 billion, with 7% of the global gold production and substantial amounts of copper, palladium, cobalt, and rare-earth elements necessary to sustain renewable technology [9]. However, below 20% of e-waste in the world is recycled in certified premises. Reasons for this gap include: Lack of collection facilities and poor rate of consumer returns; Laxity in management of international trade controls (Basel Convention); Issues with recycling technologies that require high costs; and absence of standardized information and tracking. This unrecovered material is an economic loss as well as a failure to be sustainable. The transition to a circular-economy paradigm, i.e., relying on repair, reuse, remanufacturing, and recoveries of materials, is therefore the key to achieving SDG 12 (Responsible Consumption and Production) and SDG 13 (Climate Action) objectives.

2.5 International policy and regulations

2.5.1 Basal Convention and global government

The Basel Convention on the Control of Transboundary Movement of Hazardous Wastes and their Disposal (1989) is the foundation of the world's regulation of e-waste and provides limitations to the export of hazardous e-waste between the developed and the developing world. The Basel Ban Amendment (2019) takes such exports illegally further. Nonetheless, illegal entrance and exit of e-waste continues to occur – 1.7 MT of e-waste was transferred between OECD and non-OECD countries in 2022 [10].

2.5.2 European Union (EU)

EU WEEE Directive (2012/19/EU) is the most prolific e-waste law in the world. It requires collection and recycling by the producers, provides high recovery goals (up to 85% collection of certain types of products), and advances the concept of eco-design to enhance products' longer life [11].

2.5.3 United States of America (USA)

The United States does not really have a single federal law; it is controlled in the states and thus there are fragmented systems. Nevertheless, voluntary take-back programs have been developed by the Sustainable Materials Management (SMM) initiative of EPA and National Strategy of Electronics Stewardship (NSES).

2.5.4 East Asia and Oceania

The Home Appliance Recycling Law (2001) also implements the most efficient e-waste system, with over 90% material recovery of four major home appliances in Japan. South Korea has producer take-back schemes, and Australia has the National Television and Computer Recycling Scheme (NTCRS) that covers 98% of the population in Australia [12].

2.5.5 Global South

In south Asia, Latin America, and Africa, the informal sector and weak enforcement takes over the challenge. Nevertheless, formal frameworks, based on the EPR principle, which have been recently established in several countries such as Ghana, Nigeria, and India, are indicative of a global change in the way the world is going towards producer responsibility [13].

3 SDG goals

3.1 Centrality of e-waste to global sustainable development agenda

The 17 SDGs that were adopted in 2015 are a global roadmap toward poverty eradication, ecological security, and sustainable prosperity. Although e-waste is not specifically a focus of individual goals, it touches directly on most of them (Figure 1), especially those concerning resource efficiency, well-being, sustainable industry, and climate action [15]. The digital economy is the pillar of development in all sectors around the world, but the very devices that facilitate education (SDG 4), innovations (SDG 9), and governance in general, leave an increasing poisonous footprint when discarded. As shown in the Global E-Waste Monitor 2024 released by the UNEP, the e-waste produced in 2023 amounted to close to 62 million tons of e-waste, with many material components, which had a worth of more than US $90 billion, yet only about a quarter was processed through proper recycling. The rest derails development to

Figure 1: Centrality of e-waste management to the Sustainable Development Goals (SDGs).

various SDGs, polluting ecosystems, posing a risk to the health of people, and consuming precious collateral resources [16].

3.2 SDG 3: Good Health and Well-Being

Poor management of e-waste causes the release of toxic elements like lead, mercury, cadmium, and flame retardants containing brominated compounds, which pollute the air, land, and underground water. According to the reports of World Health Organization

(WHO, 2023), almost 18 million children and adolescents on the planet are at risk due to hazardous dismantling, resulting in neurotoxicity, respiratory disease, and impaired development [1]. An example of a health–environment nexus in India is denoted by clusters of smelting of circuit boards and burning of wires in open air: the release of particulate material is much in excess of safe levels [17]. Sustainable e-waste management also helps to achieve SDG 3 by: the occupation-safety standards of recyclers; increasing formalized collection centers and dismantling centers with protective technology; and the inclusion of informal workers in form of skill development and the health insurance program under Skill India and Ayushman Bharat.

3.3 Goal 6: Clean Water and Sanitation

E-waste leakage causes heavy metals like arsenic and chromium to contaminate the rivers and water that is drained to the ground. Poor disposal sites in urban peripheries, which are usually close to drainage channels, lead to bioaccumulation of the industrial soils and aquifers in agricultural lands. Hazardous waste monitoring is now incorporated into the urban sanitation planning of India through the National Clean Energy and Environment Fund and Swachh Bharat Mission [18]. Shifting to secure landfills and creating zero-discharge recycling establishments are direct moves toward SDG 6 6.3 and 6.6, which ask for reducing water pollution and protecting aquatic ecosystems.

3.4 Goal 8: Decent Work and Economic Growth

The e-waste industry in the world has more than 20 million workers; the majority of them are informal workers. The green jobs through Producer Responsibility Organisations (PROs) and EPR-related refurbishing units formalize this workforce, which results in decent income and safe environment. The Skill India Mission of India now offers e-waste collection and recycling modules, which are to offer informal labor diversification to informal workers and self-help groups of women. Therefore, e-waste management reduces pollution as well as acts as a source of an inclusive green job, which aligns with SDG 8.3 and 8.8 [19].

3.5 Goal 9: Industry, Innovation, and Infrastructure

Sustainable management of e-waste demands technological newness and strong recycling facilities. The examples of other nations that are on the forefront of this area, such as Japan, Germany, and South Korea, prove that the incorporation of high-tech material-recovery systems enhances the competitiveness of industries, as well as environmental sustainability. E-Waste (Management) Rules: 2022 E-PR tracking systems are also encouraged in India, as manufactured products are now designed to be disassembled and recycled. The projects made to recycle waste in native technologies, like Make in India and Digital India, are turning waste into a strategic stream of resources, driving the requirements of SDG 9.4 to ensure sustainable industrialization [20].

3.6 SDG 11: Sustainable Cities and Communities

More than 70% of the e-waste produced in India is in the city, making it a hub of metropolitan agglomerations such as the Delhi-NCR, Mumbai, Bengaluru, and Chennai. The e-waste systems, sustainable collection kiosks, municipal take-back centers, and digital traceability increase the sustainability and cleanliness of cities. The application of e-waste management to Smart City Missions and waste-management bylaws in Urban Local Bodies (ULBs) directly assists the SDG 11.6 in minimizing the environmental impact of cities.

3.7 SDG 12: Responsible Consumption and Production

It is the essence of SDG that regulates e-waste policy. Responsible production requires that it is eco-designed, durable, and recyclable, whereas responsible consumption involves takeback participation. The Extended Producer Responsibility (EPR) principle, which was pioneered by Thomas Lindhqvist, is imitated across the world and attempts to hold producers responsible when it comes to the post-consumer waste. The E-Waste Rules 2022 of India make this possible via digital EPR certificates, and traceable transactions among producers, dismantlers, and recyclers [21]. The SDG 12 achievements presented include: Approval of more than 500 recyclers and refurbishers throughout the country; Creation of an integrated EPR Portal on the compliance and transparency by the CPCB; The Ministry of Environment, Forest and Climate Change's campaigns like Clean E-Waste and Green India to create awareness among the population. Still, there is the problem of customer behavior and collection logistics – formal e-waste collection in India is done in only 25% of the cases. It is essential to close this gap in order to achieve the SDG 12 goals 12.4 and 12.5.

3.8 SDG 13: Climate Action

E-waste management contributes significantly to climate mitigation. Recycling metals and plastics from discarded electronics saves energy and reduces greenhouse-gas emissions. According to UNEP, proper recycling of one ton of circuit boards can save up to 2 tons of CO_2-equivalent, compared with primary metal extraction. In India, the

National Action Plan on Climate Change (NAPCC) and National Resource Efficiency Policy (2023 draft) explicitly identify circular-economy strategies – including e-waste recycling – as climate action instruments. Hence, strengthening the e-waste recycling infrastructure is integral to achieving India's Net Zero by 2070 target and SDG 13.2 on integrating climate measures into national policies.

3.9 SDGs 14 and 15: Life Below Water and Life on Land

When dumped in open landfills, heavy metals and microplastics from e-waste seep into soils and waterways, harming biodiversity and marine life. India's coastal cities – Mumbai, Chennai, and Kochi – are particularly vulnerable, as informal dumping near estuaries and drains leads to oceanic contamination. The promotion of eco-industrial parks, closed-loop systems, and scientific landfill containment aids both SDG 14.1 (marine-pollution reduction) and SDG 15.3 (land-degradation neutrality) [22].

3.10 SDG 17: Partnerships for the Goals

Effective e-waste management relies on multi stakeholder partnerships among government, industry, academia, and civil society. India's collaboration with UNEP, WHO, and global technology firms under the Global E-Waste Statistics Partnership exemplifies SDG 17 in action. Public–private partnerships (PPP) in collection logistics and awareness campaigns – such as Karo Sambhav, Attero Recycling, and Tata Croma Take-Back Programmes – demonstrate scalable cooperation models that bridge formal and informal sectors.

4 India's e-waste landscape: trends, composition, and regional distribution

4.1 Overview of India's e-waste generation

The high rate of digitalization, growing middle class, and increased electronic use in India has transformed it to be the third-largest producer of e-waste in the world, following China and the United States. Official CPCB 2024–25 data (from the EPR Portal and Ministry of Environment) indicate India generated about 1.398 million tons of e-waste in FY 2024–25, up from 1.254 MT in FY 2023–24. Earlier years: 1.609 MT (2022–23) and 1.601 MT (2021–22). These updated numbers supersede previous model estimates based on 2019 baselines. The figures derive from producer-reported sales, product lifetimes, and registered recycler returns, giving a more realistic picture of India's formal e-waste flow. CPCB (2024–25) data show the highest generation in Maharashtra, Tamil Nadu, Uttar Pradesh, Karnataka, and Telangana, together contributing over 60% of national e-waste. States with rapidly growing ICT hubs (Delhi-NCR, Telangana, and Kerala) show the steepest year-on-year rise.

4.2 Object-wise (product-wise) composition

Based on CPCB and MeitY data (2023), India's e-waste stream can be disaggregated as follows [9]:

Table 2: Recoverable materials from e-waste.

Category / Product type	Share (%) of total 2022–23 e-waste	Major materials recoverable
IT and Telecom Equipment (computers, laptops, mobile phones, routers, and printers)	38%	Copper, gold, palladium, and plastics
Consumer electronics (TVs, audio systems, and cameras)	21%	Lead glass, indium, and aluminum
Large household appliances (refrigerators, washing machines, and AC units)	18%	Steel, refrigerants, and compressors
Small appliances (mixers, irons, and vacuum cleaners)	7%	Aluminum, copper, and mixed plastics
Lighting equipment (CFLs, LEDs, and fluorescent tubes)	6%	Mercury, glass, and phosphor dust
Others (medical, electrical tools, toys, etc.)	10%	Various alloys, plastics, and circuits

Source: Global E-Waste Monitor [5].

Mobile phones alone contribute over 12% of the IT/telecom category by weight and nearly 35% by unit count, reflecting India's fast-cycling consumer electronics market. The growing use of LED lighting and IoT devices also introduces complex composite materials that challenge existing recycling technologies.

4.3 The informal sector and its socioeconomic role

Nevertheless, with the gradual, yet incomplete regulation, approximately 85–90% of India e-waste is still being processed in the informal sector, which includes more than 300,000 workers in large hubs like Seelampur (Delhi), Moradabad (Uttar Pradesh), Saharanpur, Kodungaiyur (Chennai), and Kankurgachi (Kolkata) [24]. Commonly deployed methods are acid leaching and open burning of wires that release toxic fumes and also pollute the surrounding soil and water. In India, many informal workers belong to marginalized communities and migrant groups, lacking legal or health protection. A *social-justice-centered* e-waste policy must therefore prioritize: (a) formal inclusion through micro-licensing/cooperatives; (b) enforcement of child-labor prohibitions; (c) occupational-health support and PPE; and (d) livelihood transition under *Skill India* and *National Green Jobs* initiatives. Government efforts like: The modules of Skill India, Green Jobs., Loans of National Safai Karamchari Finance and development corporation (NSKFDC), and Aggregator models with PRO-linked aggregator models of MoEFCC, intend to fulfill the recent industry by enlisting informal workers into the formal recycling economy by training, providing credit, and covering occupational-health insurance.

4.4 Consumer Behavior and Collection Systems

A significant source of bottleneck is consumer awareness. According to a survey by the Centre for Science and Environment (CSE) in 2023, only 30% of urban consumers were aware of approved collection centers and less than 10% of consumers returned their devices to the disposal center. Mechanisms based on convenience to use are currently being tested, such as take-back kiosks in electronics outlets, tie-ups with e-commerce platforms on reverse-logistics, and online return incentives. The systems that are being integrated in municipal dashboards are mobile-app-based systems (e.g., CPCB EPR Portal, Swachhata App) to ensure traceability and citizen involvement [25].

4.5 Potential of material recovery and of resource efficiency

E-waste has abundant secondary resources, which are critical to the manufacturing economy of India: Copper (approximately 18%), aluminum (approximately 13%), iron (approximately 35%), precious metals (gold, silver, palladium approximately 0.05%), and plastics (approximately 20%). Recycled, India would record current material worth of e-waste to the tune of more than 25,000 crore/annum and would counter 3 MT CO_2-equivalent emission, in comparison with virgin extraction [26]. This possible position ensures that e-waste takes center stage in the Circular Economy Action Plan (2022–23) and Resource Efficiency Strategy, which drives its connection to SDG 12 and national Net-Zero goals.

4.6 The problems and upcoming trends

There are fundamental systemic issues: Poor collecting rates, despite being regulated; intersection of a variety of authorities (MoEFCC, CPCB, and SPCBs); Informal dominance and the creation of data gaps; and dispersed sources that involve high logistics costs. Most recyclers are technologically obsolete. However, there is a new direction: Creation of state-level structure of e-waste management cells in Gujarat, Kerala, and Karnataka; e-waste exchange portals to EPR credit B2B trade; e-waste modules in school curriculums and CSR programs; and new startups working on battery lithium retrieval, doxymagnets, and PCBs recovery.

5 Policy and regulatory framework in India: evolution, implementation, and gaps

5.1 Evolution of the legal framework

The official quest to e-waste regulation in India started in the late 2000s, as the lack of regulation and haphazard practices of dumping and recycling reached crisis levels. The E-Waste (Management and Handling) Rules, 2011 under Environment (Protection) Act, 1986 were notified by the Ministry of Environment and Forest (then, MoEFCC) in 2011. It was an innovative move by an up-and-coming country and the principle of Extended Producer Responsibility (EPR), to put the responsibility of the end-life management of products on the manufacturers and importers. Subsequent policy reforms have progressively tightened control and improved accountability.

Notable ongoing initiatives:

- Kerala – Clean Kerala Company Ltd. integrates municipal collection and recycling hubs
- Karnataka – E-Parisara demonstrates ISO-certified recycling and extended take-back drives
- Telangana – E-waste Parks & Aggregator Licensing link informal collectors to formal systems
- Delhi – Aggregator pilot enables informal recyclers' registration under the PRO framework

NGOs and PROs making impact:

- Karo Sambhav – EPR implementation and circular collection
- Toxics Link – Policy advocacy and research
- Chintan – Waste picker inclusion
- Saahas / SWaCH – urban collection cooperatives.
- Basel Action Network / WEEE Forum – Global coordination
- Ellen MacArthur Foundation – Circular economy frameworks

This steady evolution mirrors global learning – from voluntary take-back to a data-driven, compliance-verified system that reflects both EU WEEE-Directive and Basel Convention principles.

5.2 Key concepts: extended producer responsibility (EPR) and polluter pays

EPR is the ethical and operational structure of Indian policy. It requires manufacturers to maintain collection and recycling of any electrical and electronic equipment (EEE) in its final use or via designate collection centers. Key elements include: All producers and recyclers were to be registered on the national EPR portal managed by CPCB. Annual targets in terms of percentages of the sales in the previous financial year. EPR Certificates are issued to highly priced and produced-by registered recyclers on basis of proven quantities recycled, which producers can buy to satisfy targets. Digital tracking and audit systems eliminate the possibility of double-counting and allowing real-time tracking [27]. This system of credit markets makes the polluter-pays principle work in practice, whereby those introducing products to the market have to fund their sound end-of-life treatment, and stimulates efficiency and competition between recyclers.

5.3 Institutional architecture

The level of e-waste governance in India has a multitier framework:

MoEFCC- Policy development, notification of rules, and agreement with international conventions

Central Pollution Control Board (CPCB) – National regulator with the responsibility to keep the digital EPR portal, authorization issuance, checking EPR credits, and publishing data every year

State Pollution Control Boards (SPCBs) – Regional work, sites inspection, and registration of the centers of collection

Urban Local Bodies (ULBs) – E-waste integrated with the municipal solid waste systems in Smart-City and Swachh Bharat

Producer Responsibility Organisations (PROs) – The types of the intermediaries, private intermediaries, engaged in the take-back logistics, creating awareness campaigns, and aggregator training.

The architecture facilitates vertical integration, i.e., between the national strategy and the local execution, but there is a practical constraint of coordination between agencies and SPCBs.

5.4 Business highlights of the E-Waste (Management) Rules 2022

The 2022 Rules is a complete change of paradigm, shifting the manual modes of reporting to digitally traceable circular economy regime. Major innovations include: Single EPR Portal through which producers and recyclers can register in real-time, and issue and transfer credit; Uniqueness of identifiers in standardized EPR certificates; Online performance dashboard that is available to the regulatory and the public, improving transparency; Large-scale producers' annual and quarterly audits, penalty grades on noncompliance with the Environment Protection Act; Dismantler–Recycler Difference: dismantlers enable the materials to be used in their material form; those who can form EPR credits are only registered recyclers; Public awareness requirement: manufacturers should invest in the education of consumers on the disposal [28].

Through compliance of digitalization, India has developed one of the first nation-scale blockchain-ready waste-tracking systems in the world that adheres to SDG 12.6 on sustainable business reporting.

5.5 Conformance to international policy frameworks

The policy of India concerns world policies in the Basel Convention and the Strategic Approach to International Chemicals Management (SAICM). In addition, the 2022 Rules reinforce the focus of producer responsibility and producer traceability in the EU WEEE-Directive, with adjustments to the level of India development. By this, the Indian regulation would come up to the global standard of sustainability, and domestic socio economic realities in this regard would have a direct bearing on SDGs 9, 12, and 13.

5.6 Implementation progress and performance

CPCB's EPR Portal Report (2023–24) indicates measurable progress:
- Over 1,800 producers and 530 recyclers registered
- 1.2 million tons of annual processing capacity online
- EPR credits worth ₹700 crore traded in the first operational year

States such as Kerala, Karnataka, and Gujarat have achieved formal collection efficiencies above 40%, aided by local awareness drives and municipal partnerships.

Conversely, data deficits and weak enforcement persist in several northern and northeastern states.

Regular publication of state-wise dashboards and third-party audits is expected to close these gaps over time.

5.7 The road ahead: toward regulatory maturity

To consolidate gains, experts recommend:
- Integrating informal collectors through micro-licensing, cooperatives, and PRO partnerships
- Dynamic EPR targets linked to product lifecycles rather than flat percentages

- Fiscal incentives – tax rebates or green-credit schemes – for certified recyclers
- Regional treatment hubs, supported by public investment, to reduce logistics costs
- Harmonized data exchange between CPCB, MeitY, and customs to curb illegal imports

If implemented, these measures will mature India's system into a self-sustaining circular economy model, showcasing to the developing world that environmental stewardship and economic competitiveness are mutually reinforcing.

6 Praxis and the circular economy approach in India

The policy environment of e-waste in India has progressed to a market-oriented circular economy approach from a compliance-based regulation approach [30]. The E-Waste (Management) Rules 2022 created a system in which producers, recyclers, and Producer Responsibility Organisations (PROs) work together under the Extended Producer Responsibility (EPR) systems to reclaim materials and minimize landfill. The transformation is based on four pillars:
- Circular design – repairable, modular, and recyclable products
- Resource recovery – resource loops of metals and plastics
- Social inclusion – inclusion of informal collectors
- Ethical innovation – digital traceability and less impact process

Operational models: India has three hegemonic models that lead its circular e-waste model: Producer-led consortia: Companies such as HP, Dell, and Samsung have take-back programs managed by PROs (e.g., Karo Sambhav), which recycle more than a quarter-million tons of e-waste by 2024.
- Public-Private Partnerships (PPP): Municipalities partner with recyclers in states like Kerala, Karnataka, and Telangana; the Kerala model of recycling is the Clean Kerala Company, and the Karnataka model is E-Parisara, which are both ISO certified.
- Aggregator-Recycler Networks: Informal collectors deliver dismantled components to recyclers, such as Attero, Cerebra, and Ecoreco, and achieve up to 95% material recovery through digital traceability over the EPR portal.
- Technological Innovations: With the help of MeitY and CSIR, native green technologies would be bioleaching to recover metal, cryogenic dismantling, plasma-based refrigerant recovery, and lithium-ion battery recycling (Attero, Lohum).

These technologies help SDG 9 and SDG 13, as they help increase resource efficiency and decrease reliance on mining. The informal sector is considered the component of the economy that has not been integrated by the government into the formal sector.

Informal sector's integration: India is estimated to have collected nearly 80% of e-waste in the informal sector. The small aggregators are now licensed, workers trained on Skill India's Green Jobs, and health and safety is promoted through CSR grants [32], which is in line with SDG 8 on decent work.

Digital Governance: The EPR portal of India allows blockchain-advanced tracking, credit trade, geo-tagged logistics, and visible compliance [33]. Artificial intelligence-based predictive systems also streamline the waste management process and minimize emissions.

Economic and Market Growth: Recovered materials and EPR credit trading will lead to the e-waste recycling market becoming USD 9.5 billion by 2030 (CAGR ~ 14). Urban mining contributes to the vision of Atmanirbhar Bharat by recovering resources in the country [34]. Awareness and Ethics: Such campaigns as E-Waste Mukt Bharat and educational programs by Karo Sambhav facilitate responsible consumptions (SDG 12). There are courses on environmental ethics and circular economy in universities.

Persistent challenges: Problems such as the fragmented collection system, insufficient recovery of rare-earth, ineffective data verification, financing systems to SMEs, and insufficient public participation form barriers. Specific measures like micro-loans, in the form of the Green Credit Scheme of SIDBI, and the incorporation of e-waste into Swachh Bharat dashboards will be used to seal these gaps. Both ethical and cultural transition: Circularity represents intergenerational justice, worker justice, and ecological justice. These ethics, integrated in corporate, academic, and community practices, will enhance the journey of India toward Viksit Bharat @ 2047 [36].

7 India's success stories and the roadmap to Viksit Bharat @ 2047

7.1 India's emerging leadership in e-waste management

India has transformed in the past 10 years from appearing as a laggard in e-waste management globally to being a progressive example. The progression of formal recycling capacity, the establishment of the E-Waste (Management) Rules 2022, and the creation of the digital EPR Portal can be viewed as the example of a national transition of a fragmented, informal recycling to a system of data-driven, technologically advanced, and ethically inclusive recycling [37]. The change on the ground is facilitated through vibrant partnerships between industry, civil society, state governments, and grassroots organizations though policy frameworks provide the groundwork. The following case studies illustrate the way sustainable e-waste management is being put into practice and how these achievements are being matched with the vision of Viksit Bharat @ 2047, that is, a developed, circular, and equitable country.

7.2 Case Study I – Attero Recycling: the circular technology leadership

Attero Recycling Pvt. Ltd., which is located in Roorkee, Uttarakhand, and was established in 2008, is the best and modern representation of urban mining in India. Attero utilizes hydrometallurgical and electrochemical methods to extract precious and important metals, including gold, palladium, copper, cobalt, and lithium from discarded electronics and EV batteries [37]. Key features: 95% material recovery efficiency and discharge of zero liquids. Cooperation with manufacturers such as Samsung, HP, and Xiaomi in reverse-logistics. The use of recycling of lithium-ion batteries as a part of EV supply chains – compatible with the National Electric Mobility Mission Plan in India. Reduction of 0.5 MT/CO_2-equivalent emissions per year through the replacement of virgin mining. The model of Attero represents the principle of the circular economy to leave materials in the domestic cycle so that valuable metals do not leave the economy of India and provide qualified jobs [39].

7.3 Case Study II – Karo Sambhav: producer responsibility in action

Karo Sambhav is a social enterprise dedicated to operationalizing the Extended Producer Responsibility (EPR) mechanism, with the help of a multi stakeholder partnership strategy, which was established in 2017. It has partnerships with more than 60 electronics brands such as Apple, Dell, and Lenovo; it is operating in 29 states and UTs [40]. Achievements: Gathering and channelization of over 100,000 tons of e-waste to licensed recyclers, in the period between 2018 and 2024. Establishment of inversion networks between informal aggregators and formal dismantlers. School and community awareness programs to more than 3 million students and 2,000 educational institutions to be implemented. Karo Sambhav shows that by aligning ethical praxis, policy compliance, and community mobilization, it is possible to have quantifiable environmental and social impact – direct contribution to SDG 12 (Responsible Consumption and Production) and SDG 17 (Partnerships for the Goals).

7.4 Case Study III – state innovations: Kerala and Karnataka

7.4.1 Kerala: Kerala model of governance decentralization

Kerala has led localized e-waste governance. The state has the Clean Kerala Company Ltd., where district-level collection centers are formed with Smart City solid-waste centers. In 2023, Kerala was the first state to complete the formal collection coverage of 14 districts, recycling over 12,000 tons of e-waste per year [41]. The Suchitwa Mission is used to guarantee longevity of citizen involvement, whereas e-waste awareness is incorporated into schools.

7.4.2 Karnataka: IT ecosystem integration

Karnataka has E-Parisara Pvt. Ltd., which is the first ISO 14001-certified e-waste recycling plant that operates in Bengaluru, the technology hub of the country. The facility recycles about 40,000 tons annually [43] through the Karnataka State Pollution Control Board (KSPCB). The state also introduced campaigns of E-Waste Friday that made IT companies collect obsolete hardware every Friday, which can also be viewed as an example of the industry–government synergy. The two states demonstrate the role played by decentralized governments and the partnership between the government and the private sector in ensuring the shift to sustainable waste management, and is promoted in line with SDG 11 (Sustainable Cities) [43].

7.5 Case Study IV – Delhi-NCR

Aggregator licensing as a way of informal integration. India has a significant concentration of e-waste, almost 10% of which is manufactured in the Delhi-NCR region, where most has traditionally been handled by informal workers in places such as Seelampur and Mundur [43]. Granting informal collectors permission to deliver dismantled materials to licensed recyclers, Delhi has been the first to develop aggregator licensing, managed by CPCB. Results: Pilot programs registered an excess of 2,500 informal workers. Distribution of safety gear and access to microfinance loans by the cooperative societies. Acute lessening of open burning and acid-depleting activities. This model of inclusion directly promotes SDG 8 (Decent Work) and SDG 10 (Reduced Inequalities), and demonstrates how environmental policy can be used as a tool of social justice.

7.6 Roadmap to 2047: toward a circular, ethical, and technologically sovereign India

With India having the vision of having a developed nation by 2047, sustainable e-waste management will be part of Viksit Bharat story, not only as a green agenda but also as an indicator of technological independence and social accountability [44]. It is possible to imagine the roadmap to 2047 in four interconnected directions:

7.6.1 Policy deepening and harmonization

Active EPR Goals based on product lifecycles and not fixed quotas. Unified Resource Efficiency Authority (UREA) to harmonize MoEFCC, MeitY, CPCB, and state organizations. See circular economy action plan, which integrates with India National Resource Efficiency Strategy [45].

7.6.2 Infrastructure and technology acceleration

Create e-waste parks in the region, with shared effluent treatment, shredding, and smelting. Invest in local recovery technologies of rare-earths and AI material sorters. Connect the recycling of links and electronic units to the semiconductor and renewable-energy missions in India, so that no value is lost back home [46].

7.6.3 Green employment and social inclusion

By 2035, micro-license and cooperate 100% of informal collectors. Bring Green Skill Certificate to 1 million personnel under Skill India Mission. Establish collection business ventures with women heads, in collaboration with Self-Help Groups (SHGs) and Non-Governmental Organizations (NGOs) [47].

7.6.4 Education, ethics and citizen participation

Introduce e-waste literacy in National Education Policy (NEP 2020). Normalize policies on Right to Repair in order to prolong the life of products. Promote ethical consumption by labeling consumer products, offering take-back incentives, and creating awareness by the population [48]. Through a combination of these threads, policy, technology, inclusion, and ethics, India will be able to create the whole architecture of the circular economy that turns waste into the foundation of sustainable prosperity.

7.7 Critical raw materials (CRMs) recovery from e-waste: India's strategic opportunity

Critical raw materials (CRMs) such as cobalt, lithium, nickel, rare-earth elements (REEs), palladium, and indium have emerged as the strategic backbone of clean-energy technologies, semiconductors, EV batteries, telecommunications, and renewable power infrastructure. Global supply chains for these materials are highly concentrated, China controls 70–90% of REE processing, the Democratic Republic of Congo dominates cobalt mining, and geopolitical tensions have created supply instability. As highlighted in the UNEP and Global E-Waste Monitor 2024 reports, e-waste has now become the fastest-growing secondary source of CRMs, offering significantly lower extraction energy, reduced carbon footprint, and minimized land degradation, compared with primary mining.

India, which imports nearly all of its CRMs, is particularly vulnerable to supply shocks. At the same time, India's annual e-waste stream (1.39 million tons in 2024–25, CPCB) contains substantial CRM reservoirs: cobalt and lithium from discarded mobile phones and EV batteries; neodymium and dysprosium from hard-disk magnets; palladium, gold, and silver from printed circuit boards; and indium from flat-panel displays. Estimates suggest that efficient urban mining of India's e-waste could recover material worth more than ₹25,000 crore annually and reduce dependence on external supply chains by 20–25% in the next decade.

Recent initiatives indicate early progress. Attero, Lohum, and several MeitY-supported R&D programs are developing hydrometallurgical, biometallurgical, and cryogenic recovery technologies capable of isolating high-purity CRMs. India's inclusion of CRM recovery in the Draft National Resource Efficiency Policy (2023) and its alignment with the EU's 2023 Critical Raw Materials Act signal an important shift toward resource security. Embedding CRM recovery within the EPR framework can convert e-waste from a liability into a strategic economic asset, enabling India to position itself as a regional hub for secondary materials processing by 2047. Strengthening CRM recovery is, therefore, not only an environmental necessity but also a pillar of technological sovereignty, industrial competitiveness, and India's long-term vision of a circular and self-reliant economy.

7.8 Placing India in emerging global scenario

E-waste has also become a strategic security issue in the process of altering the precious raw materials (CRMs), in the world. The developed economies are also facing issues of shortages of metals and thus India, with its intensive supply of e-waste, may emerge as a hub of secondary resources recovery. India can become: A regional recycling center in South Asia, which will be founded on cross-border trading in secondary material in the SAARC in 2047. An eco-friendly technology manufacturer, with recycling services to developing countries at affordable prices, and expandable. Information economy that uses the principles of circular design and resource efficiency in the DNA of its manufacture. This would put India in the line with UNEP Global Partnership on Waste Management and would be an addition to the SDGs 9, 12, 13, and 17 and would transform India into a sustainability innovator, as opposed to a waste producer. Recent research points to three promising paths:

(a) **Biodegradable alternatives** – bioplastics and natural fiber composites replacing conventional circuit substrates;
(b) **Nanotechnological approaches** – nano-adsorbents and catalysts for high-selectivity metal recovery;
(c) **Biological processes** – microbial and enzymatic bio-leaching that can recover metals with minimal waste.

7.9 Discussion: the moral way to Viksit Bharat

Sustainable e-waste management in India is the Indian variant of the bigger moral image of Viksit Bharat 2047, a technologically sophisticated and yet socially friendly eco-competitive society. E-waste, in most of the situations, is considered merely a by-product of modernization. In this perspective it is a sort of an enlightenment, its potential to resolve consumption and conservation, advancement and forethought, invention and moral values. Nevertheless, when India takes the course it is currently taking, deepening its policy, extending circular infrastructure, ensuring to favor informal labor, assist in cultivating the sense of responsibility among the citizens, etc., not only will it become the first Global South to create a new ethos of its own, one that balances economic expansion and environmental protection, but it will also be the first to go. In such a way, the sustainable management of e-waste becomes an environmental demand to become a civilizational mission as the road to developed India is green and fair, with the principles of circularity, responsibility and shared prosperity as a guide.

References

[1] Quinto S, Law N, Fletcher C, Le J, Jose SA, Menezes PL, Exploring the e-waste crisis: Strategies for sustainable recycling and circular economy integration. Recycling, 2025, 10, 72.
[2] Sengupta D, Ilankoon IMSK, Kang KD, Chong MN, Circular economy and household e-waste management in India: Integration of formal and informal sectors. Miner Eng, 2022, 184, 107661.
[3] Bhaskar K, Turaga RM, India's e-waste rules and their impact on e-waste management practices: A case study. J Ind Ecol, 2018, 22, 930–942.
[4] Ilankoon IMSK, Dilshan RADP, Dushyantha N, Co-processing of e-waste with natural resources and their products to diversify critical metal supply chains. Miner Eng, 2024, 211, 108706.
[5] Forti V, Baldé CP, Kuehr R, Bel G, The Global E-waste Monitor 2020: Quantities, flows and the circular economy potential. United Nations University (UNU)/UNITAR–SCYCLE Programme, International Telecommunication Union (ITU) & International Solid Waste Association (ISWA, Bonn/Geneva/Rotterdam, 2020.
[6] Saha L, Kumar V, Tiwari J, Sweta, Rawat S, Singh J, Bauddh K, Electronic waste and their leachates impact on human health and environment: Global ecological threat and management. Environ Technol Innov, 2021, 24, 102049.

[7] Soetrisno F, Delgado-Saborit JM, Chronic exposure to heavy metals from informal e-waste recycling plants and children's attention, executive function and academic performance. Sci Total Environ, 2020, 717, 137099.
[8] Singh V, Bhattacharya S, Chugh P, Global perspectives on e-waste management: Challenges, innovations, and sustainable solutions, In Shukla P, Choudhary S, Ranjit PS, Verma A, Eds, E-Waste: From Generation to Management–Practices, Challenges, and Solutions. Springer, Singapore, 2025.
[9] Liu K, Tan Q, Yu J, Wang M, A global perspective on e-waste recycling. Circ Econ, 2023, 2, 100028.
[10] Mihai F, Gnoni M, Meidiana C, Schneider P, Ezeah C, Elia V, A global outlook on the implementation of the Basel Convention and the transboundary movement of e-waste, In E-waste Recycling and Management. CRC Press, Boca Raton, 2022, 49–75.
[11] Cole C, Gnanapragasam A, Cooper T, Singh J, An assessment of achievements of the WEEE Directive in promoting movement up the waste hierarchy: Experiences in the UK. Waste Manag, 2019, 87, 417–427.
[12] Lee SC, Na SI, E-waste recycling systems and sound circulative economies in East Asia: A comparative analysis of systems in Japan, South Korea, China and Taiwan. Sustainability, 2010, 2, 1632–1644.
[13] Andeobu L, Wibowo S, Grandhi S, Environmental and health consequences of e-waste dumping and recycling carried out by selected countries in Asia and Latin America. Sustainability, 2023, 15, 10405.
[14] Ravikumar D, Keoleian GA, Walzberg J, Heath G, Heller MC, Advancing environmental assessment of the circular economy: Challenges and opportunities. Resour Conserv Recycl Adv, 2024, 21, 200203.
[15] Carlsen L, Bruggemann R, The 17 United Nations' sustainable development goals: A status by 2020. Int J Sustain Dev World Ecol, 2021, 29, 1–11.
[16] Baldé CP, Kuehr R, Yamamoto T, McDonald R, D'Angelo E, Althaf S, et al, The Global E-waste Monitor 2024. International Telecommunication Union (ITU) & United Nations Institute for Training and Research (UNITAR), Geneva/Bonn, 2024.
[17] Sandwal SK, Jakhar R, Styszko K, E-waste challenges in India: Environmental and human health impacts. Appl Sci, 2025, 15, 4350.
[18] Hamdan A, Panda S, Jain MS, Raj V, Mathew S, Assessing municipal solid waste in Indian smart cities: A path towards waste-to-energy. Heliyon, 2025, 11, e42770.
[19] From waste to worth: Self-help groups in India advance circular economy and climate action. Mitigation Action Platform, 2025.
[20] Gaur TS, Yadav V, Mittal S, Singh S, Khan MA, E-waste management challenges in India from the perspective of producer responsibility organizations. IEEE Access, 2025.
[21] Faibil D, Asante R, Agyemang M, Addaney M, Baah C, Extended producer responsibility in developing economies: Assessment of promoting factors through retail electronic firms for sustainable e-waste management. Waste Manag Res, 2023, 41, 117–142.
[22] Ravish P, Sustainable management of landfill sites in India: Addressing environmental, health, and socioeconomic challenges. Curr World Environ, 2025, 20.
[23] Mohamed D, Fayad A, Mohamed AMO, Al Nahyan MT, The role of e-waste in sustainable mineral resource management. Waste, 2025, 3, 27.
[24] Rochat D, Rodrigues W, Gantenbein A, India: Including the existing informal sector in a clean e-waste channel. J Waste Manag, 2008.
[25] Nisha B, Shajil S, Dutta R, Jain T, Consumer awareness and perceptions about e-waste management in a semi-urban area of northern Tamil Nadu: A mixed-method approach. J Fam Community Med, 2022, 29, 132–137.
[26] Sharma D, Nema A, Prasad R, Sweta K, Sonaviya D, Karmakar S, Global e-waste management: Consolidated information showcasing best available practices, In Sustainable Waste Management: Policies and Practices. Elsevier, 2023, 289–314.

[27] Singh V, Pathak V, Kumari A, Roushan R, Comparative study of waste management to fulfill induced producer responsibility, In Proc 3rd Int Conf Communication, Security, and Artificial Intelligence (ICCSAI). Greater Noida, India, 2025, 710–715.
[28] Central Pollution Control Board, Frequently Asked Questions (FAQ) under the E-Waste (Management) Rules. 2022, CPCB, New Delhi, 2024.
[29] Scruggs C, Improving information flow on chemicals in electronic products and e-waste to minimize negative consequences for health and the environment. Resour Conserv Recycl, 2016, 113, 149–164.
[30] Ofori D, Mensah AO, Sustainable electronic waste management among households: A circular economy perspective from a developing economy. Manag Environ Qual Int J, 2021.
[31] Patra S, Sinha S, Mahobia GS, Kamble D, Eds, Proceedings of the International Conference on Metallurgical Engineering and Centenary Celebration (METCENT 2023). Springer, Singapore, 2024.
[32] Malik R, The alchemy of waste: How India's new E-Waste (Management) Rules 2022 are forging a digital circular economy for Viksit Bharat@2047. Chartered Secretary, 2025, 85–88.
[33] Central Pollution Control Board, Centralized EPR Portal for Plastic Packaging.
[34] Mordor Intelligence, India e-waste management market size and share analysis.
[35] Virtual Conference on "E-Waste Management in India: New Business Opportunities". Confederation of Indian Industry/Ministry of Electronics & Information Technology, Government of India, New Delhi, 2020.
[36] Prime Minister's Office, PM's 79th Independence Day address: A vision for a Viksit Bharat 2047. Government of India, New Delhi, 2025.
[37] Newar S, Deka S, From toxic trash to green governance: Legal and policy dimensions of e-waste regulation in India. Int J Environ Sci, 2025, 11, 418.
[38] Transforming waste to worth: How Attero recycles e-waste, batteries, and rare earths for India's self-sufficiency. Mint, 2025.
[39] Sobanaa M, Prathiviraj R, Prathaban M, Kiran GS, Selvin J, Closing the loop: Circular economy solutions for long-term environmental health. Evol Earth, 2025, 3, 100071.
[40] Sambhav K. Karo Sambhav: Circular economy and sustainability.
[41] Sambhav S, Waste management in Kerala: A model for success. Medium, 2025.
[42] E-Parisaraa Pvt. Ltd. About us: India's first government-approved electronic waste recycling company.
[43] Kandpal R, Okitasari M, Governance transformation towards localisation of sustainable development goal 11 in India. World Dev Sustain, 2023, 2, 100069.
[44] Dan I, Viksit Bharat @2047: A vision for India's sustainable development. Sci Temper, 2024, 15, 53–58.
[45] Jena S, Tyagi A, Kale A, Gupta S, Bassi N, Chaturvedi K, Kashyap A, Kovacs C, Khan A, Singh P, Das R, Mishra S, Baruah S, Unlocking India's circular waste economy potential for sustainability: Insights across seven key sectors. Council on Energy, Environment and Water (CEEW), New Delhi, 2024.
[46] Dey S, Ganugula TV, Anjaneya Babu PS, Manoj AVP, Recycling of e-waste materials for controlling environmental and human health degradation in India. Green Anal Chem, 2023, 7, 100085.
[47] Aayog NITI, Strategy for New India @75. Government of India, New Delhi, 2018.
[48] Government of India, National Education Policy 2020 and Mission LiFE (Lifestyle for Environment). Press Information Bureau, New Delhi, 2023.
[49] United Nations Environment Programme, The Global Waste Management Outlook 2024: Beyond an age of waste – Turning rubbish into a resource. UNEP, Nairobi, 2024.

Archana B.*, Somaraj R., Kirti

Ethical and policy frameworks for sustainable e-waste management: global challenges, circular economy pathways, and SDG alignment

Abstract: The rapid increase in the consumption of electronic devices in the world has made the issue of electronic waste more challenging, bringing about immediate environmental, social, and ethical issues. In this chapter, the author explores the development, efficiency, and lack of effectiveness of international and national policies on e-waste, in relation to frameworks such as the Basel Convention, extended producer responsibility (EPR), and circular economy strategies. It explores fundamental ethical imperatives such as precaution, polluter pays, labor rights, and transparency as well as the issues of justice that are connected to marginalization of informal sector, inequitable transboundary waste flows, and disproportionate effects on at-risk populations in developing countries. The analysis finds gaps in corporate accountability, enforcement capability, and participation of stakeholders, which impede corporate alignment with Sustainable Development Goals of responsible consumption and production (SDG 12), climate action (SDG 13), reduced inequalities (SDG 10), and good health and well-being (SDG 3) through empirical studies, comparative case studies, and global sustainability reports. The chapter goes further to discuss how digital innovations, open platforms to transparency, and collaborative governance may facilitate ethical and sustainable e-waste in a circular economy context. The synthesis of the socio-ethical, legal-political, and techno-economic approaches results in the work practicing a paradigm of the ethical stewardship and multi-level governance reforms and makes sustainable e-waste management critical to environmental integrity, equity, and human dignity in the technology-dependent world via the SDG.

Keywords: Electronic waste management, sustainable development goals, circular economy, extended producer responsibility, corporate accountability

***Corresponding author: Archana B.,** University Centre for Research and Development, Chandigarh University, Mohali 140413, Punjab, India, e-mail: archanabalachandran2000@gmail.com
Somaraj R., Kirti, University Centre for Research and Development, Chandigarh University, Mohali 140413 Punjab, India

 | https://doi.org/10.1515/9783112248072-016

1 Introduction

Acquiring digital technologies at a rapid pace and increasing the amount of global reliance on electronic goods has contributed to an unprecedented increase in electronic waste volumes (e-waste), making it the fastest-growing type of waste in the world [1]. The amount of e-waste generated by the world in 2022 was 62 million metric tons, 82% higher than in 2010, and the formal collection and recycling rates had not increased since 2010, which lost 62 billion US dollars worth materials per year. By 2030, the volume is projected to reach 82 million tons, which is 5× higher than the recycling infrastructure, due to the shorter product life cycles, society becoming electronic, and more consumption in emerging economies such as Asia. The current crisis is seen as causing severe environmental degradation and human health hazards because e-waste releases more than 1,000 substances, including lead, mercury, cadmium, and brominated flame retardants, into the soil, water, and air, bioaccumulating along food chains and ecosystems. The informal recycling that is predominant in the Global South, involves 70–90% of e-waste handling by hand, through open burning and acid baths, exposing millions of people, such as women and children to acute risks: respiratory diseases, neurotoxicity, musculoskeletal injuries, DNA damage, infertility, and cancers [2, 3]. Systematic reviews affirm a high level of blood lead, cardiopulmonary dysfunction, and hearing loss caused by noise in African and Asian hotspots, highlighting how local exposures are linked to global supply chains [2].

The failures of governance add to these damages, although landmark treaties, such as the Basel Convention (1989), which governs the movements of transboundary hazardous wastes but fails to ensure their enforcement, allows 70–80% of high-income e-waste to enter developing markets under the guise of reusable goods [4].The same can be said of regional frameworks such as the EU WEEE Directive, Bamako Convention, etc. Inadequate monitoring, corruption, and capacity gaps ensure the continuation of illegal trade, as was observed in the case of ASEAN, where Basel ratification did not change the volumes of flows considerably. In India, formal recycling has less than half the market share of E-Waste (Management) Rules (2016, 2022), with the informal sector taking more than 90% of the market share, which cannot be controlled, due to a lack of infrastructure [5]. These processes are morally visible systemic blind spots in techno-economic paradigms, which maximize recovery speed, as well as marginalizing precaution, polluter pays, procedural justice, and dignity to marginalized players. Forming a subsidizing society, informal workers have to bear the toxicity costs, which are contrary to intergenerational equity and labor rights, because the cost is projected to the Global South by rich countries [6]. This ethical imbalance requires stewardship that goes beyond regulation to accountability among the producers, consumers, and states.

The integrated framework of ethics, policy, and sustainability is urgent, which should combine the ethics of stewardship, inclusive governance, digital transparency (e.g., blockchain traceability), and SDG compliance – the SDG 12 (responsible consumption), SDG 13 (climate action), SDG 10 (reduced inequalities), and SDG 3 (health/well-being) [7, 8]. Value recovery through circular economy pathways, with focus on design-to-recycle and EPR, would help reduce inequity, yet justice precautions are needed to prevent the creation of new divides [6].

2 Conceptual foundations of e-waste management

Electronic waste (e-waste), also called waste electrical and electronic equipment (WEEE), is any discarded electrical or electronic equipment, all components, subassemblies, and consumables of the equipment at the time of disposal [9]. This definition, which is followed throughout the world by such frameworks as EU WEEE Directive (2002/96/EC), covers equipment that is dependent on electric currents or electromagnetic fields that may operate at voltages up to 1,000 V AC or 1500 V DC [10]. Depending on jurisdiction, e-waste is categorized as follows: large household appliances (e.g., refrigerators), small household appliances, IT and telecom equipment (e.g., computers and mobiles), consumer electronics (e.g., TVs), lighting, tools, toys/leisure/sports equipment, medical devices, monitoring instruments, and automatic. They are usually made of ferrous metals [50], plastics [21], nonferrous metals such as copper and precious metals [13], glass and hazardous materials such as lead, mercury, cadmium, and brominated flame retardants, which are difficult to recycle.

The rate of e-waste production has grown drastically to 53.6 million metric tons in 2019 and 57.3% increase since 2010, nearly doubling the population growth and nearly 5× formal recycling [11]. The record collection and recycling rates stood at 22.3% in 2022, and are expected to drop to 20% by 2030, even though recoverable materials were estimated to be worth $62 billion a year, which drives the idea of structural inefficiencies in collection systems [12]. There are high disparities between the different regions: Europe is the biggest per capita generation of 17.6 kg/person, with formal recycling rate (42.8%), which is followed by Oceania (16.1 kg/person) and Americas (14.1 kg/person), because of the advanced infrastructure. By contrast, Asia produces almost half the world quantity (30 million tons) but recycles very little because of the lack of legislation; Africa has the lowest rates (less than 1%), despite less output, and India doubled its production to 3.8 million tons in FY2024, making it the third-largest producer, in the face of rapid urbanization [9].

Environmentally, e-waste causes extensive pollution as the toxins seep into the land, groundwater, and air, upsetting the ecology in food chains and destroying biodiversity [9]. Open dumping and landfilling discharge intractable pollutants such as dioxins and heavy metals, and informal incineration releases volatile organic compounds and PM, which have an equal impact on climate as the 2.3 billion kg CO_2-equivalent each year of unregulated waste [12]. Health effects are very serious in informal processing facilities: through dust breathing, physical contact, and untreated water exposure, children and pregnant women are at the greatest risk because the body absorbs them more [2, 9]. Elsewhere in Africa and Asia, such as around recycling facilities, systematic reviews report high blood lead levels (as high as 10× WHO thresholds), cardiopulmonary dysfunction, and chronic kidney disease [3, 13].

E-waste has contrasting prospects and injustice socio-economically. The informal sector has employed 12–24 million individuals worldwide, mostly in the low-income region of Asia and Africa, to earn a living on low-tech dismantling and valuables recovery such as gold (15 billion potential annually), but at the price of working under hazardous conditions, without protective equipment or wages. Circular economies could be supported with the help of resource recovery, which will decrease the need for virgin materials (e.g., 100× concentration of copper vs. ore), yet the current losses sustain economic drain and reliance on mining [12]. Inequality escalates: rich countries outsource the problems to the Global South through transnational flows, creating digital rifts, whereby the marginalized groups live at the expense of high consumption societies and perpetuating a cycle of poverty [1]. Formal systems offer green jobs through the extension of producer's responsibility (EPR) but the absence of informal workers will lead to social upheaval, without involving up to transitions.

3 Ethical perspectives in e-waste governance

E-waste management is not only a technical or managerial issue; it is an ethical, philosophical issue that poses a challenge to the humanity, in relation to nature, economy, and accountability. The conceptual basis of dealing with these complexities is offered by environmental ethics that develops moral consideration outside the self-interest of immediate humanities. There are three systems of values that are inquired in this field.

Anthropocentrism assumes that the usefulness of nature is what makes it desirable, and environmental damage is seen as a threat only when it negatively affects the welfare of people. Biocentrism, in contrast, applies intrinsic value to all life forms, going on to point out that technological progression such as consumer electronics should not be done at the expense of other forms of life (Figure 1). A broader paradigm, ecocentrism, honors moral standing in the inherent processes occurring within the ecosystem and in the natural, which states that both human and environmental prosperity depend upon each other [14].

E-waste sheds light over these tensions. Endless drives toward economic growth and technological progress, which are anthropocentric in nature, have resulted in overproduction and premature obsolescence, shifting the burden of externality of en-

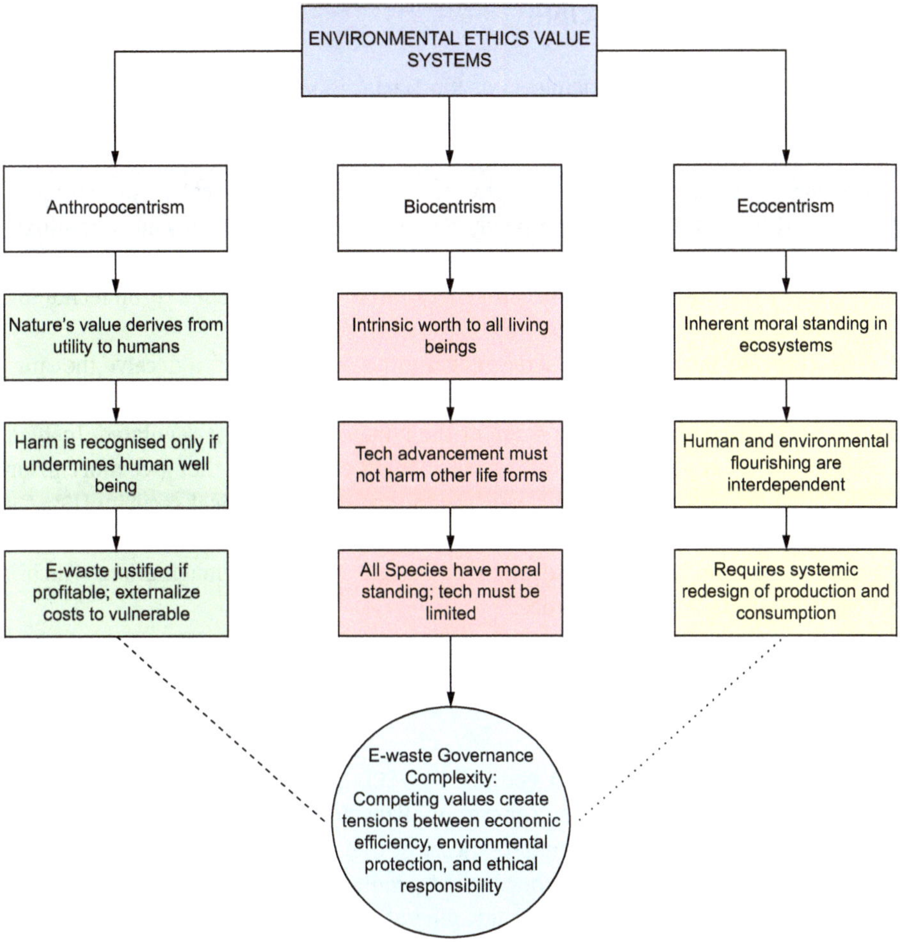

Figure 1: Schematic flowchart of environmental ethics value systems.

vironmental and human costs to the weakest. Deep ecology, as proposed by Arne Naess (1989), criticizes these trends and calls for biospheric egalitarianism, or moral equality of all life, by the use of deliberate simplicity, localized resource cycles, and by rejecting the so-called technological idolatry; deep ecology is directly contrapuntal to the consumerist culture, which creates digital detritus. In the meantime, virtue ethics provides a human dimension because it prioritizes moral qualities such as temperance, stewardship, prudence, and justice, which are to be applied to behavior in the context of ecological degradation [11, 14]. All of these ethical strategies prompt a review of e-waste governance as a moral developmental issue, institutional virtue, and intergenerational accountability, rather than policy change.

3.1 Precautionary principle

Precautionary principle, Principle 15 of Rio Declaration (1992), forms the view that precautionary action must be taken when there is reasonable suspicion of harm, even at the absence of full scientific certainty. This principle, in the context of e-waste, requires designers and manufacturers to remove harmful chemicals; lead, mercury, brominated flame retardants, and newly discovered nanomaterials, before the toxicity effect can become undeniable. This practice reflects an ecocentric ethics of moderation; it does not push the limits of self-generation of the planet but bets on technological solutions [4].

However, in the world of practice, governance systems tend to deceive the spirit of the principle. A lot of Basel Convention Annexes have been unable to keep up with technological advancement, and some of their materials are not regulated. Instead, decision-makers apply anthropocentric cost-benefit calculations that give more priority to short-term economic profits, relative to long-term ecological security [15]. The challenge ethically is not then, merely a question of implementing precautionary norms but of developing the institutional foresight and moral imagination, which is necessary to make it a strict practice.

3.2 Polluter pays principle

The polluter pays principle (PPP), codified in OECD guidelines (1972) and reflected in Article 16 of the Rio Declaration, operationalizes accountability by requiring producers to absorb the full environmental cost of their actions [16]. It is a distributive ethic of justice – distributing the burdens of production to those who reap the gains of the same. Under virtue ethics, the meaning of PPP is the moral virtue of responsibility, which discourages moral disengagement within global supply chains. EPR systems are central to e-waste policy, which strive to implement the concept of PPP, with manufacturers being charged with end-of-life collection and recycling responsibilities. This is reflected in the 85% levels of recovery in the EU WEEE Directive [10]. However, loopholes in this implementation are still existing all over the world: as a way of reducing expenses, corporations frequently outsource recycling to informal groups, which results in exploitative working environments and dumping to the environment. The poor enforcement and monitoring in India have led to less than 10% of EPR obligations being met [17]. An ethical reading of PPP, as such, is one that requires the inclusion of financial compensation alongside restorative justice, in which polluters are actively involved in the remedial efforts, help in community health recovery, and impart worldly wisdom in sustainable technologies.

3.3 Intergenerational equity

Intergenerational equity focuses on the equity between the current and future generation in the utilization of resources, and bases the concept of sustainability on moral reciprocity. Based on the robust theories of sustainability, it assumes that natural resources, as rare-earth minerals deemed indispensable in the electronics industry, cannot be replaced permanently by artificial products without the destabilization of the ecological environment [18, 19]. This is a value that is strongly echoed in the land ethic of Aldo Leopold: A thing is right when it leans toward integrity, stability, and beauty of the biotic community. The conservation of nature is therefore considered an ethical rather than a practical responsibility of maintaining the ecosystems according to ecocentrism.

This is a principle that is acutely challenged by e-waste. The current system is associated with temporal injustice by depleting finite resources by means of linear production and polluting future generations' lands. Toxic legacies of convenience in the short term, like Agbogbloshie in Ghana, where the lead concentration of the soil is hundreds of times higher than permissible limits, represent the symbol of contaminated sites [13]. The use of intergenerational ethics to regulate governance requires life cycle designing, cradle to cradle production, and ecological debt accounting to future generations.

3.4 Human dignity and labor rights

The e-waste economy provides employment to a population of 12–24 million people all over the world and mostly in the informal sectors that have no employment protective measures. The right of dignity and safe work is inherent, not based on productivity when viewed through the prism of biocentric humanism. One of the most devastating outcomes of informal recycling is the catastrophic health consequences that women, children, and migrants experience because of lead fumes and acid burns, which constitutes the rampant disregard of the ILO Convention 155 and the UN CRC Article 32 [20].

Morally, this is a form of dehumanization of human life to achieve a certain economic benefit, which is contrary to the moral imperatives of Kantian and ecofeminist ideas that gender and ecological exploitation are interconnected. Real sustainability implies the inclusion of just transition frameworks in the system of healthcare, skill training, and social security to re-establish the autonomy of workers and protect dignity [21]. Identifying these workers as planetary custodians, instead of waste pickers, would make industrial transformation conform to environmental virtue [14].

3.5 Accountability and transparency

Transparency has an expression of moral virtues of honesty, humbleness, and openness – qualities that are crucial in ethically sound governance. Information flow is not present in e-waste chains, which contributes to corruption and corporate negligence. These virtues could be maintained through blockchain-based traceability and digital product passports that will monitor the path of materials manufactured through recycling [22]. But technological openness should be accompanied by ethical uprightness, otherwise new systems will tend to be similar to current injustices, in the name of digital efficiency.

A morally consistent governance model needs to provide procedural justice as well, where actors at all levels have access to environmental information and policy affect. Lack of real-time data dashboards and third-party audit enable evasion of the system in India, compromising accountability and trust [17].

3.6 Environmental justice and the at-risk population

Environmental justice (EJ) relates distributive equity, recognition, and participation to deal with the inequitable environmental impacts. Combining Rawlsian fairness with ecocentric care, it reveals how the transboundary flows of e-waste reveal a new form of colonialism, namely, toxic colonialism, when impoverished countries are externalized by the rich ones in terms of pollution [23, 24]. On locations such as Agbogbloshie, Ghanaian and Nigerians migrants burn Western wastes on open fires and they are neurobiologically impaired and excluded socioeconomically. The ethical perspective of these practices is that they go against the principles of distributive justice (equal distribution of risks and benefits) and procedural justice (fair representation in decision-making).

Ecofeminism and virtue ethics have shared the importance of having empathy, solidarity, and humility with the affected populations. The global waste trade should recognize colonial pasts of global garbage, and should incorporate reparation policies such as capacity building funds, local empowerment, and transparency of trade into international treaties, such as the Basel Ban Amendment [14].

3.7 Current practice and ethical failures

Most e-waste governance systems are ethically thin, in spite of the move to proliferate policies. To begin with, proximity bias constrains itself to local harms that can be seen, and conceals cross-border exploitation networks. Second, utilitarian economics puts the principle of recovery efficiency and cost saving above moral principles such as equity and wellbeing. Third, North-South asymmetry is here to stay: to this day, up to 80% of e-waste is being exported by OECD to non OECD states, even with export prohibitions [6]. Lastly, techno optimism: relying on the market to provide innovations to resolve ethical crisis hides structural injustice by overlooking the underlying causes of overconsumption.

The solution to the governance of e-waste, therefore, needs what environmental thinker Holmes Rolston III might term, an enlargement in ethical space, that is, the combination of moral argument, systemic accountability, and sustainable design. Institutionalization of ethical stewardship must be in the form of cross-sector councils that comprise ethics scholars, policymakers, and unions of waste workers. Finally, genuine sustainability is a combination of ecological virtue and socioeconomic justice, where world ideals are translated into the local equity and generational sustainability

4 Global and national policy frameworks

The regulatory framework of e-waste lies between international treaties and fragmented national approaches, which points to the trends of ambition, checked by the continued lack of success. But the international tools set normative thresholds, such as prior informed consent (PIC), environmentally sound management (ESM), and transboundary controls, but still the national translations create gaps between the formalized systems of the developed economies and the capacity constraints of developing nations [25, 26].

4.1 International instruments: Basel Convention and Ban Amendment

The Basel Convention on the Control of Transboundary Movements of Hazardous Wastes (1989), which was ratified by 191 parties, is the cornerstone by which hazardous e-waste is bound to PIC procedures, and e-waste cannot be exported to non-parties or states with insufficient capacity [15]. According to the Basel Convention Secretariat, e-waste is categorized into annexes II, VIII, and IX: nonhazardous: Y46/Y48, hazardous: A1180/A1181, and new COP-15 amendments (effective January 2025) subject all e-waste, even nonhazardous, to PIC, removing the previous loopholes on the category of e-waste known as recyclables (Basel Convention Secretariat, 2024 The Ban Amendment (Article 4A), which prohibits exports of hazardous waste by OECD to non-OECD states, is still unratified by 23 major participants (including the US), allowing 70–80% illicit flows in the name of second-hand goods [15, 27, 28]. Efforts are being made to promote ESM, such as Mobile Phone Partnership (MPPI, 2002) and Partnership for Action on Computing Equipment (PACE, 2008), but they are not enforceable.

4.1.1 Connection to Stockholm and Rotterdam Conventions

The cluster of synergies of the Basel-Rotterdam-Stockholm (BRS) Conventions is holistic in e-waste. Stockholm Convention (2001) is focused on persistent organic pollutants (POPs) such as PBDEs and PBBs, which are widespread in e-waste, and it requires that these pollutants should be reduced and ESM implemented during recycling. Rotterdam (1998) regulates prior informed consent on dangerous chemicals in electronics (e.g., some flame retardants), thus the import decisions are made in accordance with capacity. These are combined in BRS joint workplans: Basel controls movements, Rotterdam governs substances, and Stockholm mandates elimination – but siloed implementation divides up oversight, because e-waste POP releases in informal dismantling cannot concentrate on action [25].

4.2 Comparative overview: developed versus developing countries

Developed countries apply EPR to the letter: the WEEE Directive (2012/19/EU) in EU requires 85% collection/65% reuse/recovery by 2025 with producer-funded take-back and national registries (Advance Recycling Fee in Switzerland is an example of 95%) [10, 26, 29]. In Japan (Home Appliance Recycling Law, 2001), South Korea (>80% rates through deposit-refund and OEM responsibility), the US has 70% population targets in its state-level laws but does not provide federal cohesion [10, 30].

The developing countries are a reflection but show deviation (Table 1): India (E-Waste Rules, 2022) sets EPR goals (60% by 2024–25) but informal recycling is done by less than 30% (formal) and 90% (informal), and the process occurs through open burning [17, 31]. China (WEEE Regulations, 2012) prohibited imports of e-wastes (2018) but has local spikes; Nigeria and Ghana regulate through NESREA/ESA but have no infrastructure, importing 70% Northern waste [32, 33]. Trends: The developed nations are focused on upstream design/collection (8 out of 10 formalization); the developing countries focus on downstream prohibition, in the absence of capacity [26].

Table 1: Comparative overview of e-waste management and EPR implementation in developed and developing countries.

Aspect	Developed countries (EU, Japan, and US states)	Developing countries (India, China, and Africa)
Core mechanism	EPR + producer take-back (85–95% compliance)	EPR mandated but <30% achieved
Collection rate	40–50% formal; infrastructure-heavy	<10% formal; informal dominance (90%)
Enforcement	Centralized registries, fines (€10k+)	Fragmented; corruption, understaffing

Table 1 (continued)

Aspect	Developed countries (EU, Japan, and US states)	Developing countries (India, China, and Africa)
Import policy	Export bans to non-OECD; traceability	Bans evaded via mislabeling
Recycling capacity	High-tech; 50–70% material recovery	Low-tech; health/environmental risks

4.3 Enforcement gaps and compliance challenges

When it comes to enforcement, it fails everywhere and unevenly. There is export leakage in developed countries, EU exports 300k tons of products to Asia every year, even with bans and free-rider issues, in which non-signatory countries undermine standards [28]. Third-world nations have weakness in terms of capacity: CPCB/SPCBs in India lack personnel (one officer/10 states), equipment to monitor, and collect in the rural areas; corruption allows 80% of illegal imports [17]. Pieces of the puzzle make it more challenging: 50+ national laws do not coexist with Basel, with no harmonized definitions or data sharing [25].

5 Extended producer responsibility (EPR) models

Extended producer responsibility (EPR) is a new way of managing waste, where producers are not only responsible to reduce their negative impact on the environment but are also responsible for other aspects such as waste collection, recycling, and disposal [34]. EPR was developed in Sweden in the 1990s as an organizing concept of shared responsibility, called product stewardship, to a binding mechanism that internalizes the environmental costs, incentivizes the creation of an eco-design, and facilitates circularity. As of 2025, EPR has been implemented by more than 30 countries in e-waste, due to synergy in the Basel Convention and requirements of the EU WEEE Directive [10].

5.1 Producer responsibility ethical justification

EPR reflects the principle of a polluter pays, the profit motive is balanced with the sense of moral responsibility: the creators of design and marketing are the most responsible in reducing toxicity and recovering resources [34]. Ecologically, it encourages virtues of stewardship, such as prudence in material choice and justice in distributing burdens, addressing anthropocentric externalization, in which consumers support corporate excreta [14]. In terms of ethics, EPR alleviates imbalances: wealthy manufacturers benefit through Global North consumption, at the expense of informal Southern laborers, which is a breach of labor and intergenerational dignity. However, there is still a tension - compliance, as a profit-maximizing box-ticking exercise risks turning into greenwashing, where shareholder returns are made to be more important than actual sustainability [17].

5.2 Comparative implementation models

EPR models vary by stringency, financing, and scope, revealing ethical trade-offs between compulsion and flexibility.

5.2.1 Collective schemes (Europe and Japan)

The EU WEEE Directive (2012/19/EU) in Europe has high collection and recovery levels of 85% (collection) and 65% (recovery) with the assistance of Producer Responsibility Organizations (PROs) that collect and pool fees of the producers [10]. An example of this practice is the EAR system in Germany, with recovery rates of up to 95% [26]. The Home Appliance Recycling Law (2001) of Japan uses a consumer-fee system, in which $Y = 2{,}000\text{–}4{,}000$ is charged on OEM take-backs and a compliance rate of approximately 80 is reached when all are required to be closely traced [30, 35]. These collective schemes allocate risk ethically and are capable of creating sector-wide innovation, although they risk undermining individual producer incentives to design environmentally good products, based on free-riding effects [26].

5.2.2 Individual producer responsibility (Switzerland and South Korea)

Switzerland has a system of Advance Recycling Fee (ARF), whereby it funds individualized take-back, frequently in conjunction with deposit refund systems. Under the control of organizations like the Sensis Foundation, the rate of recycling in this model is approximately 90% and it heavily favors design for recycling [35]. South Korea also imposes ARFs, in combination with an exemplary EPR infrastructure, such as fines on noncompliant companies, which have led to a recycle system of about 82% [36]. These systems are ethical since they offer direct and producer responsibility, to reduce evasion and match product design with end-of-life responsibility. Nonetheless, this burden of administrative complexity and compliance costs does not favor small and medium-sized enterprises [36].

5.2.3 Emerging models (India and developing economies)

The Indian E-Waste Management Rules (2022) set out slow EPR targets, and an objective of 60% collection by 2024–25, assisted with a centralized online portal, which will track down on the needs of the producers. Formal collection is yet to reach more than 30%, partly due to the important contribution of informal recycling systems [31]. The primary impediments to pilot retail-based EPR programs in Ghana and Nigeria are weak enforcement and the low presence of producers who are likely to relapse to informal e-waste channels management [37]. The novel forms are pointing out the ethical quandary on ambition to control and institutional capability, where informal economies and regulatory gaps are dooming EPR performance.

Table 2: Comparative EPR implementation models and their ethical trade-offs.

Model	Key features	Collection rate	Ethical tension
Collective (EU/Japan)	PROs, shared fees, and take-backs	80–95%	Free-riding versus scale efficiency
Individual (CH/Korea)	ARF and OEM responsibility	82–90%	SME burden versus design incentives
Emerging (India/Africa)	Portals and phased targets	<30%	Informality versus capacity gaps

5.3 Monitoring, traceability, and accountability challenges

EPR has an ethically sound principle, but its effectiveness depends on the strength of the supervision, a sphere where loopholes constantly undercut its purpose. Deficit monitoring has remained a bane of implementation: producer obligations are captured in the CPCB portal of India but it does not include real-time audit, allowing producers to select high-value waste streams and avoid low-profit and hazardous waste, like CRTs, which are avoided and filtered out. The lack of traceability still poses serious challenges, with blockchain pilot projects following at most only 20% of supply chains, and small and medium enterprises often elusive to digital compliance requirements [38].

The responsibility and accountability further divides along the profit-responsibility line, with multinational corporations assigning responsibility to PROs, which waters down the incentive to eco-design and upstream responsibility. Even though the high fines – usually more than 10,000 in the EU – can serve as a deterrent, corruption and laxity in the developing setting facilitates evading regulations. These inefficiencies are further amplified by a lack of consumer participation, due to the gap in knowledge and by cumbersome collection procedures, leading to 77.7% of all e-waste worldwide going uncollected [12]. This creates an ethical dilemma, as the ethical obligation may be diluted to a cost-center instead of being a moral obligation, which encourages the development of a compliance theatre where reporting is more of a priority than actual environmental and social change. Although such new technological approaches as RFID, AI-assisted sorting, and product passports have the potential to remedy the issue, they also present the risk of data privacy and inequity in access [25]. The compromise means multifaceted frameworks that combine moral requirements and legal measures: compulsory eco-design rating, employee inclusion capital, and transnational.

6 Circular economy pathways for sustainable e-waste management

The circular economy reimagines e-waste not as linear waste but as a regenerative resource loop, guided by Ellen MacArthur Foundation principles: eliminate waste/pollution, circulate products/materials at highest value, and regenerate nature. Applied to electronics, this demands "10R" strategies; Refuse, Rethink, Reduce, Reuse, Repair, Refurbish, Remanufacture, Repurpose, Recycle, and Recover; transcending 3 R limits to prioritize upstream prevention. Ethically, circularity embodies stewardship virtues, aligning profit with planetary limits, yet risks exclusion if rebound effects erode gains [39].

6.1 Design for durability, reuse, and recycling

The circular approach promises long lifestyles of electronics by using robust components, modular architectures, and standardized parts, which can dramatically increase the lifetime of products by 2–3× and also cut down the technological obsolescence by a fair margin. Fairphone modular smartphones are one example of how this can be a screwless practice, with users able to replace a battery and a screen without specific tools and can significantly reduce the production of e-waste by being able to self-repair and reuse. Designed for disassembly, like the use of mono-material casings, screwless, or easily dismantling joints, will also increase recyclability, which allows up to 95% of material recovery, as compared to about 20% in devices that used glued assemblies [11]. In the ecocentric view, these practices can be consistent with the land ethic, as outlined by Leopold, because they preserve biotic integrity, and ethically are a challenge to planned obsolescence as a kind of consumer fraud.

6.2 Digital tools, reverse logistics, and material tracking

The institutionalization of product take-back, various systematic sorting to refurbish or resell, and multi-cycle recycling processes, are introduced under the reverse logistics frameworks to reduce material value. By implementing AI-based software into these systems, predictive routing and real-time grading are made possible, leading to faster processing (rewarded, in turn, by a 45% cut in time) and a higher resale value (rewarded, in turn, by a 30% boost in the value). Digital innovation also enhances traceability: blockchain-based product passports like the EU Digital Product Passport should help trace the whole supply chain of a product, since its extraction to end of life, and RFID technology can be used to automatically sort materials precisely. Corporate pilots, which are also used by IBM, have shown close traceability of precious materials like gold, up to 99% which assists in decreasing illegal flows and informal leakages [38]. These tools, from a virtue ethics perspective, will institutionalize transparency and honesty in provenance, which are in opposition to practices such as ghost recycling, which tend to degrade trust [14].

6.3 Ethical implications of circularity

Circularity ethics goes further than just efficiency benefits, to issues of justice and inclusion. The modular and repair-oriented design frequently benefits urban, high-income settings where repair centers exist, and rural locations and people in the Global South experience obstacles based on a lack of tools, spare parts, and infrastructure, which further extends digital divides [24]. Formalized circular systems can also discriminate against informal recyclers, whose livelihoods are 12–24 million in the world, unless transitions are made with sufficient reskilling involving people, and that would be a breach of procedural justice [40]. Moreover, a long-term ethical dilemma: by making repairs cheaper, it will boost the rate of consumption, neutralize between 20% and 80% of the expected environmental benefits, and threaten to undermine the aims of precaution [41]. Ecofeminist critiques also warn that cyclic systems may never overcome gendered injustices, when care-based work like collection and sorting is not entirely appreciated and seen as visible.

Table 3: Ethical challenges and mitigation pathways in circular economy implementation.

Challenge	Ethical tension	Mitigation pathway
Access gaps	Urban bias excludes periphery	Community repair hubs and open-source designs
Worker displacement	Formalization erodes livelihoods	Just transitions and co-op models
Rebound effects	Efficiency fuels consumption	Sufficiency caps and degrowth ethics

These tensions demand "just circularity," fusing efficiency with equity via inclusive governance worker cooperatives, indigenous data sovereignty, and SDG-aligned metrics [6]. Digital tools must democratize access, not enclose commons, with rebound curbed by temperance virtues over endless growth.

7 Informal sector, inequality, and transboundary waste flows

The informal e-waste sector, processing 80–90% of global volumes in developing countries, sustains 12–24 million livelihoods, yet epitomizes environmental injustice, channeling Northern consumption burdens onto Southern peripheries [9, 42]. Predominant in India (Seelampur), China (Guiyu), Ghana (Agbogbloshie), and Nigeria, these operations recover valuables like gold/copper via rudimentary methods, filling formal infrastructure voids amid poverty and unemployment [43]. Ethically, this duality, economic necessity versus toxic exploitation, undermines SDG 8 (decent work), SDG 3 (health), and SDG 10 (reduced inequalities).

7.1 Role of informal sector in developing countries

Informal recyclers dominate due to formal systems' failures: consumer's reluctance to pay disposal fees, illegal imports as "second-hand goods," and absent awareness. In India, they handle 90% e-waste, despite E-Waste Rules (2022), collecting via door-to-door networks that formal systems cannot match [31]. Economically vital, recovering $15 billion metals annually, they operate in urban slums, employing migrants, women, or children, who excluded from formal labor [42]. Yet, this "paradox of necessity" traps workers: low barriers sustain entry but preclude exit via savings/education.

7.2 Occupational hazards and social marginalization

Processing unleashes horrors: open burning releases dioxins/Pb/Cd/Hg; acid leaching (HCl/H_2SO_4) generates fumes; and manual dismantling exposes skin/inhalation [9, 42]. Systematic reviews document blood Pb 10× the WHO limits, respiratory cancers, neurotoxicity, infertility, and DNA damage. Children/pregnant women are most vulnerable due to absorption/developmental windows [2]. Socially, marginalized castes and migrants face compounded stigma; women juggle caregiving and toxicity; disabled are excluded from aid; and children scavenge for nimbleness. Health literacy gaps/low-income cycles amplify: they have no PPE and strained systems prioritize acute over chronic harms.

7.3 North-South e-waste flows and ethical violations

The epitome of “toxic colonialism” is transboundary. With flows of 1.8 million tons annually, 65% of which is uncontrolled: Europe/US exports 70–80% to Africa/Asia in containers, and Basel Ban would not have been breached even with gap in ratifications. EU discards are imported to North Africa/West Africa hubs and India imports through repairable loopholes. This is ethically a breach of polluter pays/precaution/intergenerational equity. OECD subsidizes slums by imposing costs on the capacity-poor states [13]. Procedural injustice leaves out the voices of the South in trade decisions, which violate EJ.

8 Alignment with sustainable development goals

E-waste governance overlaps directly with various Sustainable Development Goals (SDGs), representing both the conjunctions and conflicts in the global context of sustainable development. Its effects, when viewed through the prism of ethics, spread throughout the sphere of health, equity, climate, and production, highlighting the interdependence of SDG 3 (Health and Well-Being), SDG 10 (Reduced Inequalities), SDG 12 (Responsible Consumption and Production), and SDG 13 (Climate Action). Meeting these goals with e-waste policy cannot simply be asked in terms of compliance with regulations; it includes redefining environmental justice, material ethics, and economic responsibility in the one and the same developmental approach [40].

It is especially true of SDG 12 – Responsible Consumption and Production. E-waste, which has surpassed 62 million tons each year and is estimated to reach 82 million tons by 2030, is indicative of the inability of global consumption models to internalize material limits (United 2559). Goals 12.5 and 12.8 require systemic minimization of waste through prevention and life-cycle consciousness, which relies on policy guidelines such as EPR and the concept of circular design. The European Union collective's EPR initiatives recycle 65–85% of e-waste, and reclaim approximately 62 billion of the material value. However, efficiency per se cannot support sustainability because there is a rebound effect, whereby lower costs of repaired or recycled goods will increase consumption by nullifying any gains made in emission reduction by 2080%. Ethically, the quest for SDG 12 to be efficient should be anchored on sufficiency: design for durability, consumer education, legislation, and legislation against planned obsolescence, which makes economic responsibility a moral one.

The health externalities of E-waste place it in the right context of SDG 3 (Health and Well Being). Formal recycling puts 1224 million workers and children (primarily in Asia and Africa) at risk of acute chemical toxicity of lead, mercury, and cadmium, increasing blood lead levels as high as ten times the levels allowed by the World Health Organization. Such practices involve violation of target 3.9 that aims at minimizing deaths as a result of dangerous chemicals. Empirical evidence has been associated with informal recycling as a cause of cancers, neurotoxicity, and congenital diseases, especially in women and children [9, 42]. Public health and waste policy integration bring synergies: models of this kind that are currently emerging in India, i.e., provision of protection gear and training to informal workers have reduced exposure by 70%. Nonetheless, tradeoffs will always exist – trade mandates may drive out unsafe jobs but also wipe out livelihoods with no fair alternative. Ethical consistency therefore demands that policies should integrate pollution control and income protection, and the human rights approach of SDG3 will come into play.

In its fundamental form, e-waste replicates structural inequalities, which SDG 10 seeks to help to decrease. The estimated 1.8 million tons of transboundary flows indicate how the prosperous economies shift the risk to the Global South, in blatant defiance of the fairness and solidarity principles. The producers of the OECD bypass impose restrictions of Basel Ban on waste, as second-hand products, at the expense of informal workers and vulnerable communities in Ghana, India, and Nigeria. In nations, caste systems and gender segregations push the e-waste livelihoods to the peripheral communities, which supports the importance of social inclusion in target 10.2. But in inclusive governance, where such communities are incorporated, whether on the basis of cooperatives, informal quotas on EPR schemes, or participatory policy formulation, e-waste is no longer a symbol of injustice, but a conveyor of empowerment. Technocratically enforced formalization and codesigned formalization strengthen inequality; SDG 10 is achieved through marginalized knowledge converted into sustainable practice [44].

There is a direct implication of e-waste to SDG 13 (Climate Action) as a challenge and opportunity. It has been estimated that the total CO_2-equivalent released by uncontrolled landfills and open burning is up to 209 million tons annually – equal to the global aviation release – and recycling and refurbishment would eliminate more than 2.3 billion kilograms of CO_2-equivalents each year [45]. The recovery of materials of critical concern through circular manufacturing and recovery will reduce the dependence on resources-intensive mining and the linking of mitigation (target 13.2) and adaptation to cleaner value chains. Nevertheless, recycling operations, particularly those that require the utilization of fossil-based transportation and thermal processing themselves, have a risk of paying off climate benefits. In achieving the real picture of coherence, green logistics and low carbon technologies and life cycle emissions accounting should be incorporated into national contributions of climate [46].

These SDGs are interdependent, which implies that the e-waste solutions cannot be thriving in isolation. The policies that will guarantee the recovery to the maximum under SDG 12 and will not safeguard workers (SDG 3) and inequitable transfer (SDG 10) will merely redistribute the damage. Instead, a virtuous nexus between the two ethical considerations will be formed: investments in cleaner recycling technologies will contribute to climate goals (SDG 13), and inclusive EPR funds will finance healthcare and training of informal collectors, which will become a combination of welfare and sustainability [44, 45]. The justice-oriented framing does not only change efficiency to empathy but to justice in the fact that there are people who need to deal with our dumped technologies [47].

In this regard, the management of electronic waste can be viewed as a sort of ethical consideration of the 2030 Agenda as such. The mechanisms of accountability may be introduced by the regularity of policies: eco-design policies (SDG12.5) match with workplace safety policies (SDG3.9); trade rules (SDG10.b) are in line with climatic policies (SDG13.2). On the international level, ethical stewardship redefines sustainability as efficiency based on consumption to one based on solidarity-based care, to ensure that the technological advancement does not continue the accumulation of environmental debt and social inequality. The future of e-waste policy is in the acknowledgement that circuits of our devices reflect circuits of the human interdependence, and the development objectives of the world cannot be fulfilled without sewing the other circuit.

9 Emerging solutions and governance innovations

New approaches to governing e-waste capitalize on digital innovation, collaborative modes, and institutional restructuring to overcome ethical ambitions and viable execution by building on evidence-based pilots that show quantifiable effectiveness. Transparency platforms are transformative tools based on blockchains, immutable end-to-end traceable ledgers to curb illicit flows, and the so-called ghost recycling fraud [48].QR codes will also be needed in the Digital Product Passport that is being implemented by the EU in 2024, by which the lifecycle data will be transferred. This program will enable the consumer to evaluate repairability ratings and access disassembly guidelines, thus creating responsibility without any concentration obstacles [22]. Apart from transparency, these platforms have the potential of expansion, though this depends on the interoperability standards and data privacy policies that are similar to GDPR in the emerging markets.

These collaborations are institutionalized through multi-stakeholder governance models which bring together producers, regulators, NGOs and workers through forums such as the multi-stakeholder policy dialogue (MSPDs) in Southeast Asia, which harmonized Basel compliance in 10 countries, cutting transboundary violations by 30% [49]. The WEEE multi-stakeholder cooperatives in China (that incorporated 50,000 informal actors through formal chains) retrieved 1.2 million tons in 2023 that combined government subsidies with producer EPR funds [50]. The Producer Responsibility Forums within the CPCB of India require quarterly meetings between NGOs and producers, with 15% EPR compliance boost rewarded; experience has shown that it helps ensuring that producers have equal power through worker quotas and veto authority over mandatory risk reduction policies.

Sustainable systems are based on capacity-building. An example is the BIASHARA project funded by the World Bank in East Africa where almost 2,000 collectors and firms received environmental management knowledge and skills that reduce health-related incidences by nearly 50% by implementing specific awareness campaigns, and microfinance to obtain personal protective equipment [51]. On the same note, the program of GIZ in India enhanced the operation of State Pollution Control Boards by introducing 100 mobile monitoring units, which helped in detecting almost three-quarters of regulation offenses. Empirical studies also suggest that dismantling practices that are evidence-based, hands-on, blockchain literacy, and EPR accounting provide a triple payback, which is based on fewer penalties and higher value recovery of high-value metals.

Those interventions can be most efficient in a hybrid governance ecosystem, where a blockchain enables the real-time sharing of data with monitoring and supervisory public departments (MSPDs), where the capacity-building centers of public-private partnerships (PPP) are financially supported by the ethical performance indicators, including the inclusion of workers and the decrease of toxic exposure, and strengthen the resilience of the whole system. Pilot projects have been documented to reduce wastes between 40% and 70%, but the problem of implementation remains. About 60% of rural participants are left out because of a lack of subsidies, the PPPs do not work well without political commitment, and the institutions cannot work effectively without a long-term funding [50]. In the end, the interactions between technology, policy, and finance need to converge so that technological innovation can be used in relation to the regulatory frameworks and equity goals instead of a limited set of efficiency benefits.

10 Discussion: toward ethical stewardship in e-waste management

The framework of e-waste sustainable governance requires an integrative ethical stewardship system, which combines normative principles and working mechanisms, beyond a disjointed approach in which ethics, law, technology, and economics operate in isolation [23, 50]. Technocentric recycling is more efficient in terms of material, but it ignores the need to demand justice, meaning that informal workers are left with toxics despite the high recovery rates of 65% by the EU under the WEEE Directive, as Northern exports perpetuate injustices [5, 23]. It is the ethics that offer the connective tissue: SDG 12 efficiency combined with SDG 3 safeguards are operationalized through precautionary design; SDG 10 reparations and SDG 13 accountability are enforced through polluter pays; SDG 3 safeguards and SDG 12 efficacy are animated through stewardship virtues; blockchain transparency is human-friendly through blockchain accountability [14].

In the present day governance there are structural constraints. The Basel Convention governs the transboundary flows without strict enforcement, and the Ban Amendment is still not enforced by the major actors to allow 70% of the flows to be illegal to the capacity-constrained nations [25]. National EPR regimes become performative compliance: the 2022 Rules document of India has obligations that bind, and a capture of less than 30% of formal collection in the process of regulatory capture [17]. Blockchain pilots have 99% supply chains, but 60% of rural actors without infrastructure cannot access digital divides [52]. It is an economically motivated selective recovery of up to US$ 62 billion recoverable value, leaving toxics to slums and continuing to perpetuate plural injustices, including postcolonial, procedural, and distributive (which requires systemic transformation) [23].

Justice-oriented stewardship makes restructuring of these silos into moral coherence. Ecocentrism dictates that the upstream toxics must be eliminated; environmental justice informal labor is the custodian rather than the casualty. The BRS Conventions have been renegotiated with local capabilities to encompass the aspects of MSPDs in polycentric models that introduce 50,000 informal recyclers in China in WEEE cooperatives [50]. The success of the repair score is shown to the communities that are exposed through the product passport, the fund of reparation is transferred by the blockchain to digital twins, and thus the technology is ethical. Just transitions are financed by virgin material taxes to meet economic pivots to make profit equal to virtue [52].

Moral opportunity thus puts the e-waste to moral stewardship instead of technical liability. The interrelatedness of ethics (precaution/justice), law (EPR enforcement), technology (traceability), and economics (sufficiency incentives) forms governance of silo-reactive to solidarity-proactive. In the absence of this synthesis, the voids of capacity and imbalances exist; justice compels to make the realization of an informal knowledge the custodian of the universe. E-waste is a problem wherein technology is being overloaded and it needs to be relapsed into intergenerational equity and dignity.

11 Policy recommendations and future directions

Ethical stewardship is operationalized in form of specific reforms that establish moral imperatives by the means of enforceable structures [5, 25]. Governments should enact laws on eco-design indexes requiring reparability 8/10, audited through blockchain passports with Basel Ban ratification by 2027 20% violation fines channeled to Southern hubs [53]. Corporate responsibility requires taxes on lifecycle resources (such as taxes on rare earths of 50 per ton) to remediate the world through global E-Waste Justice Funds and ensure ESG disclosure of 30% informal sourcing quotas to avoid greenwashing [17]. Hybrid models and informal integration scale with 1 M collector trainers, GIZ-

Chintan frameworks reduce 80% livelihoods, 70% toxics training 1 M collectors by 2030, subsidizing PPE and digital literacy, 1 million collectors collect 10TB data, subsidizing 1 M data analytics, preserving 80% livelihoods, and reducing 70% toxics training 1 M collectors by 2030 [52].

Multi-stakeholder E-Waste Stewardship Councils, which convene producers/NGOs/workers with veto power, need to institutionalize quarterly MSPDs, which reflect 30% transboundary cuts in Southeast Asia [50]. The research priorities are aimed at the rebound effects (20–80% offsets), gendered effects, blockchain equity in low-literacy contexts, piloting with North-South reparative funds tracing losses of $62 billion [11]. Such measures bring ethical change: obeying results in the good, the good results in obedience. Nexus governance that is consistent with BRS-SDGs frameworks has diversion gains between 40–70 with the results determined by the extent of political engagement and application of equity policies to avoid techno-exclusion.

12 Conclusion

Ethical stewardship in the management e-wastes incorporates law, technology, economics, and justice to fill in moral loopholes caused by efficiency-based models. In this chapter, it is contended that current systems, Basel Convention, country-level EPR regulations, and circular models, are technically successful but ethically unsound because of the externalization of risks to the informal worker and Global South. The application of ethical principles, including precaution, polluter pays, and intergenerational equity, makes the regulation be more moral and balances the efficiency focus of SDG12 with the health protection focus of SDG3 and the fairness targets of SDG10. Practical pathways enablers: blockchain-enabling transparency, multi-stakeholder governance, and inclusive cooperatives are practical pathways where the idea of just circularity can be operationalized to support the concept of environmental justice as in the Chinese models of WEEE cooperatives reclaiming more than a million tons of waste and, at the same time, providing livelihoods to local residents. The policy of the future must institutionalize the eco design standards, life cycle tax to fund the reparation, and legal obligations to safeguard the informal labor in formalized value chains. Finally, ethical stewardship reconceptualizes e-waste management as a new duty of care in the digital era: regulating the innovation with care, humility, and responsibility to make technological advancement reinforce social and ecological fairness and not recreate the toxic heritage.

References

[1] Andersen T, A comparative study of national variations of the European WEEE directive: Manufacturer's view. Environ Sci Pollut Res, 2022 Mar, 29(14), 19920–19939.

[2] Sun X, Chen X, Shi Z, Yan AF, Li Z, Chen S, et al, A comparison study of prevalence, awareness, treatment and control rates of hypertension and associated factors among adults in China and the United States based on national survey data. Glob Health J, 2023 Mar, 7(1), 24–33.

[3] Goyal S, Gupta S, A comprehensive review of current techniques, issues, and technological advancements in sustainable E-waste management. e-Prime – Advances in Electrical Engineering. Electron Energy, 2024 Sept, 9, 100702.

[4] Murthy V, Ramakrishna S, A review on global e-waste management: Urban mining towards a sustainable future and circular economy. Sustainability, 2022 Jan 7, 14(2), 647.

[5] Weiand L, Unger S, Rochette J, Müller A, Neumann B, Advancing ocean governance in marine regions through stakeholder dialogue processes. Front Mar Sci, 2021 May, 8.

[6] Basel Convention on the Control of Transboundary Movements of Hazardous Wastes and their Disposal [Internet]. [cited 2026 Jan 15]. Basel Convention E-waste Amendments. Available from: https://www.basel.int/Implementation/Ewaste/EwasteAmendments/Overview/tabid/9266/Default.aspx

[7] Basel Convention on the Control of Transboundary Movements of Hazardous Wastes and their Disposal [Internet]. [cited 2026 Jan 15]. Basel Convention on the Control of Transboundary Movements of Hazardous Wastes and their Disposal. Available from: https://www.basel.int/implementation/ewaste/overview/tabid/4063/default.aspx

[8] China WEEE | Enviliance ASIA [Internet]. [cited 2026 Jan 17]. Available from: https://enviliance.com/regions/east-asia/cn/cn-waste/cn-weee

[9] Pan X, Wong CWY, Li C, Circular economy practices in the waste electrical and electronic equipment (WEEE) industry: A systematic review and future research agendas. J Clean Prod, 2022 Sept, 365, 132671.

[10] Chen CW, Clarifying rebound effects of the circular economy in the context of sustainable cities. Sustain Cities Soc, 2021 Mar, 66, 102622.

[11] Research Scholar, Department of Education, Indira Gandhi National Tribal University, Amarkantak, Madhya Pradesh, India, Roy P, Josephine Arokia Marie S M, Assistant Professor, Department of Education, Indira Gandhi National Tribal University, Amarkantak, Madhya Pradesh, India. climate issues for e-waste management in india: The unseen consequences on ecosystem. tssr. 2025 Sept 27;(Special):160–169.

[12] Biswas S, Singh P, Singh S, Comparative analysis of formal and informal e-waste disposal methods in gurgaon: implications for environmental pollution and public health. Curr World Environ, 2025 Jan 10, 19(3), 1473–1485.

[13] Srivastav AL, Markandeya, Patel N, Pandey M, Pandey AK, Dubey AK, et al, Concepts of circular economy for sustainable management of electronic wastes: Challenges and management options. Environ Sci Pollut Res, 2023 Feb 28, 30(17), 48654–48675.

[14] Conventions, Protocols and Recommendations | International Labor Organization [Internet]. [cited 2026 Jan 17]. Available from: https://www.ilo.org/international-labour-standards/conventions-protocols-and-recommendations

[15] Borthakur A, Design, adoption and implementation of electronic waste policies in India. Environ Sci Pollut Res, 2022 Feb 9, 30(4), 8672–8681.

[16] Directive – 2002/96 – EN – EUR-Lex [Internet]. [cited 2026 Jan 17]. Available from: https://eur-lex.europa.eu/eli/dir/2002/96/oj/eng

[17] Dias P, Bernardes AM, Huda N, E-waste management and practices in developed and developing countries*, In Holuszko ME, Kumar A, Espinosa DCR, editors., Electronic Waste [Internet]. 1st ed., Wiley, 2022 [cited 2026 Jan 15], 15–32, Available from https://onlinelibrary.wiley.com/doi/10.1002/9783527816392.ch2.
[18] Agarwal A, Faizan M, Jindal Y, Iftakhar S, V S, E-waste management in India. Asian J Manage Commerce, 2024 Jan 1, 5(1), 263–267.
[19] Kumar S, Agarwal N, Anand SK, Rajak BK, E-waste management in India: A strategy for the attainment of SDGs 2030, In Materials Today: Proceedings. 2022, Vol. 60, 811–814.
[20] Turaga RMR, Bhaskar K, Sinha S, Hinchliffe D, Hemkhaus M, Arora R, et al, E-Waste management in India: issues and strategies. Vikalpa J Decis Makers, 2019 Sept, 44(3), 127–162.
[21] Noliya S, Gupta A, Kumar V, E-waste management through Blockchain Technology: Supply chain led environmental sustainability. EDPACS, 2025 July 22, 1–11.
[22] Bhagat-Ganguly V, E-Waste Management: Challenges and Opportunities in India [Internet]. 1st ed., Routledge India, London, 2021, [cited 2026 Jan 18]. Available from https://www.taylorfrancis.com/books/9780429285424.
[23] Ganguly V, E-waste management: Challenges and opportunities in India. Routledge, London, 2022, 1.
[24] Perkins N, Brune Drisse MN, Nxele T, P Sly P, E-Waste: A global hazard. Ann Glob Health, 2014 Nov 25, 80(4), 286.
[25] Electronic waste (e-waste) [Internet]. [cited 2026 Jan 15]. Available from: https://www.who.int/news-room/fact-sheets/detail/electronic-waste-(e-waste)
[26] Akese GA, Little PC, Electronic waste and the environmental justice challenge in agbogbloshie. Environ Justice, 2018 Apr, 11(2), 77–83.
[27] Singh N, Chimaniya S, Pradhan H, Singh D, Singh K, Enabling sustainable e-waste management through blockchain technology, In 2025 3rd International Conference on Disruptive Technologies (ICDT). 2025, 1–5.
[28] Environment, Climate Change and Health [Internet]. [cited 2026 Jan 15]. Available from: https://www.who.int/teams/environment-climate-change-and-health/settings-populations/children/e-waste
[29] Environmental Ethics (Stanford Encyclopedia of Philosophy) [Internet]. [cited 2026 Jan 15]. Available from: https://plato.stanford.edu/entries/ethics-environmental/
[30] Njoku A, Agbalenyo M, Laude J, Ajibola TF, Attah MA, Sarko SB, Environmental injustice and electronic waste in Ghana: Challenges and recommendations. IJERPH, 2023 Dec 23, 21(1), 25.
[31] Sandler RL, Environmental virtue ethics, In LaFollette H, editor, The International Encyclopedia of Ethics [Internet]. 1st ed, Wiley, 2013, [cited 2026 Jan 10]. Available from:, https://onlinelibrary.wiley.com/doi/10.1002/9781444367072.wbiee090.
[32] Kosarkar M, Barthwal S, Environmentally conscious but helpless: Consumer challenges for participation in Formal e-waste collection initiatives of India. Clean Waste Syst, 2025 Dec, 12, 100363.
[33] Temur GT, Bolat B, Evaluating efforts to build sustainable WEEE reverse logistics network design: Comparison of regulatory and non-regulatory approaches. Int J Sustain Eng, 2017 Sept, 10(6), 358–383.
[34] Quinto S, Law N, Fletcher C, Le J, Antony Jose S, Menezes PL, Exploring the e-waste crisis: Strategies for sustainable recycling and circular economy integration. Recycling, 2025 Apr 14, 10(2), 72.
[35] Faibil D, Asante R, Agyemang M, Addaney M, Baah C, Extended producer responsibility in developing economies: Assessment of promoting factors through retail electronic firms for sustainable e-waste management. Waste Manag Res, 2023 Jan, 41(1), 117–142.
[36] Manomaivibool P, Vassanadumrongdee S, Extended producer responsibility in Thailand. J Ind Ecol, 2011 Mar, 15(2), 185–205.

[37] Extended Producer Responsibility: Basic facts and key principles [Internet]. 2024 Apr [cited 2026 Jan 17]. (OECD Environment Policy Papers; vol. 41). Report No.: 41. Available from: https://www.oecd.org/en/publications/extended-producer-responsibility_67587b0b-en.html
[38] Choi WH, Pae KP, Kim NS, Kang HY, Hwang YW, Feasibility study of closed-loop recycling for plastic generated from Waste Electrical and Electronic Equipment (WEEE) in South Korea. Energies, 2023 Sept 1, 16(17), 6358.
[39] Tong X, Yan L, From legal transplants to sustainable transition. J Ind Ecol, 2013 Mar, 17(2), 199–212.
[40] Anghileri D, Pastori M, Marcos-Garcia P, Umlauf G, Crestaz E, Seliger R, et al, Global Water Challenges in Sub-Saharan Africa and how to strengthen science-policy dialogues on transboundary governance and cooperation. J Environ Manag, 2024 June, 365, 121417.
[41] Sui Y, Sun Q, Li S, Government mechanism in waste electrical and electronic equipment recycling among prospect theory and evolutionary game theory. Energy Environ, 2024 Feb.
[42] Sthiannopkao S, Wong MH, Handling e-waste in developed and developing countries: Initiatives, practices, and consequences. Sci Total Environ, 2013 Oct, 463–464, 1147–1153.
[43] Grant K, Goldizen FC, Sly PD, Brune MN, Neira M, Van Den Berg M, et al, Health consequences of exposure to e-waste: A systematic review. Lancet Glob Health, 2013 Dec, 1(6), e350–61.
[44] Issah I, Arko-Mensah J, Agyekum TP, Dwomoh D, Fobil JN, Health risks associated with informal electronic waste recycling in Africa: A systematic review. IJERPH, 2022 Nov 1, 19(21), 14278.
[45] Home – E-Waste Monitor [Internet]. [cited 2026 Jan 15]. Available from: https://ewastemonitor.info/
[46] Chileshe N, Kavishe N, Edwards DJ, Identification of critical capacity building challenges in public-private partnerships (PPPs) projects: The case of Tanzania. Int J Constr Manag, 2023 Feb 17, 23(3), 495–504.
[47] Joshi S, Sharma M, Barve A, Implementation challenges of blockchain technology in closed-loop supply chain: A Waste Electrical and Electronic Equipment (WEEE) management perspective in developing countries. Supply Chain Forum Int J, 2022 Nov, 24(1), 59–80.
[48] Cao J, Xu J, Wang H, Zhang X, Chen X, Zhao Y, et al, Innovating collection modes for waste electrical and electronic equipment in China. Sustainability, 2018 May, 10(5), 1446.
[49] Vojnovic I, Intergenerational and intragenerational equity requirements for sustainability. Environ Conserv, 1995, 22(3), 223–228.
[50] Becker RA, Intergenerational equity: The capital-environment trade-off. J Environ Econ Manag, 1982 June, 9(2), 165–185.
[51] Abram S, Atkins E, Dietzel A, Jenkins K, Kiamba L, Kirshner J, et al, Just Transition: A whole-systems approach to decarbonisation. Clim Policy, 2022 Sept 14, 22(8), 1033–1049.
[52] Almuwallad M, Managing E-waste in reverse logistics network through technology investment and facility effectiveness to support the electric vehicles (EV) market. Results Eng, 2026 Mar, 29, 108579.
[53] Map of the Week: Transboundary Movements of Electrical and Electronic Waste | UBIQUE [Internet]. [cited 2026 Jan 15]. Available from: https://ubiqueags.org/map-of-the-week-transboundary-movements-of-electrical-and-electronic-waste/
[54] Owusu-Sekyere K, Aladago DA, Material flow analysis and risk evaluation of informal E-waste recycling processes in Ghana: Towards sustainable management strategies. J Clean Prod, 2023 Dec, 430, 139706.
[55] New International Requirements for Electrical and Electronic Waste | US EPA [Internet]. [cited 2026 Jan 15]. Available from: https://www.epa.gov/hwgenerators/new-international-requirements-electrical-and-electronic-waste
[56] Annamalai J, Occupational health hazards related to informal recycling of E-waste in India: An overview. Indian J Occup Environ Med, 2015, 19(1), 61.
[57] OECD Legal Instruments [Internet]. [cited 2026 Jan 17]. Available from: https://legalinstruments.oecd.org/en/instruments/oecd-legal-0102

[58] Wang H, Gu Y, Li L, Liu T, Wu Y, Zuo T, Operating models and development trends in the extended producer responsibility system for waste electrical and electronic equipment. Resour Conserv Recycl, 2017 Dec, 127, 159–167.

[59] Nicol S, Thompson S, Policy options to reduce consumer waste to zero: Comparing product stewardship and extended producer responsibility for refrigerator waste. Waste Manag Res, 2007 June, 25(3), 227–233.

[60] Grandhi SP, Dagwar PP, Dutta D, Policy pathways to sustainable E-waste management: A global review. J Hazard Mater Adv, 2024 Nov, 16, 100473.

[61] Khetriwal DS, Kraeuchi P, Widmer R, Producer responsibility for e-waste management: Key issues for consideration – Learning from the Swiss experience. J Environ Manag, 2009 Jan, 90(1), 153–165.

[62] Cahill R, Grimes SM, Wilson DC, Review Article: Extended producer responsibility for packaging wastes and WEEE – A comparison of implementation and the role of local authorities across Europe. Waste Manag Res, 2010 Aug, 29(5), 455–479.

[63] Elgarahy AM, Eloffy MG, Priya AK, Hammad A, Zahran M, Maged A, et al, Revitalizing the circular economy: An exploration of e-waste recycling approaches in a technological epoch. Sustain Chem Environ, 2024 Sept, 7, 100124, Elgarahy AM, Eloffy MG, Priya AK, Hammad A, Zahran M, Maged A, et al SDG Indicators [Internet]. [cited 2026 Jan 15]. Available from, https://unstats.un.org/sdgs/report/2024/Goal-12/.

[64] Küfeoğlu S, SDG-10: Reduced inequalities, In Emerging Technologies [Internet]. Springer International Publishing, Cham, 2022 [cited 2026 Jan 18], 371–384, (Sustainable Development Goals Series). Available from https://link.springer.com/10.1007/978-3-031-07127-0_12.

[65] Küfeoğlu S, SDG-13: Climate action, In Emerging Technologies [Internet]. Springer International Publishing, Cham, 2022 [cited 2026 Jan 18], 429–451, (Sustainable Development Goals Series). Available from https://link.springer.com/10.1007/978-3-031-07127-0_15.

[66] Shokohyar S, Mansour S, Simulation-based optimisation of a sustainable recovery network for Waste from Electrical and Electronic Equipment (WEEE). Int J Comput Integr Manuf, 2013 June, 26(6), 487–503.

[67] Liu T, Zhang Q, Zheng Z, Wu S, Weng Z, Stakeholder analysis of the waste electrical and electronic equipment internet recycling industry. IJERPH, 2022 Aug 13, 19(16), 10003.

[68] Faradillah P, Sultan AAM, Samat KF, Kamarulzaman T, Strengthening e-waste governance: A decision framework for sustainable transboundary movements under the Swiss-Ghana Amendments. Chin J Popul Resour Environ, 2025 Sept, 23(3), 310–323.

[69] Ajaygopal KV, Verma R, Koul S, Sustainable and resilient circular supply chain: A remanufacturer selection. J Enterp Inf Manag, 2025 Aug, 38(4).

[70] Yu X, Sun Z, Sun D, He R, Sustainable development assessment of household e-waste reverse supply chains from an environmental ethic perspective. Humanit Soc Sci Commun, 2025 June 12, 12(1), 811.

[71] Valai Ganesh S, Suresh V, Rajakarunakaran S, Suryakumar V, Sustainable electronic waste management framework for academic institutions in India. Sci Rep, 2025 Nov 18, 15(1), 40550.

[72] Kumar RAA, Vipin B, Sridharan R, Sustainable reverse logistics network design: A case of waste electrical and electronic equipment management. Int Trans Oper Res, 2025 July.

[73] Kumar A, Gaur D, Liu Y, Sharma D, Sustainable waste electrical and electronic equipment management guide in emerging economies context: A structural model approach. J Clean Prod, 2022 Jan, 336, 130391.

[74] Santos TVD, Alvez SMD, Sehnem S, Julkovski DJ, The circularity paradox: The rebound effect, governance, and the limits of corporate sustainability. Corp Soc Responsib Environ, 2026 Jan, 33(1), 616–646.

[75] Lin CH, Chen Y, Chang KI, Chi PY, The evaluation of economic impact of the Basel Convention on the trade of waste electrical and electronic equipment. J Asian Econ, 2025 Oct, 100, 101996.

[76] The Growing Environmental Risks of E-Waste – Geneva Environment Network [Internet]. [cited 2026 Jan 15]. Available from: https://www.genevaenvironmentnetwork.org/resources/updates/the-growing-environmental-risks-of-e-waste/
[77] Castro CG, Trevisan AH, Pigosso DCA, Mascarenhas J, The rebound effect of circular economy: Definitions, mechanisms and a research agenda. J Clean Prod, 2022 Apr, 345, 131136.
[78] Pansera M, Barca S, Martinez Alvarez B, Leonardi E, D'Alisa G, Meira T, et al, Toward a just circular economy: Conceptualizing environmental labor and gender justice in circularity studies. Sustain Sci Pract Policy, 2024 Dec 2, 20(1), 2338592.
[79] Daum K, Stoler J, Grant R, Toward a more sustainable trajectory for e-waste policy: A review of a decade of e-waste research in Accra, Ghana. IJERPH, 2017 Jan 29, 14(2), 135.
[80] Cao D, Puntaier E, Gillani F, Chapman D, Dewitt S, Towards integrative multi-stakeholder responsibility for net zero in e-waste: A systematic literature review. Bus Strat Environ, 2024 Dec, 33(8), 8994–9014.
[81] Thapa K, Vermeulen WJ, Deutz P, Olayide OE, Transboundary movement of waste review: From binary towards a contextual framing. Waste Manag Res, 2023 Jan, 41(1), 52–67.
[82] Tsai WH, Hung SJ, Treatment and recycling system optimisation with activity-based costing in WEEE reverse logistics management: An environmental supply chain perspective. Int J Prod Res, 2009 July, 47(19), 5391–5420.
[83] Maphosa M, Maphosa V, Using the circular economy to mitigate the global electronic waste challenge: A systematic literature review approach. Eur J Soc Impact Circ Econ, 2024 Dec 20, 5(3), 28–54.

Index

 | https://doi.org/10.1515/9783112248072-017